Constructing the Universe

CONSTRUCTING THE UNIVERSE

David Layzer

**SCIENTIFIC
AMERICAN
LIBRARY**

An imprint of Scientific American Books, Inc.
New York

Library of Congress Cataloging in Publication Data

Layzer, David.
  Constructing the universe.

  Bibliography: p.
  Includes index.
    1. Cosmology. I. Title.
  QB981.L33   1984   523.1   84-5351
  ISBN 0-7167-5003-1 (W.H. Freeman)

Printed in the United States of America
Book design by Malcolm Grear Designers

Scientific American Library is published by
Scientific American Books, Inc., a subsidiary
of Scientific American, Inc.

Distributed by W. H. Freeman and Company,
41 Madison Avenue, New York, New York 10010.

1 2 3 4 5 6 7 8 9 0 KP 2 1 0 8 9 8 7 6 5 4

To Jean with love

CONTENTS

PREFACE

This book is about the two great modern theories of space, time, and gravity—Newton's and Einstein's—and about the theories of cosmic structure and evolution that have been built on them. It is also about the process by which these theories came into being. Like living organisms, scientific theories are the outcomes of an evolutionary process and cannot be fully understood except in the light of that process.

Four distinct but intertwined activities have shaped modern ideas about the universe and its underlying laws: the development of experimental and observational instruments and techniques; the telescopic exploration of space; the construction of basic physical theories; and the construction and testing of cosmological hypotheses. These four activities enjoy a symbiotic relationship: each nourishes the others. Although this book is mainly about the construction of theories and hypotheses, I hope it will also help its readers gain insight into how theory and observation interact, and into how, at critical stages in the development of our present world-picture, technological advances, from the invention of the telescope at the beginning of the seventeenth century to the development of low-noise microwave radiometers in the 1950s, have made it possible for us to see deeper, not only with the eye of the body but, in Plato's phrase, with the eye of the mind. If instruments and observations play less conspicuous roles in my story than do theories and hypotheses, it is not because I think they are less important. It is not even because I know less about them, though that is certainly true. It is because a single medium-sized book (I had hoped it would be shorter!) cannot treat both the practical and the theoretical aspects of the story I set out to tell in the depth they deserve.

In *The Character of Physical Law*, Richard Feynman writes, "It is impossible to explain honestly the beauties of the laws of nature in a way that people can feel, without their having some deep understanding of mathematics. I am sorry, but this seems to be the case." Feynman then goes on to explain, simply and honestly, some of the deep mathematical ideas that underlie modern physical theories. I have tried to follow his example. Instead of making detours around crucial mathematical arguments, I have tried to make the arguments understandable to readers who have only a modest mathematical background (high-school geometry, algebra, and analytic geometry), but do have a hearty appetite for mathematical reasoning. At the same time, I have tried to tell my story in a way that will allow readers with a more delicate mathematical appetite to skip the heavier passages without losing the thread of the narrative.

Feynman makes the same point as Galileo (see the epigraph to Chapter 1): The book of nature is written in the language of mathematics. But we must bear in mind that Galileo's mathematical background was that of a modern high-school student; analytic geometry had yet to be invented. Newton himself knew nothing of vectors, rates of change, or integrals when he began to puzzle over Kepler's laws of planetary motion. He invented these concepts in order to describe more general kinds of motion than could be described with the mathematics of his day. In the process, he created the mathematical language that he needed to formulate his laws of motion and gravitation and to apply them to astronomical phenomena. Newton's mathematical, physical, and astronomical inventions and discoveries were parts of a single project, and that is how I have dealt with them in this book. The reader who has not previously encountered vectors and rates of change will encounter them here in the contexts in and for which Newton invented them.

Historians of science usually view the theories they are writing about in a different light from the people who made these theories. For example, historians writing about the astronomical revolution of the sixteenth and seventeenth centuries rarely focus their attention on the mathematical structure of the theories of Kepler, Galileo, and Newton. Instead, they focus on the theories' "underlying concepts," treating mathematics as a language for lending precision to these concepts. The scientists themselves took a different view. For them the mathematics came first. Verbal concepts, in their view, could be helpful or even necessary, but they were not the essence.

The people who made the astronomical revolution looked upon science as a growing body of truth about the world. This view still prevails among working scientists. It is not widely held, however, by contemporary historians of science. Some of them regard theories as nothing more than convenient but ephemeral devices for making predictions. Others reject the notion of scientific objectivity, arguing that all so-called facts are "theory laden" and denying that any intelligible meaning can be given to the notion of scientific truth.

It may be that the historians of science are right and the scientists are wrong about these issues; a hen may not be the best judge of an egg. But that would not make the way scientists talk about them less interesting or important, because the philosophical attitudes of Kepler, Galileo, and Newton, however naive they may seem to today's historians of science, were inextricably bound up with their achievements. So this book is not only about scientific theories and their evolution but also about how the people who built the theories viewed scientific knowledge and its relation to truth.

The book's final chapter deals with the origin and evolution of the astronomical universe. These are unquestionably the most exciting topics in modern cosmology, and the most speculative and controversial. There are two ways of presenting such material. One is to minimize the distinction between strongly confirmed theories and widely accepted (but not yet strongly supported) hypotheses; the other is to emphasize the distinction. I have chosen the second way. I have also included a fairly detailed account of my own views about the early universe, the origin of astronomical systems, and the origin of the cosmic radiation background. My aim is not to persuade the reader that my theory is better than its rivals. Rather, I hope that my discussion of competing cosmogonic hypotheses will help readers understand how and why scientists who are working within essentially the same theoretical framework can interpret a given body of observational evidence in radically different ways. I hope it will also convey a sense of the excitement and uncertainty of science in the making.

I have used parts of this book in an undergraduate course attended mainly by nonscientists, and more of it in courses attended mainly by concentrators in mathematics, physics, and astronomy. The intelligence and curiosity of the students and discussion leaders in these courses have been a constant source of stimulation.

A *Scientific American Library* book, I have come to understand, is the outcome of a team effort in which the author has a disproportionate share of the fun and the other members of the team have a disproportionate share of the hard work. Part of the fun is writing these acknowledgements.

In the beginning there was Gerard Piel's vision of a thin, beautifully illustrated book struggling to emerge from two fat volumes of typescript. "I'll send Linda Chaput to see you," he said; and she ran the project with such energy, enthusiasm, and intelligence that we all came to believe that it could be done, and done well, in the allotted time. James Maurer supported, encouraged, and coordinated our efforts with great skill, tact, and imagination, ably assisted by design coordinator Lisa Douglis, production coordinator Sarah Segal, layout designer Daniel Earl Thaxton, art coordinator Jill Feldheim, and indexer Julia McVaugh.

I owe a special debt of gratitude to Aidan Kelly, my editor, who also served as my navigator, helping me to steer between the shoals of oversimplification and the maelstrom of abstract mathematical discourse.

I had hoped to be able to tell an important part of my story with pictures. For making this possible, I am indebted to many people: to George Kelvin, who transformed my crude sketches into the clear and handsome illustrations that adorn the following pages; to photo researcher David Barkan, who ransacked Manhattan's photo archives for beautiful and appropriate images; and to my colleagues at the Harvard-Smithsonian Center for Astrophysics: Owen Gingerich, John Huchra, Robert Stachnik, Robert F. C. Vessot, Martha Liller, and, especially, Christine Jones and William Forman, who supplied and helped me select the X-ray images that appear in Chapter 8.

Last but not least, I thank Joan Thompson, who, with supreme competence and unflappability, not only prepared and kept organized a seemingly endless series of versions and revisions, but managed to greet each one of them with a smile.

David Layzer
Harvard University
February 8, 1984

Constructing the Universe

CHAPTER ONE

COSMOLOGY AND SCIENTIFIC TRUTH

Philosophy is written in this grand book, the universe, which stands continually open to our gaze. But the book cannot be understood unless one first learns to comprehend the language and read the letters in which it is composed. It is written in the language of mathematics, and its characters are triangles, circles, and other geometric figures without which it is humanly impossible to understand a single word of it; without these, one wanders about in a dark labyrinth.

GALILEO, in *The Assayer*

The night sky looks much the same today as it did 400 years ago, but we see it with different eyes. This is how it seemed to Lorenzo in *The Merchant of Venice* (1596):

> *Sit, Jessica. Look how the floor of heaven*
> *Is thick inlaid with patines of bright gold:*
> *There's not the smallest orb which thou behold'st*
> *But in his motion like an angel sings,*
> *Still quiring to the young-eyed cherubins.*
> *Such harmony is in immortal souls;*
> *But whilst this muddy vesture of decay*
> *Doth grossly close it in, we cannot hear it.*

A modern Lorenzo would explain to his Jessica, or she to him, that the stars are spheres of gas held together by the mutual gravitation of their atoms; that all the stars we can see are themselves atoms in a vast system shaped like the solar system, but a million million times as massive and a hundred million times as large; that this system is itself an atom among innumerable similar atoms in a "gas" that extends endlessly in all directions; that this "gas" and the space it fills are expanding at a rate that will double the distance between any two "atoms" in about 20 billion years; and that the universe began its existence 10 billion years ago as pure, undifferentiated energy at an unimaginably high density.

Left, God as the Orderer of the (Ptolemaic) universe, with the Earth at its center: a colored woodcut from Martin Luther's Bible (1543).

Although modern scientific cosmology began with the work of Nicolaus Copernicus (1473–1543) and Johannes Kepler (1571–1630), their pictures of the world seem to have more in common with Shakespeare's picture than with the modern one. In fact, the earliest scientific cosmologies embody many of the same cosmological motifs as the world view they replaced. All of these motifs go back to the Greeks; some of them, to still older civilizations. Let us take a closer look at a few of them.

Recurrent Cosmological Themes

Shakespeare wove four ancient cosmological motifs into Lorenzo's speech. The first is that of the *firmament*. In ancient Babylonian, Hebrew, Egyptian, and Greek cosmologies, the Earth was at the center of the universe, and a solid vault supported the stars and divided Heaven from Earth. Most stars were fixed to this vault, but the Sun, the Moon, Mercury, Venus, Mars, Jupiter, and Saturn wandered against the background of the fixed stars (our word planet comes from the Greek word for wanderer). By the time of Plato (427–347 B.C.), Greek astronomers had assigned each wanderer its own solid, transparent sphere, whose rotation accounted for most of the wanderer's motion relative to the firmament. Eudoxus (408–355 B.C.), one of the most original mathematicians of his day, was able to represent the wandering motions more accurately by giving each wanderer several spinning, concentric spheres ingeniously coupled together. His pupil Callipus added more spheres, and Plato's pupil Aristotle (384–322 B.C.) added still more, bringing the total to 55.

The firmament or "sphere of the fixed stars" still appears in Copernicus' picture of the world (opposite, on right), though he shifted its center from the Earth to the Sun and attributed its daily revolution to the rotation of the Earth itself. A century later, Kepler dispensed with a solid vault, but he never relinquished the idea that all the stars are contained in a thin shell centered on the Sun.

The second cosmological motif in Lorenzo's speech is the idea that *celestial bodies have souls*. Plato believed that stars, like other self-moving bodies, have souls. Aristotle seems to have shared this belief, but in his mature doctrine the ultimate cause of the motions of stars and planets is an Unmoved Mover. Copernicus took a less abstract view in *On the Revolutions of the Heavenly Spheres* (p. 169):

> *In the middle of all sits Sun enthroned. In this most beautiful temple could we place this luminary in any better position from which he can illuminate the whole at once? He is rightly called the Lamp, the Mind, the Ruler of the Universe; Hermes Trismegistus names him the Visible God, Sophocles' Electra calls him the All-seeing. So the Sun sits as upon a royal throne ruling his children, the planets, which circle round him. The Earth has the Moon at her service. As Aristotle says, in his de Animalibus, the Moon has the closest relationship with the Earth. Meanwhile the Earth conceives by the Sun, and becomes pregnant with an annual rebirth.*

Kepler, as a young man, considered the planets to be animate beings. Later he became convinced that the cause of their motions is purely physical and resides in the Sun's rotation.

The third cosmological motif is that of *celestial perfection*. Plato and Aristotle, like other philosophers of their time, believed that the heavens are perfect in every way. From this doctrine they deduced that the heavenly bodies and their supporting structures must be composed of an

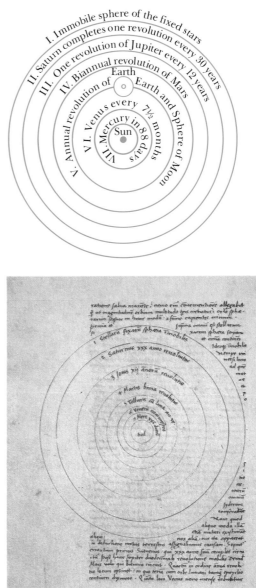

Plato and Aristotle, as portrayed in Raphael's fresco *The School of Athens* (1509–1510). Galileo and Kepler saw their own view that mathematical laws express the deep structure of reality as a continuation of the Platonic tradition in Greek philosophy, which, in their eyes, included Pythagoras and Archimedes. Aristotle, in contrast, stood for the opposing view that the deepest truths about the world are logical refinements of commonsense judgments, expressed in nonmathematical languages.

A reproduction of Copernicus' own illustration in *De Revolutionibus;* the Latin labels are translated into English in the upper drawing.

The principle of uniform circular motion. A planet Q moves with constant speed around a small circle (the *epicycle*) whose center P moves with constant speed around a larger circle (the *deferent*) centered on the stationary Earth at O.

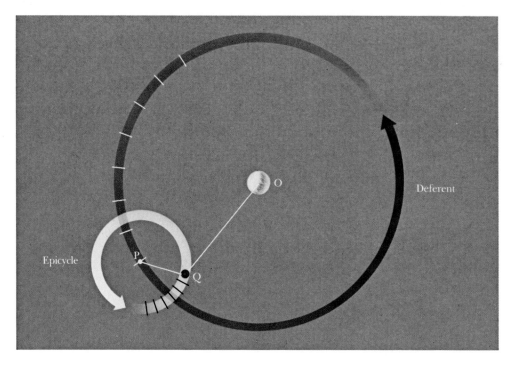

imperishable substance (the "ether") and must be spherical in shape, for the sphere is the only solid whose surface is everywhere equidistant from its center. The same doctrine led Greek mathematical astronomers of Plato's day to formulate the *dogma of uniform circular motion:* Either celestial bodies move in circles at constant speed (this is called uniform circular motion), or else their motions are a combination of two or more uniform circular motions. For example, a planet Q may travel with constant speed along a circle centered on a point P that travels with constant speed on a larger circle centered on the Earth at O (see figure above). This dogma dominated mathematical astronomy for 2,000 years. Copernicus was one of its most outspoken and inflexible adherents. Perhaps the greatest of Kepler's achievements was his invention of an alternative and, as it turned out, better way of describing planetary motions, as we will see in Chapter 2.

The fourth motif is *the music of the spheres*. Lorenzo's notion of a celestial choir in which each orb "like an angel sings" came from Plato's *Timaeus,* a late cosmological dialogue that was translated into Latin by Cicero. Timaeus was a follower of Pythagoras, who flourished during the second half of the sixth century B.C., and the *Timaeus* is largely an exposition of the cosmological ideas of the Pythagoreans, the religious sect founded by Pythagoras. Among these ideas is the notion that musical harmony and the motions of the planets are governed by closely related mathematical laws.

Pythagoras (*ca.* 580–500 B.C.), here portrayed in a French engraving of 1584 holding a magical talisman, not only proved the famous theorem that bears his name and discovered the mathematical basis of musical harmony, but also invented the doctrine that the deep structure of reality resides in mathematical relations, and founded an influential religious society based in part on that doctrine. The figure on the right shows Al-Tusi's Arabic rendering (in A.D. 1258) of Euclids proof of the Pythagorean Theorem.

Pythagoras discovered a remarkable connection between numbers and musical harmony. He found that the pitch of a vibrating string at fixed tension depends in a simple way on its length. Halving the length of the vibrating part of a violin string raises its pitch by an octave. Reducing the length by one third raises the pitch by a perfect fifth; by one quarter, a perfect fourth; by one fifth, a major third; by one sixth, a minor third. (See the figure on p. 6.) From such discoveries Pythagoras and his followers developed an elaborate theory of musical scales and harmony, whose success strengthened their belief that mathematical regularities underlie the flux of experience. The observation that there are eight heavenly spheres (one each for the Sun, the Moon, and the five bright planets, and one for the fixed stars), just as there are eight tones in the diatonic scale, seemed to confirm this belief. In the *Republic* Plato tells the story of Er, a fallen soldier restored to life by the gods after a tour of the heavens. "On each of the circles," Er reports, "stood a Siren who was carried along with it and uttered one note, and the eight notes made up a concordance." The *Timaeus* presents a theory of celestial harmony that includes an algorithm for constructing the twelve-tone scale.

The cosmological motifs popular in Shakespeare's time did not long survive the astronomical revolution. Although Copernicus himself clung to the notion of a solid firmament, his followers recognized that it was a superfluous feature of his world model. The sudden appearance and subsequent fading away of bright new stars in 1572 (Tycho's supernova) and 1604 (Kepler's supernova), along with Galileo's telescopic discovery that large black patches appear and disappear on the face of the Sun, made it clear that heavenly bodies are not changeless and perfect. The new physics pioneered by Kepler and Galileo gradually took over the explanatory role of

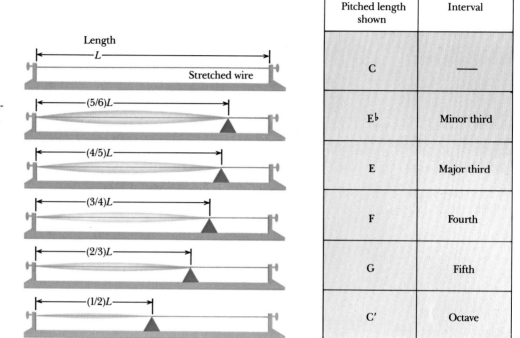

Pythagoras discovered the relation illustrated in this diagram between consonant musical intervals and the lengths of vibrating strings.

Pitched length shown	Interval
C	—
E♭	Minor third
E	Major third
F	Fourth
G	Fifth
C′	Octave

Plato's doctrine of souls. And although the notion of a permanent mathematical harmony underlying the flux of experience continued to inspire scientists, as it does to this day, no great scientist after Kepler looked for musical meanings in the motions of celestial bodies.

Not all of the ancient cosmological motifs were made obsolete by the astronomical revolution, however, and some of them play as prominent a part in present-day cosmology as they did in Greek philosophy. One of these persistent motifs is that of *Void versus Plenum.* Greek philosophers speculated about the ultimate constituents of the world and, characteristically, divided into two opposing schools. Leucippus and Democritus, who flourished at the end of the fifth century B.C., held that matter consists of tiny, invisible atoms moving in otherwise empty space. They attributed differences in the sensible properties of objects (color, taste, texture, hardness, temperature) to differences in atomic composition. The opposing view, much more widely held, was that the world is filled with one or more continuous substances. Aristotle, for example, believed that the part of the world contained within the sphere of the Moon is made up of earth, water, air, and fire, but that the heavens are made of a fifth substance, ether. According to this view, the ether is unchanging and indestructible, but the other four substances can be transformed into one another. For example, water, which is cold and moist, is transformed into air, which is hot and moist, by heating, and air is transformed into fire by drying.

The scientific revolution revived rather than resolved the conflict between those who saw the world as a Void thinly populated by invisible atoms and those who saw it as a Plenum filled with continuous, indestructible fluids. Isaac Newton (1642–1727) was a thoroughgoing atomist. He believed that both light and matter consist of invisible particles, and that heat consists in their motion. He also believed in an all-pervading ether consisting of exceptionally light and fast-moving particles. René Descartes (1596–1650), on the other hand, believed in a continuous ether that completely fills the space not occupied by solid bodies and mediates their interactions by means of a system of vortices.

For Descartes, as for Aristotle, the doctrine of the Plenum rested on two self-evident truths: Nature abhors a vacuum, and there can be no action at a distance. During the second half of the seventeenth century and well into the eighteenth, the followers of Descartes defended these and other tenets of his cosmology against the "mathematical philosophy" of Newton. Descartes' cosmology was qualitative, complex, and full of *ad hoc* explanations. In contrast, Newton's of-

Hellenistic bronze bust of Democritus (*ca.* 460–370 B.C.), who developed Leucippus' (fifth century B.C.) doctrine that the world is made up of indestructible, unchanging atoms moving in a Void. This was also Newton's picture of the world.

"Having darkened my chamber, and made a small hole in my window-shuts, to let in a convenient quantity of the Sun's light, I placed my Prism at his entrance, that it might thereby be refracted to the opposite wall." By this and other experiments, Isaac Newton demonstrated in 1666 that white light is made up of differently colored rays, each with its own "refrangability." This engraving was made in the nineteenth century.

René Descartes. He and Pierre de Fermat (1601–1665) independently invented analytic geometry. Descartes' cosmology, which rivaled Newton's in popularity, rested on the premises that the world is a Plenum, that there is no action at a distance, and that all physical properties are reducible to extension and motion.

fered a complete and exact description of the world based on a few simple mathematical laws, and the agreement between its predictions and observation was nothing short of spectacular; yet Newton himself found the concept of action at a distance hard to accept.

In the nineteenth century new physical entities came to fill the Void: the gravitational field, light waves, electric and magnetic fields, electromagnetic radiation, and a new ether to support electromagnetic waves. Twentieth-century physics abolished the ether but filled space with brand-new fields and endowed it with new kinds of structure. The "vacuum" of present-day physics is so complex that a complete, self-consistent description of it still eludes theorists.

Another ancient theme that remains as fundamental to modern science as it was to the cosmologies of the Greek atomists and of Aristotle is the idea that *permanence lies at the heart of change.* According to modern physics, neither atoms nor subatomic particles are indestructible. An electron and its antiparticle, the positron, may come together and disappear in a flash of light, but the *properties* of an electron, which constitute what Plato would have called its Form or Idea, are universal and unchanging. The modern analogues of earth, air, fire, and water are *conserved quantities,* of which energy and electric charge are the most familiar.

Epicurus (*ca.* 340–270 B.C.), who carried on the atomistic philosophy of Leucippus and Democritus, developing it into both an ethical theory and a cyclic cosmology. Epicurus' philosophy is set out in Lucretius' (*ca.* 96–95 B.C.) poem *De Rerum Natura* (*On the Nature of Things*).

Finally, the conflict between *centrism* and *uniformity* figures in both ancient and modern cosmologies. Are we at the center of the universe? The world of Plato and Aristotle was like an onion, with the Earth at its core and the sphere of the fixed stars as its skin. Leucippus and Democritus, on the other hand, held that the stars are distant suns, scattered throughout an infinite space. The historian of science Alexandre Koyré has argued, in *From the Closed World to the Infinite Universe,* that a transition between these two world views was a key element in the scientific revolution. But did a transition actually take place?

The writings of Saint Thomas Aquinas in the thirteenth century removed Christian theology from its Platonic setting, which had evolved during the preceding millennium, and based it instead on Aristotle's philosophy, which thus took on religious overtones. The resulting cosmotheology is enshrined in Dante's *Inferno* and Milton's *Paradise Lost.* The rival cosmological doctrine did not become known in Europe till the fifteenth century, and then only at third hand. Lucretius, a Roman philosophical poet who flourished in the first half of the first century B.C., was a disciple of the philosopher Epicurus (*fl. ca.* 300 B.C.), who borrowed his physical and cosmological ideas from Leucippus and Democritus. The manuscript of Lucretius' long philosophical poem *De Rerum Natura* (*On the Nature of Things*) turned up in 1417. Lucretius described the universe as a Void, thinly but uniformly populated by "completely solid indestructible particles of matter flying about through all eternity." The universe, he said, has no center and contains infinitely many populated worlds. Moreover, "nature is free and uncontrolled by proud masters, and runs the universe by herself without the aid of gods."

Lucretius' most influential disciple was Giordano Bruno (1548–1600), whose checkered career was marked, and eventually terminated, by conflict with ecclesiastical authorities. As a Dominican friar he was accused of heresy and forced to leave the order. In the course of his subsequent wanderings he was imprisoned by the Calvinist authorities in Geneva, "excommunicated" by the Lutherans in Germany, and finally, in February 1600, burned alive by the Inquisition after an imprisonment of seven years. Bruno's cosmotheology was eclectic rather than original: It combined Lucretius' views on the infinity of space and the multiplicity of inhabited worlds with mystical, magical, and animistic doctrines drawn from the occult tradition attributed to Hermes Trismegistus. Bruno believed that stars and planets have souls, and he ridiculed the notion that mathematical reasoning could contribute to an understanding of the heavens.

As we have seen, Copernicus believed that the universe is a sphere with the Sun at its center, and Kepler believed that the stars are confined to a thin spherical shell centered on the Sun. Galileo was prudently noncommittal on the size and shape of the stellar universe. In his *Dialogue on the Two Chief World Systems* he tactfully allowed Simplicio, the interlocutor who speaks for Aristotle and the Church, to have the decisive say on the subject:

> SALVIATI. *Now what shall we do, Simplicio, with the fixed stars? Do we want to sprinkle them through the immense abyss of the universe, at various distances from any predetermined point, or place them on a spherical surface extending around a center of their own so that each of them will be the same distance from that center?*
>
> SIMPLICIO. *I had rather take a middle course, and assign to them an orb described around a definite center and included between two spherical surfaces—a very distant concave one, and another closer and convex, between which are placed at various altitudes the innumerable host of stars. This might be called the universal sphere, containing within it the spheres of the planets which we have already designated.*
>
> SALVIATI. *Well, Simplicio, what we have been doing all this while is arranging the world bodies according to the Copernican distribution, and this has now been done by your own hand.*

Giordano Bruno (1548–1600), who antagonized religious and civil authorities in Europe and Britain by preaching the infinity of space and of inhabited planets.

In a letter written in 1640, two years before his death, Galileo refers to the question whether the world is finite or infinite as "one of those questions haply inexplicable by human reason, and similar perchance to predestination, free will, and such others in which only Holy Writ and divine revelation can give an answer to our reverent demand." Descartes believed that the world is *indefinite* (or, to use the modern term, *unbounded*) but not infinite. Newton believed that the stars are suns, and that they condensed from matter that was initially "evenly disposed throughout an infinite space." Newton's universe is not only unbounded but uniform. The average properties of Newton's world are the same for every observer, no matter where situated: At every point they are the same in all directions. Newton's universe is also the universe of Lucretius and of Bruno, who wrote that "the center is everywhere, the circumference nowhere." Descartes accepted the second part of this aphorism but stopped short of affirming the first part. To that extent his cosmology remained Aristotelean.

Although Descartes' physics was no match for Newton's, Aristotelean centrism did not give way quickly or easily to the picture of the world that Newton had inherited from the Greek atomists. During the eighteenth and nineteenth centuries and well into the twentieth, most practical astronomers—those who actually looked through telescopes and counted and measured—took their ideas about the structure of the world from Copernicus, Kepler, and Galileo rather than from Newton. They soon learned that the stars are suns; yet they continued to believe (for what at the time seemed excellent observational reasons) that the Sun is at or close to the center of the world. When in 1918 Harlow Shapley discovered that the Sun is close to the edge of a vast sphere outlined by globular star clusters, he moved the center of the universe to the center of this sphere. The first observational astronomer to make a convincing case for Newton's postulate of cosmic uniformity was Edwin Hubble, in 1929. But the opposite hypothesis—that the world has a center, and that we are close to it—was defended by some cosmologists until the late 1960s.

Description *Versus* Explanation

If the scientific revolution of the seventeenth century did not simply replace ancient cosmological themes by more "scientific" ideas, what *did* it accomplish? How does modern scientific cosmology differ from older mythical descriptions of the world and its origin? Most physicists and astronomers would probably answer this question in roughly the following way.

Scientific cosmology and its underlying physical theories are not merely more or less plausible stories about the world. They are *true* stories. Of course, not everything scientists now believe about the world will turn out to be true. History shows that scientific ideas about the world are continually changing. Their underlying physical theories—the theory of gravitation, for example—change too, though not as quickly. But both physical and cosmological theories have a core of truth. That core is continually growing, and there is no limit to how much we can learn about the physical universe or to how certain that knowledge can become.

This view of scientific knowledge was enunciated very clearly and forcefully by Galileo and Descartes early in the seventeenth century, and most working scientists have accepted it more or less uncritically ever since. It has been less well received by historians and philosophers.

The Scottish philosopher David Hume (1711–1766) constructed a powerful logical argument demonstrating that knowledge of causal connections between observed events is impossible if, as he postulated, all knowledge derives from experience. (The gist of the argument is that experience can never tell us that event *A* is the cause of event *B*; we can observe only that *A* habitually precedes *B*.) The German philosopher Immanuel Kant (1724–1804) tried to rescue science from Hume's critique by arguing that scientific knowledge does have universally valid features, but that these do not derive from experience. We do not abstract our notions of space and time from experience (as Hume and Aristotle believed). Rather, said Kant, we cannot avoid experiencing the outer world "in space" and the inner world "in time." Space and time are "forms" of outer and inner experience, respectively—filters built into the human perceptual apparatus. Along with these "forms of sensibility," Kant posited "categories of understanding" (cognitive filters, as it were) that are responsible for at least the deep structure of what we call physical laws. Pythagoras' theorem and Newton's law of gravitation are truths that structure experience but are not, according to Kant, derived from experience. They are truths not about the world, but about the way we interact with an essentially unknowable world.

Ernst Mach (1838–1916), a physicist, historian of science, and philosopher in the empiricist tradition of Locke, Berkeley, and Hume, regarded physical laws as nothing more than economical ways of representing the outcomes of past experiments and predicting future outcomes.

In modern times the most influential view of science has been that of Ernst Mach (1838–1916), who also made important contributions to physics and psychology. Mach was a positivist; that is, he believed, like Francis Bacon (1561–1626) and August Comte (1798–1857), that science consists in the methodical gathering and arrangement of facts, and that "scientific method" is the only source of genuine knowledge. Mach argued that laws are nothing more than concise summaries of past experience, useful in predicting future observations. Theories, according to Mach, do not *explain* phenomena; they merely *describe* them. To say that a successful theory expresses some kind of truth behind phenomena is to indulge in metaphysical speculation, for such a statement cannot be tested by observation or experiment.

Mach's refusal to credit scientific theories with any deep meaning is echoed by some contemporary historians and philosophers of science who in other respects have broken with the positivist tradition. For example, in *The Structure of Scientific Revolutions*, the historian Thomas Kuhn writes (pp. 206–207):

> One often hears that successive theories grow ever closer to, or approximate more and more closely to, the truth. . . . I do not doubt . . . that Newton's mechanics improves on Aristotle's and that Einstein's improves on Newton's as instruments for puzzle-solving. But I can see in their succession no coherent direction of ontological development. On the contrary, in some important respects, though by no means in all, Einstein's general theory of relativity is closer to Aristotle's than either of them is to Newton's.

The conflict between the naive view of science as a growing body of truth and the positivist view of science as mere description did not begin with Mach in the nineteenth century, or even with Berkeley and Hume in the eighteenth. It goes back to the dawn of modern science. Copernicus' great book *De Revolutionibus Orbium Caelestium* (*On the Revolutions of the Heavenly Spheres*), published in 1543, a few months after its author's death, opens with a message "To the Reader Concerning the Hypotheses of This Work," which explains the positivist view of mathematical astronomy and instructs the reader to view the work that follows in that light:

> It is the job of the astronomer to use painstaking and skilled observation in gathering together the history of the celestial movements, and then—since he cannot by any line of reasoning reach the true causes of these movements—to think up or construct whatever causes or hypotheses he pleases such that, by the assumption of these causes, those same movements can be calculated from the principles of geometry for the past and for the future too. . . . And if [mathematical astronomy] constructs and thinks up causes—and it has certainly thought up a good many—nevertheless it does not think them up in order to persuade anyone of their truth but only in order that they may provide a correct basis for calculation. . . . And as far as hypotheses go, let no one expect anything in the way of certainty from astronomy, since astronomy can offer us nothing certain.

This passage was written not by Copernicus but by his friend Andrew Osiander, a Lutheran clergyman—a fact uncovered by Kepler half a century later. Osiander's motives are clear. Luther had already denounced Copernicus' ideas as contrary to Scripture, observing that

"Joshua commanded the Sun to stand still, and not the Earth." Osiander's preface was intended to reassure Copernicus' readers that mathematical astronomy is not concerned with truth, that its sole purpose is to describe—not explain—the motions of the Sun, the Moon, and the planets.

Several modern historians and philosophers of science have endorsed Osiander's view of *De Revolutionibus*. But Copernicus himself viewed the heliocentric theory in a different light. In his dedication to Pope Paul III, he explains why he became dissatisfied with geocentric world models. They are unsatisfactory, he says, because different models use different computational devices to achieve the same end, and none of them adheres strictly to the principle of uniform circular motion.

Moreover, they have not been able to discover or to infer the chief point of all, i.e., the form of the world and the certain commensurability of its parts. But they are in exactly the same fix as someone taking from different places hands, feet, head, and the other limbs—shaped very beautifully but not with reference to one body and without correspondence to one another—so that such parts make up a monster rather than a man.

Having discovered that the Pythagorean school of Greek astronomers had attributed motions to the Earth, he was encouraged to try out the same idea, and so

. . . finally discovered by the help of long and numerous observations that if the movements of the other wandering stars are correlated with the circular movement of the Earth, and if the movements are computed in accordance with the revolution of each planet, not only do all their phenomena follow from that but also this correlation binds together so closely the order and magnitudes of all the planets and of their spheres or orbital circles and the heavens themselves that nothing can be shifted around in any part of them without disrupting the remaining parts and the universe as a whole. . . . in this ordering we find that the world has a wonderful commensurability and that there is a sure bond of harmony for the movement and magnitude of the orbital circles such as cannot be found in any other way.

Thomas Kuhn in *The Copernican Revolution* recognizes that Copernicus' arguments for the truth of his system hinge on mathematical harmony. But, he writes,

"Harmony" seems a strange basis on which to argue for the earth's motion, particularly since the harmony is so obscured by the complex multitude of circles that make up the full Copernican system. Copernicus' arguments are not pragmatic. They appeal, if at all, not to the utilitarian sense of the practicing astronomer but to his aesthetic sense and to that alone. . . . [They appealed] primarily to that limited and perhaps irrational subgroup of mathematical astronomers whose Neoplatonic ear for mathematical harmonies could not be obstructed by page after page of complex mathematics leading finally to numerical predictions scarcely better than those they had known before.

But is truth simply an illusion in the minds of mathematicians who happen to have certain aesthetic sensibilities? Or is harmony, as mathematicians perceive it, a mark of truth?

Mathematical Harmony as a Criterion for Scientific Truth

Harmony and coherence are criteria for truth in everyday life as well as in science. I look across a crowded room and say to myself, "That's Jane." I have formulated a hypothesis that unites a few fragmentary impressions with a "bond of harmony." My hypothesis may suggest new observations. I recall that Jane has a penchant for chartreuse; so when I notice that the figure across the room is wearing a chartreuse scarf, I regard my hypothesis as confirmed.

Scientific generalizations have a similar character. I observe that a flash of lightning is always followed—sometimes immediately, sometimes after an interval of several seconds—by a thunderclap. But I am not content to infer from repeated observations of this kind the general rule "Thunder always follows lightning." Knowing that light travels faster than sound, I formulate the unifying hypothesis that the lightning flash and thunderclap are caused by a single atmospheric event, and I check this hypothesis by noticing how the interval between the lightning and the thunder is related to the elevation of the flash and to the loudness of the clap.

A good scientific theory or hypothesis, like the denouement of a well-constructed detective story, not only fits all the facts but also reveals unexpected connections between apparently unrelated phenomena. A scientific theory that manages to do this may carry more conviction than individual facts. "Theories are like balloons floating on the surface of the sea, and facts are like battleships," said Sir Arthur Stanley Eddington. "Occasionally a balloon collides with a battleship—and the battleship goes down."

From the time of Hipparchus in the second century B.C. to Tycho Brahe at the end of the sixteenth century, the angular diameters of stars (their apparent sizes on the bowl of the night sky) were regarded as firm data. For example, Tycho gave the angular diameter of a first-magnitude star as two minutes of arc—one fifteenth the angular diameter of the Sun or Moon. This venerable battleship collided first with Aristarchus' theory and then with Copernicus', but managed to stay afloat until Galileo looked at the stars through his telescope and found that even their *magnified* images were much smaller than two minutes of arc.

The harmony that scientists perceive in all modern physical theories has a special character: it is mathematical. Pythagoras' vision of an unchanging mathematical order underlying the flux of experience is an important element of Plato's philosophy, and may even have inspired his belief in a world of eternal, unchanging Forms, the objects of true knowledge. But the Pythagorean view of the world found its most direct expression in the mathematical world models of such astronomers as Philolaus, Ecphantus, Heraclides, and Aristarchus, and in the mathematical and physical theories of such mathematicians as Euclid and Archimedes. As we will see in Chapters 2 and 3, Copernicus, Kepler, Galileo, Huygens, and Newton regarded themselves as the inheritors of the scientific tradition founded by these men, a tradition in which mathematical regularities were not, as Aristotle believed, abstractions from the surface appearances of things but the very heart of reality.

Thus, in his preface to *De Revolutionibus*, Copernicus dismisses in advance the criticisms of "idle babblers, ignorant of mathematics," but expresses confidence that "gifted and learned

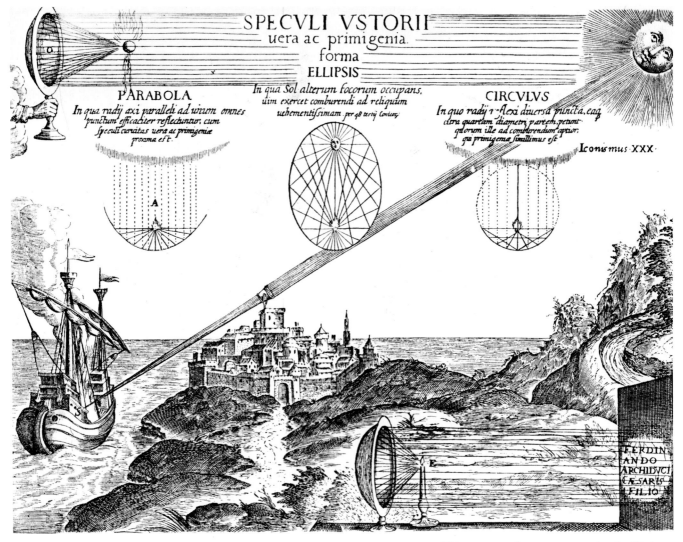

This seventeenth-century engraving by Kircher illustrates Archimedes' invention of a system of mirrors to focus the Sun's rays on a hostile ship (and thus set it afire). A mirror whose surface is formed by rotating a parabola about its symmetry axis focuses parallel rays. A spherical mirror brings a bundle of parallel rays coming from any direction to an approximate focus. An ellipsoidal mirror causes rays produced at one focus to converge on the other focus.

Gottfried Wilhelm von Leibnitz (1646–1716), Newton's contemporary, independently invented the differential calculus at the same time as Newton. He differed sharply with Newton on the nature of space and time, which he held to be "relational" rather than absolute.

mathematicians will agree" with the theory if they take the trouble to study it thoroughly. "Mathematics," he remarks, "is written for mathematicians." Galileo, replying to exactly the kind of criticism that Copernicus had anticipated, made the same point more elegantly in the passage quoted on the title page of this chapter.

Galileo took his inspiration directly from Archimedes' work. In Alexandre Koyré's apt phrase, he apprenticed himself to Archimedes. Galileo's "two new sciences," like Archimedes' statics and hydrostatics, are addenda to Euclid's *Elements* and Apollonius' *Conics*. Although more than eighteen centuries separate Archimedes from Galileo, there is an obvious continuity of spirit, style, and substance between Archimedes' *On Floating Bodies* and Galileo's *Dialogues on Two New Sciences*.

Newton, too, believed that physical laws are irreducibly mathematical. Replying to critics who complained that his theory of gravitation contradicted philosophical principles laid down by Descartes and Leibniz—in particular, the principle that two bodies can act on one another only if they are in contact—Newton said, "To us it is enough that gravity does really exist, and acts according to the laws which we have explained, and abundantly serves to account for all the motions of the celestial bodies, and of our sea."

The Growth of Science

Human culture grows by a process that both resembles and extends organic evolution. The ability to invent and transmit culture is no less—and no less important—an evolutionary "adaptation" than upright posture or binocular vision. Of course, the languages, customs, social organizations, tools, and works of art that constitute culture are not, strictly speaking, evolutionary adaptations; after all, they are not encoded in DNA. Yet, as many evolutionary biologists (Ernst Mayr, Theodosius Dobzhansky, and G. Ledyard Stebbins among them) have emphasized, there is a deep parallel between organic and cultural evolution. Natural science is analogous to a specialized organ in a complex organ system—human culture as a whole.

Three key aspects of organic evolution have more or less precise counterparts in the growth of science: the conservation of successful structures and functions; the rapid evolution of qualitatively new structures and functions that make possible a new mode of interaction between the population and its environment; and the evolutionary dynamics of variation and selection.

Consider first the conservation of successful structures and functions. New biological structures do not as a rule replace the structures that preceded and prepared the way for them; instead, they assimilate them. Some adaptations present in the earliest known forms of life have persisted in nearly all their descendants. For example, all known forms of life from the amoeba to the higher plants and animals use the same basic chemical strategy: The information needed to specify the development of an organism is encoded in one form of nucleic acid (DNA), transcribed into another form (RNA), and translated, according to a universal code, into highly specific organic catalysts (enzymes). The chemical machinery involved is essentially identical in all living organisms. So too are the molecules and the complex cycles and chains of chemical reactions through which foods are broken down and their usable energy is extracted and stored. Other invariant structures emerge at higher levels of biological organization, as described by G. Ledyard Stebbins (p. 126):

> *the muscular foot of the molluscs; the segmentation patterns of higher arthropods, consisting of three body segments and six legs in insects and two segments with eight legs in spiders and their relatives; two pairs of limbs with comparable bones in all land vertebrates; seven neck vertebrae in all mammals, including the long necked giraffe as well as the almost neckless whale.*

Likewise, successful features of science's theoretical framework may be modified as the framework evolves, but they are rarely, if ever, discarded. At the deepest level, the mathematical language of physical laws is analogous to the chemical language of basic biological processes. And the difference between the Aristotelean view of natural philosophy and the Archimedean view (philosophical principles *versus* mathematical axioms) is as profound as the difference between the view that life is the manifestation of an irreducible vital force and the view that life at its most elementary level is chemistry. This is not to say that life is *nothing but* chemistry. From the physicist's standpoint as well as from the biologist's, there is more to, say, seeing than the sequence of chemical reactions that begins when a rhodopsin molecule absorbs a photon.

The conservative aspect of science is as evident in the substance of physical laws as in their language. Euclidean geometry, itself an accretion of conserved structures and methods, remains central to present-day physical theories, even though its immediate context has undergone several radical changes. Galileo and Newton made time a fourth dimension of the world, embedding Euclidean space in Galilean spacetime (as we will see in Chapter 3). Einstein and Minkowski modified the geometry of spacetime, but Galilean spacetime survived as a limiting case of Minkowskian spacetime (Chapter 5). Einstein's general theory of relativity reduced Minkowskian spacetime itself to a limiting case, valid locally in a freely falling reference frame (Chapter 6). Finally, relativistic cosmology (Chapter 7) revealed the "global" properties of Einsteinian spacetime.

Much the same story can be told about theories of gravitation, whose development is, indeed, inseparably linked to that of theories of space and time. Archimedes' principles of statics and hydrostatics and Galileo's theory of motion became limiting cases of Newton's theory of gravitation, which in turn became a limiting case of Einstein's theory. The modern mechanical engineering curriculum owes a great deal to Archimedes nothing to Einstein.

Next, let us consider the role of innovation in evolution. Major evolutionary advances such as temperature regulation in mammals and the biological capacity for language in humans involve many intricately coordinated changes. We might therefore expect them to have taken place slowly, over long periods of time. In fact, the fossil record shows that exactly the opposite is true. Major adaptive shifts occurred very rapidly, during such short periods of time that transitional forms are usually scarce or absent in the fossil record. By contrast, relatively minor evolutionary changes—for example, increases in body size—often occurred very slowly, over long periods of time.

The growth of science shows a similar alternation of tempos. Long periods during which a theory is elaborated and used to interpret experiments and observations are punctuated by brief periods during which the theory undergoes a major qualitative change. It is fashionable to refer to major qualitative changes in the theoretical framework as "revolutions," but the distinction between revolution and evolutionary innovation is worth preserving. A successful scientific theory like Archimedes', Galileo's, or Newton's is not overthrown by a competing theory in the way that a government or constitution is overthrown in a political revolution. It lives on, more or less altered, within a new, more inclusive theoretical framework. On the other hand, the scientific revolution of the sixteenth and seventeenth centuries *was* a genuine revolution. It deposed Aristotelean natural philosophy and put in its place the radically different natural philosophy of Galileo and Newton. It replaced philosophical explanations by mathematical laws. Copernicus was seen by Kepler and Galileo as a revolutionary, but not as an innovator.

Finally, let us consider the dynamics of the evolutionary process. Evolutionary innovations result from an interplay between two processes: genetic variation and differential reproduction (or natural selection, as Darwin called it). Random processes associated with reproduction constantly create new genes and gene combinations, enabling a biological population to explore a variety of potential evolutionary pathways. Usually this process goes unrewarded, but occasionally there is a breakthrough. Under the combined action of undirected genetic variation and natural selection, an evolutionary pathway may open out into a broad thoroughfare leading to a new adaptation.

The growth of science may be described in roughly similar terms. By means of experiment, observation, and theoretical construction, scientists continually search for new evolutionary pathways. The search is certainly not aimless, but in a certain sense it is blind, for no one can know beforehand which paths will lead to new knowledge and which will terminate in dead ends. Scientists rarely agree about which directions ought to be explored, and their preferences often reflect philosophic, aesthetic, or merely personal prejudices. As the historian of science Paul Feyerabend has emphasized, such prejudices promote the growth of science. They motivate the individual scientist as no "correct" methodology could do, and they ensure a diversity of exploratory activity that is as essential to scientific progress as it is to organic evolution.

The *fitness* of a gene or cluster of linked genes is measured by its contribution to the reproductive success of its carriers in a biological population. Thus, the fitness of a gene relative to that of its variants in a given population is a measure of its long-term prospect for survival. What determines the equivalent prospect for a scientific theory? The fitness of a scientific theory seems to depend mainly on three closely related factors: its accuracy, its scope, and the extent to which it is overdetermined by experience. In the short term, of course, a theory's

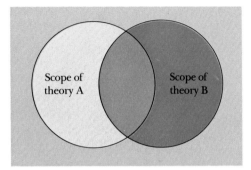

Highly fit theories have always been related in one of the three ways illustrated below, never in the way illustrated at the right.

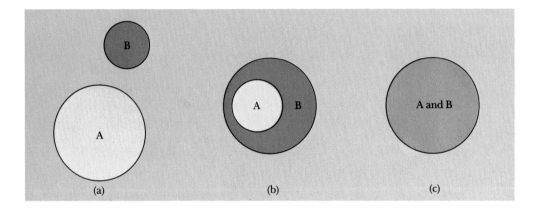

(a)   (b)   (c)

reception may be more strongly influenced by other factors: fashion, ideology, or general obtuseness. And the short term may be of long duration. Aristarchus' model of the solar system escaped the notice of astronomers for seventeen centuries, and Kant's theory of the Milky Way (Chapter 4) failed to capture their attention for nearly two centuries.

The *scope* of a theory is the range of phenomena about which it makes predictions. In principle, two theories could overlap in scope (see above, top): here theory A and theory B both make predictions about certain phenomena, but each also makes predictions about phenomena that lie outside the other's scope. Historically, highly fit theories have never overlapped in this fashion. Instead, they have always been related in one of the three ways shown above. In situation (a), theory A and theory B refer to distinct phenomena; Archimedes' theory of statics and Galileo's theory of motion illustrate this case. A later theory, such as Newton's theory, may include A and B in their entirety as limiting cases. In situation (b), theory B may include theory A, just as Newton's theory includes Galileo's, and Einstein's includes Newton's. In situation (c), two theories may have exactly the same scope. For example, a theory of gravitation with exactly the same scope as Newton's can be constructed by replacing his inverse-square law of gravitation by some other law. (Such a modification was seriously considered by nineteenth-century mathematical astronomers.)

The final factor that affects the fitness of a theory is *overdetermination*. Every theory has a certain number of adjustable parameters, some of which define initial conditions; others, properties of the system under consideration. For example, the adjustable parameters in Newton's theory of the solar system are the masses of the Sun, the planets, and their moons, and the position and velocity of each of these bodies at some moment in time. The values of these adjustable parameters are *determined* by an equal number of independent data, for example, measurements of position on the celestial sphere. When the number of independent data exceeds the number of adjustable parameters, the theory is said to be overdetermined by the data. Overdetermination has a qualitative as well as a quantitative aspect, and the qualitative aspect is the more important. Newton's theory is strongly overdetermined by experiment and observation, not only because it furnishes very accurate predictions of planetary motions but also because (as we will see in Chapter 3) it explains, at no extra cost, a host of qualitatively different phenomena, including the ocean tides, the precession of the equinoxes, and the rings of Saturn.

(Economy and simplicity contribute to overdetermination, but these criteria do not always agree. As we will see, Einstein's theory of gravitation is more strongly overdetermined than Newton's, but it is certainly not simpler, nor can it be said to provide a more economical representation of astronomical observations.)

The three factors of scientific fitness—accuracy, scope, and overdetermination—are closely related. The progression Galileo–Newton–Einstein is a progression in all three. Galileo's theory applies to freely falling bodies and projectiles near the surface of the Earth. Its accuracy is limited by its neglect of air resistance, the curvature of the Earth's surface, and the variation of gravity with height. Newton's theory not only removes these limitations but also embraces a much wider range of phenomena, from the fall of an apple to the dynamics of astronomical systems a hundred million million times as massive as the Sun. Yet Newton's theory rests on very simple mathematical laws, which, accordingly, are strongly overdetermined by the phenomena they account for. Einstein's theory is significantly more overdetermined than Newton's because it fuses three distinct Newtonian concepts—inertial mass, gravitational mass, and energy—into a single concept. Experiment and observation have shown that Einstein's theory is also more accurate than Newton's. Finally, Einstein's theory is significantly broader in scope than Newton's. Newton's law of universal gravitation applies to all self-gravitating systems, from planets and their moons to clusters of galaxies, but it does not apply to the universe as a whole (Chapter 4); Einstein's theory does.

So far we have been considering similarities between the growth of science and biological evolution. We now come to an important difference. Evolution is a multiply branching process, and so is science in the making, but there is a striking difference between the outcomes of the

The shell of theory. Its expanding outer boundary represents
predictions; its contracting
inner boundary, the theory's
most basic principles, which
communicate with the outer
boundary via intermediate theories derived from the basic
principles.

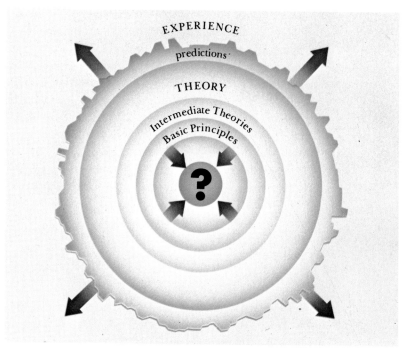

two processes. From simple beginnings in a highly specific environment, evolution has created
an enormous diversity of environmental niches and filled them with millions upon millions of
distinct species. From equally simple beginnings—Thales' use in the sixth century B.C. of similar
triangles to measure the distance of a ship at sea, Pythagoras' arithmetical laws of musical
harmony—science has woven a web of explanation that encompasses the ultimate constituents
of matter, the astronomical universe, and the phenomenon of life. As science grows, theories
proliferate. But those that survive—those that prove to be strongly overdetermined by the
phenomena—are related not only by descent from common ancestors but also structurally and
functionally, like organs in an organism. There are no sharp dividing lines between biology and
chemistry, between chemistry and physics, or between physics and cosmology. And as scientific
theories grow in scope and accuracy, they tend to become more tightly and more deeply connected. Newton's theory exhibited Archimedes' statics and Galileo's theory of motion as details
in a larger pattern. Einstein, in his special theory of relativity, welded space and time into a
single four-dimensional continuum, and merged mass and energy. His general theory merged
spacetime with matter/energy.

   We may think of theories as occupying an expanding spherical shell embedded in the sea of
experience. As the fundamental laws become fewer and deeper, the shell's inner surface con-

tracts—perhaps toward a core of ultimate scientific truth. Does such a core—Plato's "first principle"—really exist? No one knows, but a belief in its existence has undoubtedly motivated most theory-builders from Pythagoras' time to the present day.

CHAPTER TWO

ARISTARCHUS, COPERNICUS AND KEPLER

All nature and the graceful sky are symbolized in the art of Geometria.
KEPLER, *Tertius Interveniens*

To a naive observer unencumbered by scientific knowledge, the sky seems to be an inverted bowl resting on a flat plate. Sprinkled over the inner surface of the bowl are the stars, arranged in fixed, identifiable patterns that do not change noticeably from day to day, year to year, or even century to century. During the course of a single night, the stars wheel around a fixed point, as if the bowl were spinning on an axis passing through this point and the observer's position. On a time-exposure photograph made with a camera pointed toward the fixed point, each star traces out an arc of a perfect circle (top left figure on p. 24). Careful observation shows that the bowl rotates at a uniform rate, completing one revolution in just under a day—23 hours and 56 minutes, to be exact.

The bowl's rotation causes some stars to sink below the western horizon as others rise in the east, suggesting that the bowl is part of a complete sphere covered all over with stars. This sphere is called the *celestial sphere* (top right figure on p. 24). The two fixed points where it is attached to the rotation axis are called the *celestial poles*. The imaginary circle whose points are equidistant from the two poles is called the *celestial equator*. The celestial sphere is thus a greatly expanded and idealized model of the surface of the Earth. It is concentric with the Earth; its rotation axis is an extension of the Earth's polar axis; and the celestial equator is the circle in which the Earth's equatorial plane, extended, meets the celestial sphere.

But the celestial sphere came first. It is much easier to "deduce" its existence from the daily motions of the stars than to arrive at the notion of a round Earth. Greek philosophers formulated the hypothesis of a round Earth at the beginning of the fifth century B.C. They were led to it by reports of travelers who noticed that the elevation of the North Celestial Pole increased as they traveled northward and decreased as they traveled southward. The bottom figure on p. 24 shows how the hypothesis of a round Earth explains this observation.

The hypothesis of a round Earth is strongly overdetermined by observation, for it contains a single adjustable parameter—the Earth's radius—and makes two strong and logically independent predictions: that the change in elevation of the celestial pole will be strictly proportional to the distance traveled northward or southward; and that the elevation of the pole will not change at all if one travels eastward or westward.

Time-exposure photograph of the southern sky, made at the Anglo-Australian Observatory in Australia. The circular star trails are centered on the South Celestial Pole.

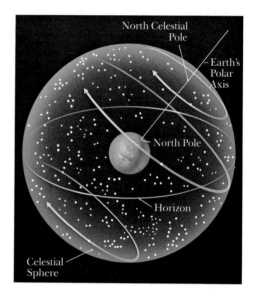

(*Above right*) The diurnal motions of the stars along circles concentric with the North Celestial Pole are explained by the hypothesis that the stars are sprinkled on the surface of a transparent sphere rotating about an axis passing through the center of the Earth and piercing the celestial sphere in the North Celestial Pole.

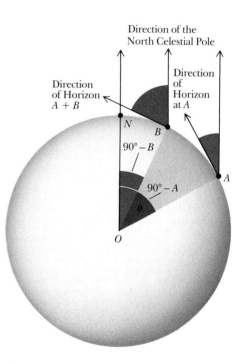

One can estimate the Earth's circumference by measuring the change in elevation of the North Celestial Pole between two points. O is the center of the Earth, N the north pole, A and B two points on the same meridian. The difference between the (marked) elevation angles at A and B is equal to the angle $AOB = \theta$; $AB/OA = 2\pi\,(\theta/360°)$.

The Sun's daily and annual motions. The horizon is drawn for an observer in a northern midlatitude.

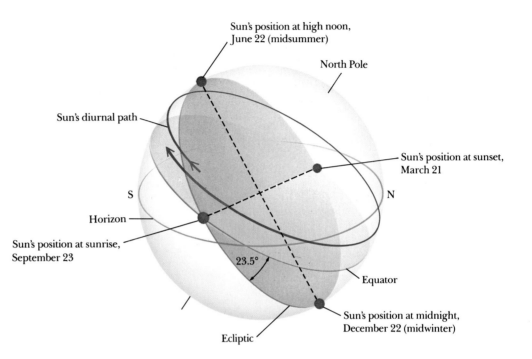

The hypothesis also enabled Greek mathematicians to estimate the radius of the Earth. Using the method explained in the bottom legend on p. 24, the Greeks obtained progressively more accurate estimates of the Earth's radius and circumference. Aristotle quotes a value 60 percent larger than the modern one, and Eratosthenes' estimate in the following century had an error of only about 1 percent.

The fact that the celestial sphere completes a revolution in slightly less than one day implies that the Sun's position on the sphere is not fixed. If it were, the interval between two successive risings of the Sun would be the same as the interval between two successive risings of a star. The fact that the first interval is on average four minutes longer than the second implies that the Sun moves along the celestial sphere in a direction opposite to the direction of rotation, so that its rising is delayed each day relative to the rising of the stars. By noticing the Sun's position among the stars at sunrise (or sunset) each day, we can plot its path on the celestial sphere. This path turns out to be another circle, centered on the center of Earth, and inclined at a moderate angle (23.5°) to the celestial equator (see above). On this circle, called the *ecliptic*, the Sun creeps from west to east at a nearly uniform rate of about one degree (twice its own angular diameter) per day, taking 365 days and almost six hours to complete the circuit.

Using the figure above, we can make predictions about seasonal changes in the Sun's elevation at noon and in the relative lengths of day and night. When the Sun is at either of the two

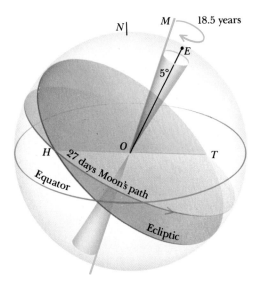

The Moon's path on the celestial sphere is a great circle inclined 5° to the Sun's path (the ecliptic). The polar axis *OM* of the Moon's path describes a cone about the (nearly) fixed axis *OE* of the Sun's path, so that the line *HT*, in which the planes of the Moon's and the Sun's paths meet, rotates from east to west with a period of 18.6 years.

points (*V, A* in the figure on p. 25) where the ecliptic meets the celestial equator, day and night are equal in length. On these days, the *equinoxes,* the Sun's elevation at high noon is equal to the observer's *colatitude* (90° minus the latitude). For example, in Boston (latitude 42°) the Sun's elevation at high noon on March 21 is 48°. As the Sun moves north of the equator, the intervals between sunrise and sunset lengthen in the northern hemisphere, shorten in the southern hemisphere. At the summer solstice (June 22 in the northern hemisphere), the Sun reaches its maximum noontime elevation, equal to the observer's colatitude plus 23.5° (the angle that the ecliptic makes with the equator). This is the longest day of the year in the northern hemisphere and the shortest in the southern hemisphere (at the equator, the days and nights are always of equal length). At the solstices the Sun "stands still": its elevation at high noon has stopped increasing and has not yet begun to decrease (or has stopped decreasing and has not yet begun to increase).

The figure on p. 25 predicts different seasonal variations at colatitudes less than 23.5°. For example, at the north and south poles the Sun's daily path is a circle parallel to the horizon. On midsummer day this circle has an elevation of 23.5°. Thereafter its elevation decreases, and at the fall equinox the path of the Sun coincides exactly with the horizon (though, owing to the bending of light rays in the atmosphere, the Sun appears to be above the horizon). For the next six months the Sun's daily circle lies below the horizon.

The Moon also moves among the stars from day to day. Its path is also a circle centered on the Earth; the plane in which the path lies makes an angle of 5° with the ecliptic (see above). The Moon moves at a nearly uniform rate along this path, in the same direction as the Sun (that is, opposite to the daily rotation of the celestial sphere), completing its circuit in a little over 27 days. Just as the Sun's motion lengthens the interval between its successive risings (relative to the stars) by about four minutes, so the Moon's motion increases the interval between its successive risings (relative to the stars) by nearly a full hour.

The Moon happens to have almost precisely the same angular diameter as the Sun. When the Sun and the Moon are both at one of the two points (*H, T* in the figure opposite) where their great circles intersect, the Sun, the Moon, and the Earth lie in a straight line, and the Moon may eclipse the Sun. When the Sun is at one of these points and the Moon at the other, the Earth may eclipse the Moon.

The five bright planets—Mercury, Venus, Mars, Jupiter, and Saturn—also move among the stars, but their motions are much less regular than those of the Sun and the Moon. Accounting for these apparently irregular motions became the central problem of Greek astronomy.

Explaining the Daily Motion of the Stars

Greek philosophers began to construct geometric representations of celestial motions early in the sixth century B.C. Characteristically, as we saw in Chapter 1, these founders of theoretical astronomy split into two opposing schools. The members of both schools strove for accuracy and simplicity, but they held different views about the nature of mathematics and mathematical models.

Members of the first school, whose major spokesman was Aristotle, regarded mathematics as the handmaiden of philosophy and common sense. Mathematical models, they believed, could usefully describe phenomena but could not express their deeper meaning. The achievements of this school culminated around A.D. 150 in Claudius Ptolemy's *Mathematical Collection,* which Moslem astronomers of the Middle Ages labeled *Al-Magest,* "the greatest." His astronomical observations and his geometric representation of them stood unrivaled in accuracy and completeness until the work of Tycho Brahe and Johannes Kepler fourteen centuries later.

Members of the second school, the Pythagoreans, believed that the reality beneath appearance is essentially and irreducibly mathematical. They held that mathematical harmony is a more reliable guide to an understanding of the heavens than common sense. When mathematical harmony came into conflict with common sense, they scandalized right-thinking Greeks by cheerfully abandoning common sense. The achievements of the Pythagoreans culminated in the heliocentric model of Aristarchus during the third century B.C. Copernicus revived Aristarchus' model in the sixteenth century, and in the opening years of the following century Kepler, aided by a splendid catalog of planetary positions inherited from Tycho Brahe, achieved the Pythagorean vision of a harmonious mathematical model of the solar system. The rest of this chapter traces the main steps in the construction of Kepler's model.

The Pythagoreans began by questioning the prevailing view that the stars are embedded in a transparent spinning sphere. It seemed simpler to them to attribute the daily motions of the stars to an unperceived, oppositely directed circular motion of the Earth itself. At the center of the circle they postulated a "central fire" to which the Earth in its revolution always presented the same hemisphere, as the Moon does to the Earth. Greece was assumed to lie in the opposite

The Pythagoreans speculated that the Earth and the counter-Earth revolve synchronously around a central fire, with a period of 24 hours. Inhabitants of the part of the Earth opposite Greece are shielded from the central fire by the counter-Earth.

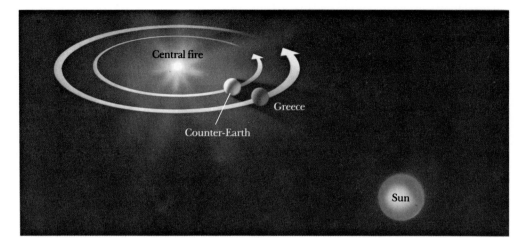

hemisphere. To shield the inhabitants of the Antipodes from the central fire, the Pythagoreans postulated a synchronously rotating "counter-Earth" that permanently eclipsed the central fire. This hypothesis was refuted near the beginning of the fourth century B.C. by reports from explorers who had penetrated westward beyond the Straits of Gibraltar and eastward as far as India but had failed to observe anything resembling a central fire or a counter-Earth.

Ecphantus and Heraclides thereupon put forward the still simpler but intuitively more difficult hypothesis that the Earth spins about an axis passing through its own geometric center "like a cartwheel on its axle." This was a novel idea, difficult to grasp and still more difficult to believe, and apparently was rejected by all but the small group of philosophers who carried on the Pythagorean mathematical tradition. Yet it greatly simplified the description of celestial motions: it brought the stars to rest, and accounted for a major component of the motions of the Sun, the Moon, and the planets. And although the hypothesis of a spinning Earth seemed to contradict experience and common sense, none of its *mathematical* consequences conflicted with any *measurable* aspect of experience.

In accepting the hypothesis of a spinning Earth, the Pythagoreans, pursuing their vision of mathematical harmony, left the mainstream of Greek astronomy. Eventually the mainstream dried up, and the tributary, gradually broadening, led to what Newton in his old age described as "the great ocean of truth."

Representing the Motions of the Planets: Genesis of an Overdetermined Model

The five bright planets move among the stars in complicated ways. To begin with, their paths—unlike those of the Sun and Moon—are not great circles centered on the Earth but are much more complicated curves. What is worse, a planet does not return to its starting point after making a complete circuit of the heavens; rather, it traces out different curves in successive circuits. Finally, each planet moves in a highly nonuniform way along its path. Mercury and Venus stay close to the Sun, sometimes rushing ahead, sometimes falling behind, like puppies walking with their master. Mars, Jupiter, and Saturn move steadily westward relative to the Sun, but at highly variable rates, moving most rapidly relative to the Sun when they are in the opposite part of the sky, and most slowly when in the same part of the sky. Near opposition, Mars' westward motion relative to the Sun becomes faster than the Sun's eastward motion relative to the stars. Mars, which had been moving steadily eastward among the stars, slows down, stops, and reverses direction. A few days later it slows down, stops, and reverses direction again, resuming its normal eastward progression. This part of its path thus resembles a flattened letter S, as shown at the left. Jupiter and Saturn perform the same trick. Understanding these complex and apparently erratic motions presented a formidable challenge.

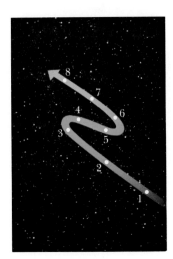

Retrograde motion of Mars against the stars (schematic).

One of the scientist's basic strategies is to divide a complex phenomenon into two parts, one of which is large and simple, the other small and complex—and then ignore the second part. This strategy succeeds brilliantly for the planets. The complicated curves that the planets trace out on the celestial sphere are all contained within a narrow strip, the Zodiac, which is centered on the ecliptic (the Sun's path). We may think of a planet as having two separate motions: one parallel to the ecliptic, the other at right angles to the ecliptic. Because each planet stays close to the ecliptic as it wanders among the stars, its motion perpendicular to the ecliptic is small compared with its motion parallel to the ecliptic. This small perpendicular motion is also very complicated. In fact, it turns out to be the main source of complexity.

This insight marked the beginning of a successful assault on the problem. It was not an easy insight. It eluded Eudoxus (408–355 B.C.), the greatest mathematician of his day and the chief founder of Greek mathematical astronomy.

When we ignore the perpendicular motions of the planets, the problem simplifies in two important ways. First, the paths of all the planets coalesce with the Sun's path, the ecliptic. Second, the motion of each planet becomes *periodic*. Each planet returns to an arbitrarily chosen starting point on the ecliptic after a fixed period of time.

We are left with two related problems: understanding why the planet's motions are so variable; and constructing a mathematical model that accurately represents the observed variability. For the Pythagoreans, the key problem was the first. Once the *why* of variable motion was understood, they expected the *how* to present little difficulty. For the Aristoteleans the first question did not arise. The purpose of mathematical astronomy, in their view, was to describe, not to explain.

Why do the planets move at such variable rates along the ecliptic? Clearly (as we can now say in retrospect; at the time it was clear to very few), the Sun has something to do with it. Imagine the plane of the Sun's path to be the floor of a merry-go-round with the Earth at its center, rotating at the same rate as the Sun. In this rotating frame of reference the Sun's direction is fixed, and a line drawn through the centers of the Earth and the Sun is an axis of symmetry for

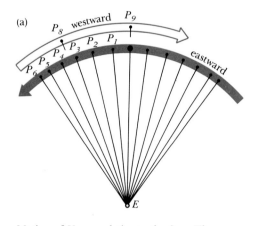

Motion of Venus relative to the Sun. The apparent positions P_0–P_9 correspond to equal intervals of time. Thus the eastward motion from P (the position of the Sun) to P_0 takes about 5 times as long as the westward motion between the same two positions.

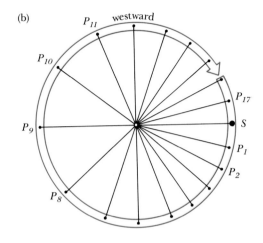

Apparent motion of Mars relative to the Sun. P_1, P_2, . . . are the apparent positions of Mars on the celestial sphere at equal intervals of time.

"Venus changes shape precisely as does the Moon; and if Apelles will now look through his telescope he will see Venus to be perfectly circular in shape and very small (though indeed it was smaller yet when it [recently] emerged as an evening star). He may then go on observing it, and he will see that as it reaches its maximum departure from the Sun it will be semicircular. Thence it will pass into a horned shape, gradually becoming thinner as it once more approaches the Sun. Around conjunction it will appear as does the Moon when two or three days old, but the size of its visible circle will have much increased. Indeed, when Venus emerges [from behind the Sun] to appear as an evening star, its apparent diameter is only one-sixth as great as at its evening disappearance [in front of the Sun] or its emergence as a morning star [several days thereafter], and hence its disk appears forty times as large on the latter occasions."
GALILEO, *Letters on Sunspots* (1613).

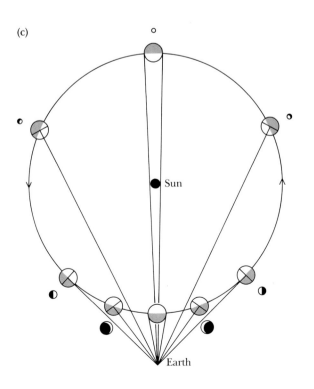

(a) *P* (representing Mercury or Venus) moves uniformly on a circle centered on the Sun at *S*. As viewed from the Earth at *E*, the motion is nonuniform. The apparent motion, relative to the Sun, is eastward from *Q* to P_7 (in the direction of the arrow) and westward from P_7 to *Q*.
(b) *P* (representing Mars, Jupiter, or Saturn) moves uniformly on a circle centered on the Sun at *S*. As viewed from the Earth at *E* the motion is nonuniform, being fastest at *A*, slowest at *B*. These models explain the motions represented by the corresponding diagrams on the facing page.

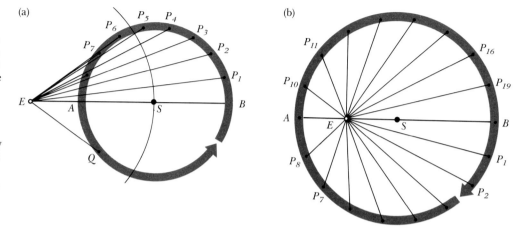

the motion of each planet. That is, if two positions of a planet are symmetric relative to this line (when lines connecting the Earth to each of the two positions make equal angles with the Earth–Sun line), then their rates of motion along the ecliptic at these two positions will be equal (see figures on p. 30). For example, if Jupiter moves *x* degrees per day when it is *y* degrees east of the Sun, then it also moves *x* degrees per day when it is *y* degrees west of the Sun; the back-and-forth motions of Venus and Mercury are centered on the Sun, and so are the retrograde sections of Mars', Jupiter's, and Saturn's paths. So our problem is now to understand why a line connecting the Earth to a planet rotates at a rate that is variable, but variable in a way that is symmetric with respect to the direction of the Sun.

The dominant school of Greek mathematical astronomers imposed a further condition on the problem: that each planet be attached to the surface of a sphere centered on the Earth. This condition seems bizarre to us, but it did not seem so to Greek philosophers in the fourth century B.C. Aristotle never questioned it. However, Heraclides of Pontus (*ca.* 388–310 B.C.), a contemporary of Aristotle and a younger contemporary of Eudoxus, decided that a planet need not maintain a fixed distance from the Earth. He proposed that the orbits of Mercury and Venus are circles centered on the Sun, and that each planet moves along its circle at a constant rate as seen from the Sun. This simple model represents the observed motions of Venus and Mercury very well. It also correctly predicts variations in the two planets' brightness, as well as the phases of Venus (see the bottom figure on p. 30). In short, Heraclides' model is strongly overdetermined by the phenomena.

The same idea works for Mars, Jupiter, and Saturn, provided the radii of their heliocentric orbits are assumed to be greater than the Sun's distance from the Earth (see above). This model immediately explains the main qualitative features of the outer planets' variations in speed and brightness: the symmetry of the variations about the Earth–Sun line; and the observation that

Tycho Brahe's (and perhaps
Heraclides) model of the planetary system. The five planets
move in circular orbits around
the Sun (*S*), which moves in a
circular orbit about the Earth
(*E*). The orbital periods are
given in days.

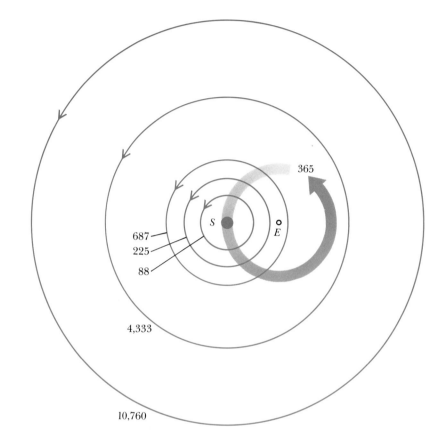

each outer planet is brightest and moves most rapidly when it is opposite to the Sun in the sky,
and is faintest and moves most slowly when it is in the same part of the sky.

Each circular orbit has two adjustable parameters: its period, and the ratio between its radius
and the distance from the Earth to the Sun. The periods are easily measured. We may infer the
ratio between the radius of the orbit of Venus and the Earth–Sun distance (and the same ratio
for Mercury) from the measured angle that each orbit subtends at the Earth (see above). The
radius of an outer planet's orbit relative to the Earth–Sun distance is proportional to the ratio
between the planet's apparent speeds at opposition and conjunction (see the figure on p. 31).

After the period and radius of each planetary orbit have been calculated as described, no
further adjustments can be made in the model. It must now submit to the test of experience: it
must reproduce the observed day-to-day variations in each planet's motion among the stars.
And it does so, astonishingly well by the standards of Aristotle's day, tolerably today.

The complete model with its five heliocentric planetary orbits is shown above. The model
is not drawn to scale; the radii of the orbits (in units of the Earth–Sun distance) are shown

The Earth–Sun line rotates relative to the stars, but the effect is the same whether it rotates about the Earth (top), the Sun (middle), or any other point P on the Earth–Sun line (bottom).

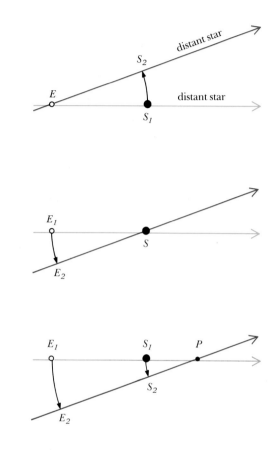

in the plot on p. 35. The orbital periods are periods relative to the stars, as viewed from the Sun.

Aristarchus and the Dimensions of the World

Observation tells us that the line that passes through the Earth and the Sun rotates relative to the stars. It does not tell us about which point the line rotates. The figure above shows that the Sun will appear to revolve around the Earth whether the Earth or the Sun or some other point on the Earth–Sun line is taken to be at rest relative to the stars. How can we decide?

If you hold your finger in front of your face, close one eye, and move your head slowly from side to side, your finger will appear to move in the opposite direction against its background. This motion is called a *parallactic shift*. If the Earth went around the Sun, the stars would show annual parallactic shifts—if they were close enough. Since no such shifts were observed (before the nineteenth century), it followed that either the Earth was at rest relative to the stars or the stars were too distant to have measurable parallactic shifts. Thus, observation alone could not settle the question.

But agreement with observation was not the only criterion for a theoretical model in the Pythagorean tradition. Which hypothesis—a stationary Earth or a stationary Sun—makes the data more harmonious? On physical grounds it makes more sense to have the Earth go around the Sun like the planets than to have the Sun with its retinue of planets go around the Earth. The Sun is the source of light; the Earth, the Moon, and the planets all borrow their light from the Sun.

But there is a more compelling reason for regarding the Earth as the Sun's third planet. Consider the sequence of planetary periods in the figure on p. 32. The periods increase steadily with orbital radius. If we regard the Earth as a planet, its period of one year seems to fit neatly into the sequence.

To study this question more closely, let us plot the period of each planet against its orbital radius. The relation between these two quantities stands out most clearly if we plot the logarithm of the period against the logarithm of the orbital radius, as in the facing figure (left). The plotted points for the five planets lie on a straight line. And if we regard the Earth as a planet, its point, too, falls on the line. Of course, if the Sun were the Earth's satellite, *its* point would fall in the same place. But that would make no sense. The line is defined by the periods and orbital radii of satellites of the Sun. We would not expect the relation between period and orbital radius for satellites of the Sun to be satisfied by satellites of the Earth. And, in fact, the point defined by the Moon's period and orbital radius lies nowhere near the line.

It is not known whether Aristarchus used such an argument to support his heliocentric model of the planetary system. The book in which he discussed the model was lost. We know about it from a single reference by Archimedes (as translated by Heath, p. 302):

> But Aristarchus brought out a book consisting of certain hypotheses wherein it follows, as a consequence of the assumptions made, that the universe is many times greater than the 'universe' just mentioned [a sphere of radius equal to the Earth's orbital radius]. His hypotheses are that the fixed stars and the Sun remain unmoved, that the Earth revolves about the Sun in the circumference of a circle, the Sun lying in the middle of the orbit, and that the sphere of the fixed stars, situated about the same center as the Sun, is so great that the circle in which he supposes the Earth to revolve bears such a proportion to the distance of the fixed stars as the center of the sphere bears to its surface.

The last of these hypotheses ("that the circle in which he supposes the Earth to revolve . . .") is probably just a concise statement of the conclusion that Aristarchus drew from his contemporaries' failure to observe annual parallactic shifts of the stars. As shown on the right on the facing page, the Earth's annual motion would cause a star to trace out a tiny ellipse on the celestial sphere. This ellipse would look exactly like the Earth's orbit viewed from the star; it would be a tiny circle viewed from directly above the Earth's orbital plane, a tiny straight-line segment viewed from a point in the plane. Thus the absence of observed parallactic shifts meant that the Earth's orbit, viewed from a bright star, would be indistinguishable from a point.

(*Left*) A logarithmic plot of sidereal period T against orbital semi-diameter a for the five bright planets and the Earth. The straight line represents the relation $T \propto a^{3/2}$. (*Right*) A star's parallactic ellipse looks exactly like the Earth's orbit viewed from the star.

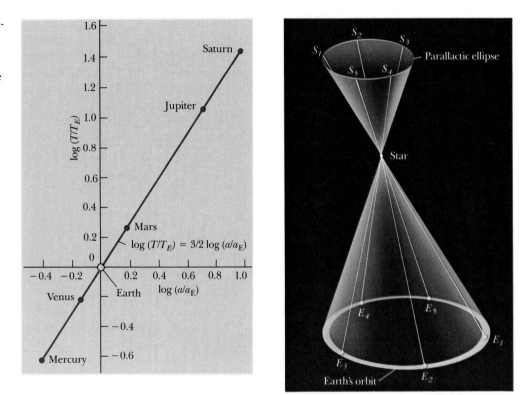

Aristarchus' only surviving work is the monograph *On the Sizes and Distances of the Sun and Moon*. This work makes it clear that Aristarchus' model of the planetary system was part of a grand project to establish the geometry of the universe. As we have seen, Aristarchus' model of the planetary system would have allowed him to deduce from astronomical observations the radii of the five planetary orbits, measured in units of the Earth's distance from the Sun, as well as the five planetary periods, measured in years. The lost book mentioned by Archimedes presumably elaborates this part of the project. Next, Aristarchus needed a method for finding the ratio between the orbital radii of the Sun and the Moon, and, finally, a method for finding the ratio between the radius of the Moon's orbit (or of the Earth's orbit) and the radius of the Earth itself, which, as mentioned earlier, had recently been accurately measured by Eratosthenes.

Aristarchus' method for estimating the ratio of the distances of the Sun and Moon is shown in the top figure on p. 36. It hinges on an accurate measurement of the angle between the center of the Moon and the center of the Sun when the Moon is exactly half full. The method is sound in principle, but the difference between the angle just mentioned and 90° is too small to measure accurately. Aristarchus estimated the difference to be 3°, which would have made the

When the Moon is exactly half full, the cosine of the angle *MOS* (the angle between the center of the Moon and the center of the Sun) = *OM/OS*, the ratio of the two distances. (In reality *OM/OS* ≃ $\frac{1}{400}$, and the difference between angle *MOS* and a right angle is too small to measure.)

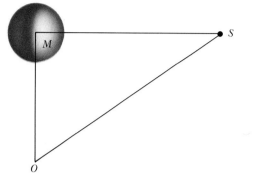

From the average duration of lunar eclipses and the length of the month, Aristarchus calculated the angle $\delta = \angle MET$ (S is the center of the Sun, E the center of the Earth, M the center of the Moon, MT the half-width of the part of the Moon's orbit eclipsed by the Earth). Since $\alpha + \beta + \angle AET = 180° = \gamma + \delta + \angle AET$, $\alpha + \beta = \gamma + \delta$; γ is the angular radius of the Sun, which is measurable. Hence $\alpha + \beta$ is measurable. But $\alpha \simeq \sin \alpha = EB/ES$, and $\beta \simeq \sin \beta = EB/ET$. So $1/R_S + 1/R_M \simeq \gamma + \delta$, where $1R_M$ is the Moon's distance measured in units of the Earth's radius, and R_S is the Sun's distance in the same units. In reality $R_S \gg R_M$; so the method yields an accurate estimate of R_M, namely, $R_M \simeq 60$ Earth radii.

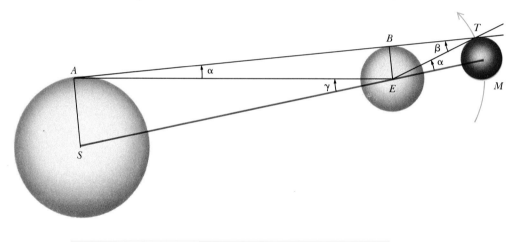

Sun's distance 19 times as great as the Moon's. The true value of the difference is about one-sixth of a degree, corresponding to a distance ratio of about 400.

Aristarchus devised an ingenious method for deducing the ratio between the Moon's distance and the radius of the Earth from the average duration of a lunar eclipse and the length of the month (see the figure directly above). His estimate of 60 Earth radii for the Moon's distance is in fact reasonably accurate.

Finally, Aristarchus deduced the diameters of the Sun and Moon from their measured angular diameters (about half a degree) and their estimated distances.

Aristarchus' project was now complete: he had constructed a coherent geometric model that represented what was known about the apparent sizes, motions, and brightnesses of the Sun, the Moon, and the planets with tolerable accuracy, and that was strongly overdetermined by these data. The first qualitative improvements in Aristarchus' model were embodied in Kepler's three laws of planetary motion, more than eighteen centuries later.

Thales' method of triangulation for measuring the distance of a ship at sea:

1. Measure angles *a* and *b* and distance *AB*.
2. Construct triangle *A'B'S'* with angles *a* and *b* at *A'*, *B'*.
3. Use *AS*/*AB* = *A'S'*/*A'B'* to find *AS*.

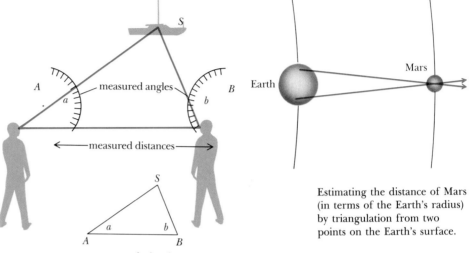

constructed triangle

Estimating the distance of Mars (in terms of the Earth's radius) by triangulation from two points on the Earth's surface.

A medieval astronomer using Thales' method of triagulation to calculate the distance of a ship at sea.

The only serious quantitative flaw in Aristarchus' model was his estimate of the Sun's distance. This was not corrected until 1672, when French astronomers measured the distance of Mars during its closest approach to Earth by the same method—triangulation—that Thales had used to measure the distance of a ship at sea in the sixth century B.C. Teams at Paris and Cayenne in French Guiana carried out simultaneous measurements of Mars' position in the sky. The difference between the two measured positions yielded the angle at the apex of the triangle formed by the center of Mars' disc and the two observing sites. This single measurement established the linear scale of the entire planetary model.

The work of Aristarchus completed the grand program of scientific cosmology initiated three centuries earlier by Pythagoras and his followers. It is true that some of Aristarchus' estimates of size and distance were vitiated by systematic errors of which he was unaware, but such errors are almost invariably present in pioneering work. In the 1920s Harlow Shapley overestimated the distance of the Sun from the center of the Galaxy by a factor of 10, and Edwin Hubble underestimated the distances of external galaxies by a factor of 5. The pioneer does not yet know the right answer, even approximately, because his is the first measurement and the only test he can apply is the test of internal consistency. That Aristarchus' distance scale turned out to be inaccurate is unimportant; what matters is that he and those whose work he built upon created, along with an elegant and strongly overdetermined model of the planetary system, the modern scientific approach to cosmology.

Copernicus

Nicolaus Copernicus (1473–1543). This idealized but contemporary portrait shows how he may have looked as a young man.

Aristarchus' magnificent achievement seems to have made little impression on his fellow mathematical astronomers, who directed their efforts toward constructing increasingly accurate geometric descriptions of the observed motions of the Sun, the Moon, and the planets. Their models were intended to represent as accurately as possible the *appearance* of the heavens, not the structure of the world; so they held fast to the intuitive idea that the Earth is at rest at the center of the heavens, and constructed increasingly elaborate, unrealistic, and underdetermined models. The orthodox tradition reached full flower in the work of Claudius Ptolemy, but then gradually ran to seed.

Copernicus revived the Pythagorean tradition. He began by reconstructing Aristarchus' model but then proceeded to complicate it with epicycles and offsets. The figures on pp. 40 and 41 show his final version, along with Ptolemy's model. Copernicus' model certainly does not look any simpler than Ptolemy's. Indeed, as Arthur Koestler pointed out in *The Sleepwalkers*, it uses 48 circles to Ptolemy's 40. Nor did it represent astronomical observations more accurately; it was more accurate in some ways, less in others. And in one crucial respect it blatantly contradicted well-established evidence: it predicted that the stars would show annual parallactic shifts. Neither Copernicus nor any of his predecessors had ever been able to detect such shifts. Copernicus argued that the stars must be distant enough to make their parallactic shifts unobservably small. But if this were the case, the diameters they would need to have to appear as large as they do would be greater than the diameter of the Earth's orbit! This conclusion affronted common sense and undermined Copernicus' claim that the Sun, who "is rightly called the Lamp, the Mind, the Ruler of the Universe . . . sits . . . enthroned" at the center of the universe.

In short, Copernicus' heliocentric model of the planetary system was as clumsy and complicated as the competing geocentric model; it was not notably more accurate; and its implications about the sizes of stars seemed to be absurd and inconsistent with his own central thesis. It was not even particularly original, as Copernicus himself was careful to point out (originality being a dubious virtue in a provincial canon). The Greeks, as usual, had gotten there first. All Copernicus had to do was work out the heliocentric translation of Ptolemy's geocentric model, a task that would not have presented a serious challenge to any competent mathematician of his day.

Why, then, did Kepler and Galileo, men not given to idle praise, hail Copernicus as the founder of the New Astronomy? Why did Martin Luther and John Donne (in *Ignatius his Conclave*) berate him with such vehemence?

In constructing the complicated model illustrated on p. 41, Copernicus began with the simplest model that incorporates the three essential features of Pythagorean/Aristarchan astronomy: a spinning Earth, the Sun at the center of the planetary orbits, and the Earth as a planet, circled by the Moon. Let us call this model, shown on p. 42, Model *A* (for Aristarchus). It represents the sound core of Copernicus' theory. For Galileo and Kepler it *was* Copernicus' theory. In *De Revolutionibus*, Copernicus tried to construct a heliocentric model of the world that would represent astronomical observations better than Ptolemy's geocentric model, but he succeeded only in constructing a mathematical thicket within which Model *A*, the simple and beautiful idea that inspired his life work, slept for nearly a century.

For Copernicus, Kepler, and Galileo, Model *A* was more than a device for representing celestial motions in a rough first approximation. It had something that was lacking in the geocentric models that represented astronomical observations far more accurately: a kernel of truth.

Aristotle said that a statement is true if what it asserts is the case. "Rain fell on the Boston Common this morning" is true if and only if rain did fall on the Boston Common this morning; and we can find out whether it did by asking someone who was there. Does Aristotle's definition apply to scientific theories? May we say that a theory is true if and only if what it asserts is the case?

At first sight this may seem an attractive possibility. The statement "Model *A* is true" would then mean that a hypothetical observer, suitably placed and equipped with telescopic vision, would be able to look down on the planetary system and report, "Yes, the planets really do go around the Sun in nearly circular orbits whose planes nearly coincide." Spacecraft sent to explore the solar system have sent back pictures assuring us that the planets are indeed exactly where our vastly improved version of Model *A* says they ought to be; so Model *A* is true in the everyday, Aristotelean sense.

But for Copernicus, Kepler, and Galileo, Model *A* was already true in some less direct but equally compelling sense. What is that sense? What are the marks of scientific truth?

One of them is surely agreement with experience. A true theory must "save the phenomena." But that is not enough. Ptolemy's planetary model agreed very well with experience, but no one regarded it as more than a convenient device for predicting the apparent positions of celestial objects. On the other hand, Model *A*, which represented astronomical observations far less accurately, had—for Copernicus, Kepler, and Galileo—the unmistakable ring of truth. Why? Because Model *A* revealed, in the words of Copernicus, "a wonderful commensurability and . . . a sure bond of harmony for the movement and magnitude of the orbital circles such as cannot be found in any other way."

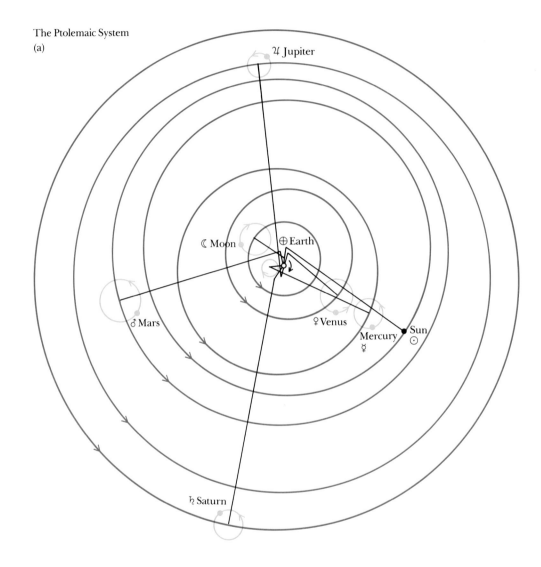

The Ptolemaic System
(a)

The Copernican System
(b)

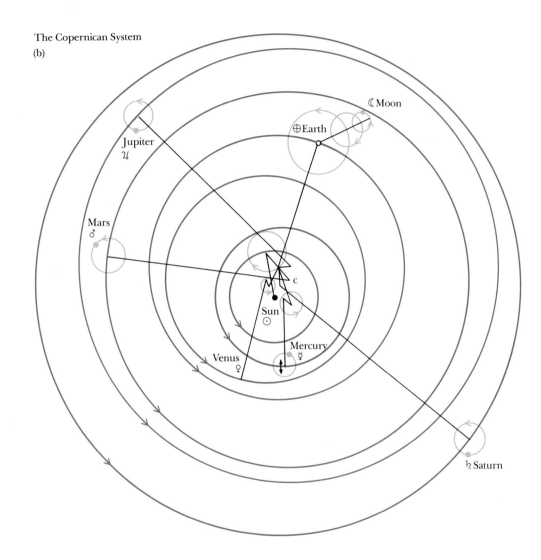

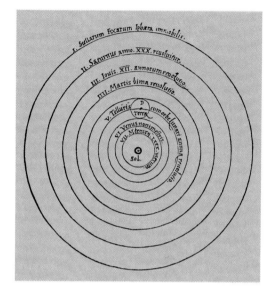

A plate from Copernicus' *De Revolutionibus* showing the Copernican system, with the Sun in the center.

Johannes Kepler (1571–1630), in an engraving by MacKenzie.

Kepler

The theme of celestial harmony, as developed by Plato in his cosmological dialogue *Timaeus*, inspired and unified the life work of Johannes Kepler who completed the astronomical revolution launched by Copernicus. In this dialogue, Timaeus, a Pythagorean, identifies four of the five Platonic solids with the four classical elements: cube = earth, tetrahedron = fire, octahedron = air, icosahedron = water; he identifies the fifth Platonic solid, the dodecahedron, with the universe. And he relates the construction of the Heavens to the construction of the diatonic scale. Kepler, in the *Mysterium Cosmographicum (The Cosmographic Mystery)*, published in 1596, elaborated a different cosmological interpretation of the Platonic solids (here as translated by Koyré, *The Astronomical Revolution,* p. 146):

> *The Earth [the sphere of the Earth] is the measure for all the other spheres. Circumscribe a Dodecahedron about it, then the surrounding sphere will be that of Mars; circumscribe a Tetrahedron about the sphere of Mars, then the surrounding sphere will be that of Jupiter; circumscribe a Cube about the sphere of Jupiter, then the surrounding sphere will be that of Saturn. Now place an Icosahedron within the sphere of the Earth, then the sphere which is inscribed is that of Venus; place an Octahedron within the sphere of Venus, and the sphere which is inscribed is that of Mercury.*

Having explained why there are just six planets (including the Earth), Kepler set out to construct a geometric model that would represent the extraordinarily complete and precise observations of the planet Mars bequeathed him by the great Danish astronomer Tycho Brahe, whose assistant he became in 1601. In the *Astronomia Nova (New Astronomy)*, completed in 1607 and published two years later, Kepler announced two of his famous three laws of planetary

Tycho Brahe (1546–1601). An engraving by J. de Gheyn made from a portrait done by Tobias Gemperlin in 1586 and first published in Brahe's *Astronomical Letters* in 1596.

A diagram from Kepler's first-published book, *The Cosmographic Mystery* (1596), showing that the five regular solids can be nested in a set of concentric spheres in a way that reproduces the relative radii of the planetary orbits in the Copernican system.

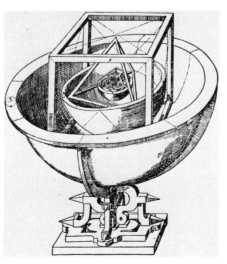

Brahe's model of the solar system, in which the Sun revolves around the Earth, but the other planets revolve around the Sun. Brahe may have been anticipated by Heraclides; see the figure on p. 32.

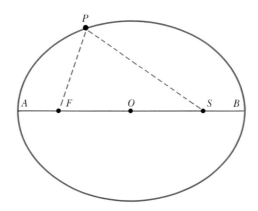

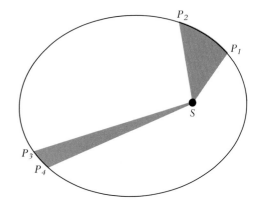

Kepler's first law: The orbit of each planet is an ellipse with the Sun at one focus. The sum (*PF* + *PS*) of the distances from any point *P* of an ellipse to the two foci *F,S* is equal to the major diameter *AB*. The eccentricity of the ellipse is the ratio *OS/OB*.

Kepler's law of areas: The Sun–planet line *SP* sweeps out equal areas in equal times. P_1, P_2 and P_3, P_4 are pairs of positions of the planet separated by short, equal time-intervals.

motion: that the planetary orbits are ellipses with the Sun at one focus, and that a line drawn from the Sun to a moving planet sweeps out equal areas in equal times.

Kepler was the first astronomer to advance Aristarchus' program of constructing an overdetermined geometric model of the cosmos. How can we account for the gap of 1,800 years that separates the two astronomers? Let us consider, with the benefit of hindsight, the improvements that needed to be made in Aristarchus' model.

(1) The six planetary orbits, which in Aristarchus' model all lie in the same plane, needed to be put into distinct planes, each passing through the Sun.

(2) The principle of uniform circular motion, the fixed idea of mathematical astronomy from its inception to the end of the sixteenth century, needed to be replaced by a new principle: that the line joining a planet to the Sun sweeps out equal areas in equal times.

(3) The orbits of the planets needed to be changed from circles or figures resulting from combinations of circular motions to ellipses with the Sun at one focus.

These steps were all taken by Kepler. Historically there were no intermediate models, and even with hindsight it is not easy to construct models that could have served as stepping stones. Kepler's achievement required physical insight and mathematical genius, fueled by enormous energy and dedication. It also required unprecedentedly accurate and extensive observational data. In short, it required a Kepler supplied with the observational material of a Tycho Brahe.

EXPLAINING THE MOTIONS OF THE PLANETS AT RIGHT ANGLES TO THE ECLIPTIC In Aristarchus' model all the planets move in the same plane; so their paths on the sky coincide with the Sun's path

If the Sun lies in the orbital planes of both the Earth and Mars, the two planes intersect in a line *NSN* that passes through the Sun (*S*). The nodal points *N* are the intersections of Mars' path on the celestial sphere with the ecliptic. By assuming that Mars' orbital plane passes through the Sun (rather than through the center of the Earth's orbit), Kepler was able to explain the component of Mars' motion on the celestial sphere at right angles to the ecliptic (motion in latitude).

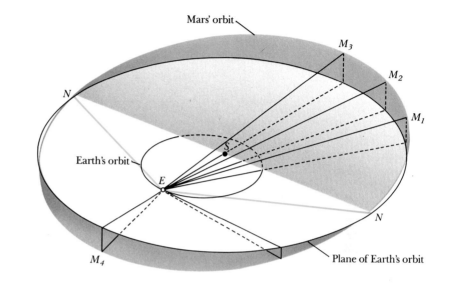

(the ecliptic). Aristarchus made no attempt, so far as we know, to understand the small and apparently irregular motions of the planets at right angles to the ecliptic. Over the centuries these small motions challenged the ingenuity of mathematical astronomers. From Eudoxus to Copernicus, they all tried and failed to represent them in a reasonably simple way. Ptolemy assumed correctly that each planetary orbit lies in a plane and that these planes are all distinct. But he assumed that they all pass through the center of the Earth. Copernicus modified Ptolemy's model by moving the common point of the orbital planes—not, as a modern reader might expect, to the Sun, but to a mathematical point close to the Sun, the center of the Earth's orbit. (We will see soon why Copernicus did not put the Sun at the center of the Earth's orbit.) The resulting predictions of planetary motions at right angles to the ecliptic agreed better with observation than Ptolemy's, but still did not represent the data accurately.

Kepler took what seemed to him the obvious step of making all the orbital planes pass through the Sun. This hypothesis has a simple testable implication. Consider the line of sight from Earth to a planet. It will be inclined to the Earth's orbital plane—and hence the planet's position in the sky will be above or below the ecliptic—except when the planet is at one of the two points where its orbit meets the Earth's orbital plane (see above). At each of these points, the planet will appear to lie exactly on the ecliptic. But it is clear from the figure that the line joining these two points, the line in which the planet's and the Earth's orbital planes intersect, *passes through the Sun*. Using Tycho's observations of Mars, Kepler was able to confirm this prediction. He could then adjust the angle between the two planes to represent with tolerable accuracy the planet's (angular) distance from the ecliptic at other times. As we can see from the figure above, the planet's angular distance from the ecliptic depends not only on its position in its orbit but also on its distance from the Earth. The interplay between these two factors is responsible for the complexity of the apparent motions at right angles to the ecliptic.

By taking the Copernican hypothesis seriously, and with the help of Tycho's excellent observations, Kepler had made the first substantial advance in theoretical astronomy since Aristarchus. Commenting on this achievement, he remarked that Copernicus had not known the value of his own treasure.

THE LAW OF UNIFORM AREAL VELOCITY   Although Aristarchus' model explained all the qualitative aspects of planetary motions parallel to the ecliptic, it did not represent these motions as accurately as even his contemporaries could have wished. As the observations accumulated and became more accurate, it became evident that some of the discrepancies between the predictions of Aristarchus' model and observation had a simple, systematic character.

Consider, for example, the apparent motion of the Sun. Even in Aristarchus' day astronomers knew that the Sun does not move along the ecliptic at a perfectly uniform rate. But the deviations from uniformity, as inferred from careful measurements of the length of the day (see the box on the bottom of the facing page), are themselves simple and regular. Ptolemy found that he could represent the observed variability of the Sun's motion fairly accurately by allowing the Sun to move at a constant rate around an *eccentric* circle—that is, a circle whose center does not coincide with the Earth (see the top right figure on p. 31). Similarly, in Copernicus' heliocentric adaptation of Ptolemy's model, the Earth's circular orbit is centered on a mathematical point near the Sun. This point, though without physical significance, played a central role (literally) in Copernicus' construction: the deferents of all the planetary orbits were centered on it (see figure on p. 41).

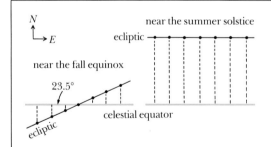

The length of the solar day is proportional to the eastward component of the Sun's daily motion along the ecliptic. It is greatest at the solstices, smallest at the equinoxes.

We may consider the Sun's daily displacement along the ecliptic to consist of two separate displacements in perpendicular directions: an eastward displacement parallel to the celestial equator (i.e., along a parallel of celestial latitude), and a northward or southward displacement, along a circle of celestial longitude. Only the eastward component affects the interval between successive risings or settings of the Sun. The ratio between the eastward component and the total daily displacement along the ecliptic varies seasonally. It is equal to unity at the solstices, when the Sun is moving parallel to the celestial equator, and has its smallest value at the equinoxes, where the Sun's path cuts the celestial equator at an angle of 23.5°. The resulting seasonal variation in the length of the day is easy to calculate. When we calculate the Sun's daily displacements along the ecliptic, taking this effect into account, we find that they are not exactly the same throughout the year but show the systematic variation discussed in the text.

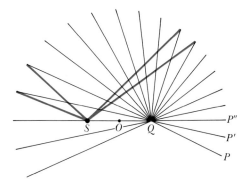

Ptolemy's equant construction, adapted to the heliocentric hypothesis. The planet P moves on a circle whose center O is displaced from the Sun S. As seen from the equant Q, with $OQ = OS$, the planet sweeps out equal angles in equal times. As seen from S (or O), its angular motion is nonuniform.

Astronomical observations show that the other five planets also move around the Sun at variable rates. Offsetting their circular orbits by appropriate amounts in appropriate directions reduces but does not eliminate the systematic discrepancies between the model and the observations. Now, Ptolemy had met and solved essentially the same problem fifteen centuries earlier. Ptolemy's solution, adapted to the heliocentric hypothesis, is shown above. The planet moves in a circular orbit whose center is offset from the Sun. The planet moves around this circle in such a way that a line drawn from the point Q to the planet sweeps out equal angles in equal times. The point Q, called the *equant*, is also offset from the center of the circle. As a result the planet speeds up as its distance from the equant increases and slows down as its distance from the equant decreases.

By trial and error Ptolemy discovered that the best place for the equant in the models for Mars, Jupiter, and Saturn is the one illustrated above: the equant's offset is equal and opposite to the Sun's (the Earth's, in Ptolemy's models). In other words, the center of the circular orbit bisects the line joining the equant to the Sun.

Copernicus rejected this ingenious device because it violated the principle of uniform circular motion. He replaced Ptolemy's equants by extra epicycles; this is one reason why his model uses more circles than Ptolemy's. Kepler saw the equant construction in an entirely different light. He recognized that it must express a genuine regularity of planetary motion, and that it should therefore apply to all planetary orbits, including the Earth's. Ptolemy had used a simple

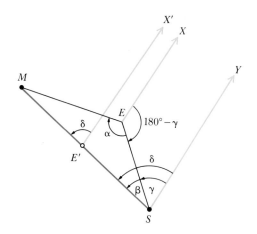

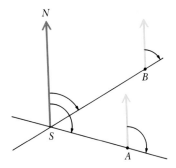

S, *E*, and *M* represent the Sun, the Earth, and Mars, respectively, and *SX* and *EX* are parallel to the direction of an arbitrary fixed star. We neglect the small inclination between the orbital planes of the Earth and Mars. The angles at *E* (α, $180° - \gamma$) are measurable. When Mars is at opposition (*i.e.*, when the Earth is at *E'*) the angle δ is measurable. Taking *SM* as a fixed reference segment, draw a line through *S* making an angle $\beta = \delta - \gamma$ with *SM*, and a line through *M*, making an angle $180° - \alpha - \beta$ with *SM*. These two lines intersect at *E*, the position of the Earth associated with the measured directions of Mars and the Sun. One Martian year later, Mars must again be at the point *M* (wherever that may be), and observations of the directions of Mars and the Sun will yield the Earth's new position in the diagram. In this way, by means of observations of Mars and the Sun at intervals separated by one Martian year, Kepler calculated the shape of the Earth's orbit and its dimension relative to the reference distance *SM*.

Finding a ship's position relative to fixed beacons (*A* and *B*) from measurements of the directions of the beacons relative to a fixed reference direction (*N*). If *A* and *B* are marked on a chart, the ship's position on the chart is the point where lines through *A* and *B*, making the measured angles with the reference direction, intersect.

eccentric circle to represent the Sun's Earth-centered orbit, and Copernicus used the same eccentric circle to represent the Earth's orbit. Now, Ptolemy's representation of the Sun's orbit did not directly affect his models for the five planets, each of which contained its own hidden representation of the Earth's motion. But in any heliocentric model the representation of the Earth's motion is of crucial importance, because it enters directly into the model's predictions for the apparent motions of the planets. Kepler realized that, to construct an accurate theory of planetary motion, he would need to begin by testing the equant hypothesis for the Earth's orbit.

The equant construction and the simple eccentric circle each have a single adjustable parameter, the ratio between the Sun's offset and the radius of its orbit, whose value, called the *eccentricity*, is determined by the observed ratio between the largest and smallest values of the Sun's daily dispacement along the ecliptic. But the eccentricity or offset needed to produce a given value of the ratio between the two extreme daily displacements is smaller for the equant construction than for the simple eccentric circle. In fact, it is only half as large. To decide between the two constructions, Kepler needed to find a way to measure the Sun's distance from the center of its orbit.

He devised a simple and beautiful way of carrying out this measurement. Think of the Earth as a ship, and the Sun as a fixed beacon. To plot a ship's position, the navigator needs a fixed reference direction (which could be supplied by a magnetic compass) and two fixed beacons (see figure above, left). The stars provide the fixed reference direction; the Sun, one of the fixed beacons. The problem is to find a second beacon, a fixed object whose direction changes notice-

Fragment of Kepler's Model *B*. Each planet moves in an eccentric circular orbit, its motion governed by Ptolemy's equant construction. The orbits lie in distinct planes, all passing through the Sun (*S*).

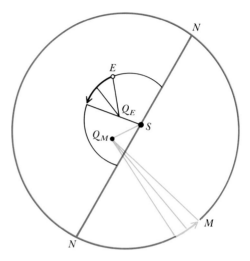

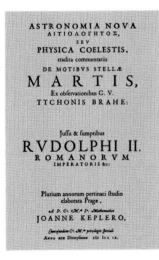

The title page of Kepler's *New Astronomy*.

ably as the Earth moves in its orbit. But the Solar System contains only one fixed object, the Sun. Kepler found a simple way to "manufacture" a second beacon. He reasoned that Mars returns to the same point in space at intervals of one Martian year. The length of a Martian year was accurately known. By measuring the directions of Mars and the Sun at intervals of one Martian year, he could plot a series of points on the Earth's orbit. Details of the construction are shown on the facing page, right.

In this way Kepler discovered that the eccentricity of the Earth's orbit is exactly that predicted by the equant construction, as he had hoped it would be. He could now construct the model shown above; let us call this Model *B*. In this model, as in Model *A*, the orbits of the planets are still circles, but they lie in distinct planes, all of which pass through the Sun. In each orbit the Sun is eccentrically situated, and the motion of each planet is represented by Ptolemy's equant construction. This model represents a significant advance over Model *A*; but Kepler did not stop here. The model was not yet accurate enough to fit Tycho's measurements as well as they deserved to be fitted. Moreover, it had, in Kepler's eyes, a serious *qualitative* flaw.

Kepler's astronomical studies were motivated by two main ideas: the Pythagorean notion of an underlying mathematical harmony, and the conviction that celestial bodies are physical objects whose motions are produced by natural causes. The title of his most important book is also a succinct statement of his philosophy of science:

A NEW ASTRONOMY Based on Causation
or A PHYSICS OF THE SKY
derived from Investigations of the
MOTIONS OF THE STAR MARS
Founded on Observations of
THE NOBLE TYCHO BRAHE

(a) Heliocentric counterpart of Ptolemy's equant construction. A planet P sweeps out equal angles in equal times as viewed from Q, situated so that $QO = OS$ (O the center of the circle, S the Sun).

(b) The same diagram with points Q and O deleted and lines drawn from S. By virtue of the equality $QO = OS$, the areas of the two shaded triangles are equal.

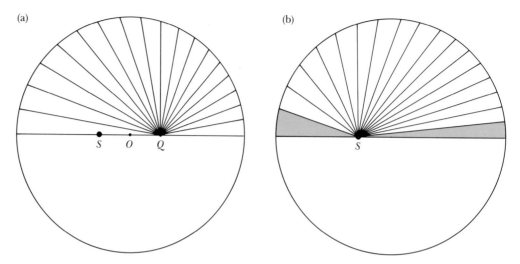

Kepler believed not only that the planets move around the Sun but also that the Sun *causes* their motion; his astronomy was *heliarchic* as well as heliocentric. As Newton was to show later in the century, this view is not quite correct. The Sun does not cause a planet's motion; it causes *changes* in the speed and direction of the motion. The mistaken idea that the Sun moves the planets did, however, suggest to Kepler that each planet's orbital plane contains the Sun. And it made him dissatisfied with Ptolemy's equant construction.

Although the equant construction made good mathematical sense, it made poor physical sense. It vested control of a planet's motion in a mathematical point (the point Q in the figure above, left), not in the Sun. Kepler perceived that the solution to this physical difficulty lay in the mathematics. Since the equant construction correctly describes the phenomena, Kepler may have reasoned, there must be a mathematically equivalent construction that makes better physical sense. Suppose we redraw the figure (above, left) to eliminate the equant and the center of the circular orbit, neither of which has any physical meaning. The resulting diagram (above, right) is mathematically equivalent to the original diagram. But how can we specify its construction without reintroducing the deleted equant? Kepler noticed that the *area* swept out by a line joining the Sun to the planet during a short fixed time-interval is the same at perihelion (when the planet is closest to the Sun and moving fastest) as at aphelion (when it is farthest from the Sun and moving most slowly). This observation suggested the rule now known as Kepler's second law, or law of areas: *A line drawn from the Sun to a planet sweeps out equal areas in equal times.*

This rule is not, in fact, equivalent to Ptolemy's equant construction. The two prescriptions agree in the neighborhoods of aphelion and perihelion, but not elsewhere. Although the law of areas did not initially represent the observations better than the equant construction, Kepler adopted it straightaway and used it exclusively in the long series of intricate calculations that culminated in the rule (which we will come to presently) that is now called his first law of planetary motion. In the end, the two laws turned out to represent Tycho's and Kepler's observations far better than any previous model, but it is worth noting that Kepler began to use the law of areas in his calculations *before* it had strong observational support. He was prepared to wager thousands of hours of laborious calculation on its validity, not because he knew it to be better than the models of Ptolemy and Copernicus, but because it conformed more closely to his notions of mathematical harmony and physical causality.

Kepler's law of areas is the first mathematical description of planetary motions that does not use uniform circular motions as building blocks. Moreover, unlike all earlier descriptions, it expresses a relation between the *instantaneous* values of *continuously variable* quantities (the planet's angular velocity with respect to the Sun and its distance from the Sun). This "instantaneous" mode of description, which Kepler handled with complete competence during his subsequent analysis of the motion of Mars, is one of the conceptual breakthroughs of seventeenth-century science. Its implications were first fully understood by Newton, who invented the differential calculus partly to express and to develop them.

BENDING THE CIRCLE   Kepler had now, in effect, constructed a third model of the planetary system, Model *C*, in which, as in Model *B*, the planetary orbits are circles lying in distinct planes that all contain the Sun but in which the angular rate of each planet's motion as seen from the Sun obeys the law of areas. Kepler's final discovery depended on a piece of luck. When Kepler arrived at Benatek Castle near Prague to take up his duties as one of Tycho's assistants, Longomontanus, Tycho's senior assistant, was engaged in a study of Mars' orbit. Kepler took over this work. "Had [Longomontanus] been occupied with another planet," Kepler wrote in Chapter 7 of the *New Astronomy,* "I would have started with that one. That is why I consider it an act of divine providence that I arrived when he was studying Mars; because to arrive at the secret knowledge of astronomy, it is absolutely necessary to use the motion of Mars; otherwise it would remain eternally hidden." The reason that Mars was essential is that Mars' orbit has a relatively large eccentricity (nearly 0.1, as compared with 0.02 for the Earth, and about 0.05 for Jupiter and Saturn); so its distance from the Sun changes by 20 percent between aphelion and perihelion. Model *C* would have fit Tycho's observations of Venus, Jupiter, and Saturn tolerably well, but it did not fit his observations of Mars well enough; so Kepler felt obliged to improve it.

Fruitless attempts to make it fit finally led Kepler to conclude that the fault lay not with the law of areas, but with the assumption that Mars' orbit is a circle. "Under the spell of common opinion," he writes, "I had bound them [the planets] into circles like donkeys in a mill." Free from this spell, he quickly discovered that Mars' orbit must be some kind of oval figure with its major diameter along the line joining Mars to the Sun (as shown here on the left).

From a purely theoretical point of view, the problem of finding the shape of Mars' orbit was now straightforward. Knowing the Earth's orbit and orbital motion, and using the law of areas, one can in principle derive the distance of Mars at closely spaced points on its orbit from observations of its direction as seen from the Earth. But the problem presented great practical difficulties. Kepler had neither a modern computer nor graduate students. He did have enor-

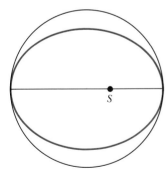

Kepler's oval orbit for Mars compared with a circular orbit.

Tycho Brahe's observatory at Uraniborg; engraving made in 1598.

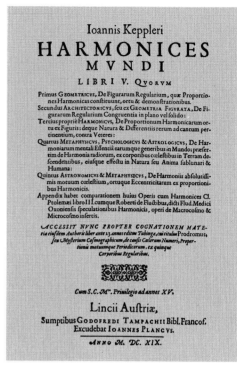

The title page of Kepler's *Harmonics of the World*.

mous mathematical skill and endless patience, and he knew that the problem was soluble, provided the law of areas was correct. After many conjectures and trials he found that he could fit the data by giving Mars an elliptical orbit with the Sun at one focus (as shown in the lefthand figure on p. 44). Later he found that the same law, now known as his first law of planetary motion, accurately describes the orbits of the other planets.

The fact that the Sun is at a focus of the ellipse introduces a new "bond of harmony" into Kepler's model. The shape of an ellipse is completely determined by its eccentricity; ellipses with the same eccentricity differ only in scale. The law of areas tells us that the variations in a planet's angular velocity as seen from the Sun are determined by the shape of the orbit. Hence variations in both the planet's distance from the Sun and its angular velocity as seen from the Sun are determined by the eccentricity of its orbit—that is, by the ratio between the Sun's offset and the orbit's semidiameter.

Finally, in 1618 Kepler announced his third law of planetary motion, the relation between period and orbital semidiameter illustrated in the lefthand figure on p. 35: *The squares of the orbital periods are proportional to the cubes of the orbital semidiameters.*

Jupiter and its four bright moons, discovered by Galileo with his new telescope while Kepler was working out the orbits of Mars, resemble a miniature solar system. Kepler found that Jupiter's moons satisfy his three laws of planetary motion, a beautiful and unexpected confirmation of their validity.

Kepler had at last constructed a strongly overdetermined model of the solar system that, with few exceptions, represented the motions of the planets and their satellites to within the accuracy of Tycho's observations. He had completed the scientific program initiated by the followers of Pythagoras, and he had constructed one of the two great pillars (Galileo's theory of motion is the second) on which Newton later built the theoretical structure that provided the framework of physics and cosmology for nearly three centuries.

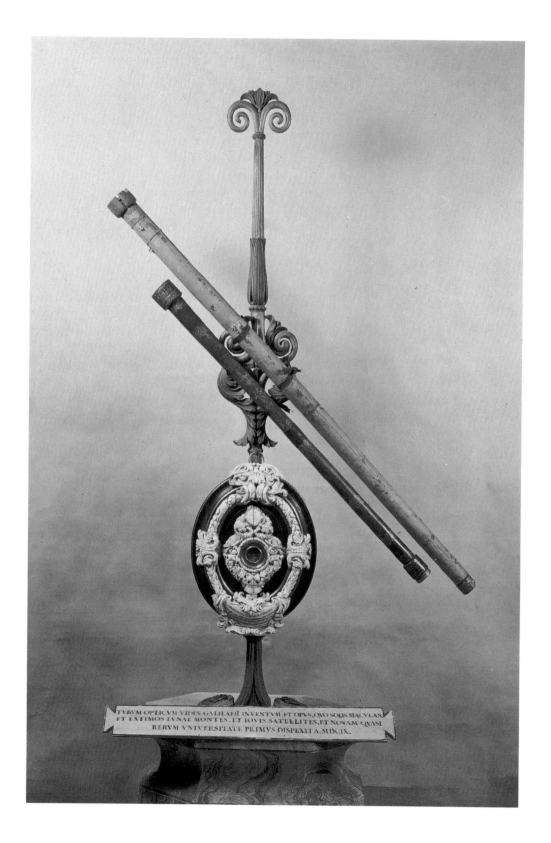

TVBVM OPTICVM VIDES GALILAEI INVENTVM, ET OPVS, QVO SOLIS MACVLAS
ET EXTIMOS LVNAE MONTES, ET IOVIS SATELLITES, ET NOVAM QVASI
RERVM VNIVERSITATE PRIMVS DISPEXIT A. MDCIX.

CHAPTER THREE

ARCHIMEDES, GALILEO

AND NEWTON

I do not know what I may appear to the world; but to myself I seem to have been only like a boy playing on the seashore, and diverting myself in now and then finding a smoother pebble or a prettier shell than ordinary, whilst the great ocean of truth lay all undiscovered before me.

SIR ISAAC NEWTON

Solutions to important scientific problems usually bring to light new and deeper problems. Kepler's laws of planetary motion explained not only the observed motions of the planets and their satellites, but also their variations in brightness and apparent size. But why should the planetary orbits be ellipses rather than some other curve? Why should the Sun lie at a common focus of these ellipses (rather than, say, at a common center)? Why should an arrow drawn from the Sun to a planet sweep out equal areas (rather than, say, equal angles) in equal times? Before we can tackle questions like these, we need an *explanatory framework* to define the kinds of questions we can meaningfully ask and the kinds of answers we can expect to receive.

Archimedean Science

The natural philosophers of the seventeenth century inherited from the Greeks two related but distinct explanatory frameworks, exemplified by the theories of Aristotle on the one hand and by those of Archimedes and the Pythagoreans on the other.

Aristotle's discussion of time and motion (in the *Physics*) is full of subtle and sophisticated mathematics; yet for Aristotle mathematics was merely a tool. His basic categories of explanation—cause and effect, form and substance, potentiality and actuality, being and becoming, rest and motion, continuity and contiguity, and so on—were drawn from common sense and intuition. Archimedes' theories, by contrast, are essentially and irreducibly mathematical. Their underlying axioms are justified not by their plausibility, as Aristotle's are, but by the scope and accuracy of their verifiable implications.

From our vantage point in history, the superiority of Archimedes' approach seems obvious. But at the beginning of the seventeenth century, the domain of Archimedean science was still small and fragmentary. Besides geometry, it included the Pythagorean theory of harmony; optics, which dealt with the apparent sizes and shapes of geometric objects and with reflection from plane and curved mirrors; Archimedes' theory of simple machines (devices such as levers

(Left) Galileo's telescope.

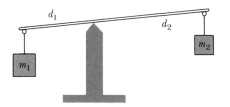

The unequal-arm balance.

and pulleys for transmitting and multiplying force); Archimedean statics, which dealt with the equilibrium of geometric objects endowed with weight; and hydrostatics, the theory that enabled Archimedes to find the proportions of gold and silver in a crown commissioned by the ruler of Syracuse, the principal Greek city of Sicily. By contrast, Aristotelean explanations, besides being easier to understand than mathematical theories, seemed to have unlimited scope. All natural phenomena came within their purview. Aristotle's own theories dominated European natural philosophy for 2,000 years.

Archimedean theories do have one advantage over Aristotelean theories, however: within their domains, narrow though these may be, they can become overdetermined by relevant data (see Chapter 1). This is illustrated by the simplest of all Archimedes' theories, the theory of the unequal-arm (Archimedean) balance.

Consider two weights hanging from a weightless rod that rests on a knife edge (see above). The theory asserts that the weights balance if and only if they are on opposite sides of the knife edge and their masses are in inverse ratio to their distances from the knife edge: symbolically, $m_1/m_2 = d_2/d_1$ or $m_1d_1 = m_2d_2$. Suppose we had an ideal balance and a collection of weights, and wished to test this theory. We might begin by selecting two weights and finding a pair of distances d_1, d_2 at which they balance. The theory then tells us the ratio of their masses: $m_1/m_2 = d_2/d_1$. Of course, this experiment does not *test* the theory; it merely tells us how to assign values to ratios of masses. But now we may vary the distances d_1, d_2, and notice whether their ratio always has the same value when two given weights balance. There is no *logical* reason why it should. (If the experiment is performed with a real balance, the ratio varies, owing to the weight of the rod.) The first experiment determines the mass ratio m_1/m_2; subsequent experiments *overdetermine* it: the mass ratio must satisfy more independent conditions than are needed to determine its value. The mutual consistency of all these independent conditions corroborates the theory.

The theory is overdetermined in another way. Suppose we select a third weight, and measure the mass ratios m_3/m_1 and m_3/m_2. There is no *logical* reason why the three measured ratios should satisfy the relation $m_3/m_2 = (m_3/m_1)/(m_2/m_1)$, even if each of the measured mass ratios has been found to depend only on the relevant distance ratio. For example, suppose that the true law was not $m_2/m_1 = d_1/d_2$ but the more complicated law $d_1/d_2 = (m_2/m_1)(1 + 0.001\ m_2/m_1)$. Then the mass ratios inferred from the simpler law and measurements of the distance ratios would pass the first test (two weights would balance whenever their distance ratio had a given value) but would fail the second. Thus, Archimedes' law is overdetermined in two qualitatively distinct ways.

The theory of the unequal-arm balance illustrates the features that characterize all modern physical theories and that distinguish them not only from Aristotelean theories but also from mathematical theories in the social sciences. An Archimedean theory has: (1) a set of mathematical axioms or laws that implicitly or explicitly define the theory's basic notions, such as mass and equilibrium in the theory of the Archimedean balance; (2) a set of well-defined procedures for testing the theory's predictions; and (3) the ability to become strongly overdetermined in some well-defined domain.

A theory that could become strongly overdetermined in a well-defined domain evidently could be *falsified* in that domain. Falsifiability is Karl Popper's well-known criterion for distinguishing scientific from nonscientific theories. A falsifiable theory, however, need not be capable of becoming strongly overdetermined, for it may not be able to generate enough testable predictions. Economic theories—at least those that are testable—belong to this category. By Popper's criterion, they are scientific; but they are not Archimedean.

Galileo's Philosophy of Science

Archimedes (*ca.* 287–212 B.C.), as envisioned by Domenico Fetti in the early seventeenth century.

The works of Archimedes became widely known in Europe during the second half of the sixteenth century through a Latin translation by the Italian mathematician Tartaglia. As the French historian Alexandre Koyré comments in his *Études Galiléenes* (p.76), Galileo became Archimedes' disciple:

> *Galileo made his scientific début with the* Bilancetta, *a treatise on the hydrostatic balance; he owed his first chair in mathematics at the University of Pisa to a work on the centers of gravity of solids, a work thoroughly Archimedean in inspiration and technique; and it was by apprenticing himself, consciously and resolutely, in the school of Archimedes, by adhering to the tradition of thought that Archimedes represented—the "ancients" [i.e., the Greek atomists] as against Aristotle—that Galileo was able to go beyond the [contemporary] physics of impetus, to attain to a level of mathematical physics that is nothing else than an Archimedean dynamics.*

Galileo was the first modern writer to articulate the double aspect of Archimedean physics: its quest for simple and general mathematical laws, and its reliance on experience to validate these laws. In *Dialogues Concerning Two New Sciences*, Galileo explains (p. 160) how he was led to postulate that a freely moving object falls at a uniformly accelerated rate:

> *Finally, in the investigation of naturally accelerated motion we were led, by hand as it were, in following the habit and custom of nature herself, in all her various other processes, to employ only those means which are most common, simple and easy.*
>
> *For I think no one believes that swimming or flying can be accomplished in a manner simpler or easier than that instinctively employed by fishes and birds.*

Galileo Galilei (1564–1642), after Susterman.

Title page of Galileo's *Dialogues on Two New Sciences,* in the first edition, 1638.

When, therefore, I observe a stone initially at rest falling from an elevated position and continually acquiring new increments of speed, why should I not believe that such increases take place in a manner which is exceedingly simple and rather obvious to everybody?

But mathematical simplicity is merely a guide; experience is the final arbiter:

Let us then, for the present, take this [his proposed law of motion] as a postulate, the absolute truth of which will be established when we find that the inferences from it correspond to and agree perfectly with experiment.

Galileo was keenly aware of the difference in outlook between Aristotelean philosophy and Archimedean physics. The Aristotelean searches for causes, and causes behind causes; the Archimedean searches for simple mathematical axioms whose consequences agree with experience:

SALVIATI. *The present does not seem to be the proper time to investigate the cause of the acceleration of natural motion, concerning which various opinions have been expressed by various philosophers . . . . Now, all these fantasies, and others too, ought to be examined; but it is not really worth while. At present it is [our] purpose . . . merely to investigate and to demonstrate some of the properties of*

Milton writes in the *Areopagitica* that he "found and visited the famous Galileo, grown old a prisoner to the Inquisition." Marjorie Nicolson has argued in *Milton and the Telescope* that looking at the heavens through a telescope "was an experience he (Milton) never forgot; it is reflected again and again in his mature work; it stimulated him to reading and to thought; and it made *Paradise Lost* the first modern cosmic poem in which a drama is played out against a background of interstellar space."

accelerated motion (whatever the cause of this acceleration may be) . . . and if we find the properties which will be demonstrated later are realized in freely falling and accelerated bodies, we may conclude that the assumed [law of free fall was correct].

Galileo points out another important difference between science and philosophy. In the opening paragraphs of the second of the *Dialogues,* he contrasts assertions about motion in "books written by philosophers" with his own mathematical results. He then continues (p. 153):

But this and other facts, not few in number or less worth knowing, I have succeeded in proving; and what I consider more important, there have been opened up by this vast and most excellent science, of which my work is merely the beginning, ways and means by which other minds more acute than mine will explore its remote corners.

Galileo is not here displaying false modesty—a vice to which he was a stranger. Almost every page communicates to the reader a sense of science as a perpetually unfinished enterprise whose answers always raise fresh questions.

Galileo's Theory of Motion

Galileo's theory offers an idealized description of the motions of objects near the surface of the Earth, in that it neglects air resistance and friction, the curvature of the Earth's surface, and the dependence of gravitational acceleration on height. The theory rests on four simple axioms, which Galileo did not state explicitly but which are implicit in his discussion. The first is a special case of what is nowadays called the law of inertia, or Newton's first law of motion. The second is Galileo's law of free fall. The third governs the motion of objects sliding without friction on inclined planes, and the fourth the motion of projectiles. Let us consider these axioms in some detail.

1. *Free horizontal motion is constant in speed and direction.*
   According to this law, an object sliding without friction on a horizontal table would never slow down, speed up, or swerve. Now, this is not an empirical generalization suggested by experience. If it were, it would read, "An object moving freely on a horizontal surface slows down and eventually stops." Instead, it refers to motion of a kind that has never been, and presumably never can be, actually observed.
   Galileo had learned from Archimedes that a physical law is more like a geometric axiom (perfect triangles and circles don't exist either!) than like an empirical generalization. But he did not simply ignore complications like friction and air resistance, since he would then have had no way to confront his theory with experience. Instead he devised experiments in which such effects were *demonstrably* negligible. For example, he dropped two balls of the same size but made of different materials "from an elevation of 150 or 200 cubits. . . . Experiment shows that they will reach the Earth with slight difference in speed, showing us that in both cases the retardation caused by the air is small."

The motion of a gliding skater illustrates Galileo's law of horizontal inertia.

Galileo deduced the law of free horizontal motion not from experience but from a *thought experiment*. Imagine an object sliding without friction down an inclined plane. It seems obvious that the object must gather speed no matter how little the plane is inclined to the horizontal. Similarly, an object sliding up an inclined plane must slow down no matter how little the plane is inclined. By symmetry, an object sliding on a perfectly horizontal plane can neither gain nor lose speed.

2. *A freely falling body is uniformly accelerated.*
A uniformly accelerated body, by definition, acquires equal increments of speed in equal increments of time. In algebraic language this law is expressed by the equation $\Delta v = g \Delta t$, where Δv is the velocity gained during the time-interval Δt, and g, a constant, represents the acceleration. Thus v, the velocity of a freely falling body at time t, is given by

$$v = g(t - t_0) + v_0, \tag{3.1}$$

where v_0 is the speed at some arbitrarily chosen time t_0.

The left figure on page 61 is a plot of vertical velocity v against time t in free fall. To estimate the distance covered between time t_0 and time t, Galileo argued as follows. Consider an object moving with constant speed $\bar{v} = \frac{1}{2}(v_0 + v)$, midway between its initial speed v_0 and its final speed v (see facing figure, on left). During this time-interval, the object moving at speed $\bar{v}$ would cover the same distance as the freely falling object. This distance is

$$d = \frac{1}{2}(v_0 + v)(t - t_0)$$

or, substituting for v its value given by Equation 3.1,

$$d = v_0(t - t_0) + \frac{1}{2}g(t - t_0)^2. \tag{3.2}$$

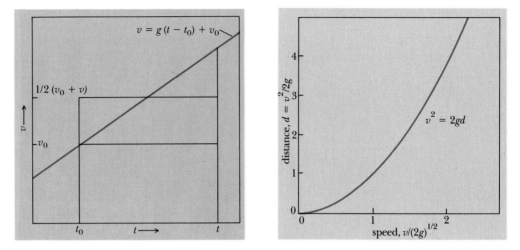

Left, a plot of vertical velocity v against time t in free fall.

Right, Galileo's relation between terminal speed v and distance d for a body falling freely from rest; g is the gravitational acceleration.

For a body falling from rest, $v_0 = 0$; so Galileo's formula becomes

$$d = \frac{1}{2}g(t - t_0)^2. \tag{3.3}$$

Using the relation $v = g(t - t_0)$ for free fall from rest, we obtain from Equation 3.3 Galileo's relation between the terminal velocity of a body falling from rest and the distance through which it has fallen:

$$v^2 = 2gd. \tag{3.4}$$

This relation is plotted in the righthand figure, above.

How did Galileo arrive at the law of free fall? His writings suggest three distinct stages:

a. *He realized that the velocity of an object falling from an initial state of rest increases smoothly from its initial value $v = 0$.* Nowadays this seems obvious, but Galileo's contemporaries believed that when a weight is released, it immediately attains a finite velocity (whose value is greater the heavier the weight), which it then maintains until it strikes the ground. Galileo invented ingenious thought experiments to demonstrate that a weight falling from rest must move very slowly immediately after it is released, and must continue to pick up speed as it descends.

b. *The choice of a specific law.* Galileo believed that the motion of falling bodies must be governed by a simple law, because simplicity is the "habit and custom" of nature. For some time he pinned his hopes on the law $\Delta v = C \Delta x$: equal increments of speed in equal incre-

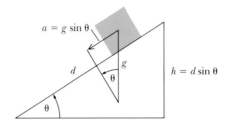

$a = g \sin \theta$

d

g

θ

θ

$h = d \sin \theta$

Galileo's axiom of acceleration on an inclined plane. In Newton's theory this axiom is a consequence of the fact that acceleration is a vector and hence can be split into components parallel to and perpendicular to the inclined plane.

ments of distance (instead of time). He abandoned this law when he realized that if it were true, a body initially at rest would have to remain at rest forever.

c. *Testing the law $v = gt$.* As we have seen, this law predicts that the distance covered in free fall from rest is proportional to the square of the elapsed time. In Galileo's day this prediction was not easy to test. An accurate clock had yet to be invented, and Galileo usually used his own pulse to measure time-intervals. Thus the shortest time-interval he could hope to measure with an accuracy of, say, 10 percent was around 10 seconds. But in 10 seconds an object falling from rest covers nearly half a kilometer. Galileo circumvented the practical difficulties of measuring large distances and short time-intervals by using inclined planes. To interpret his measurement, he needed the following postulate.

3. *A body sliding without friction down an inclined plane experiences the uniform acceleration $g \sin \theta$, where g is the acceleration in free fall and θ is the plane's inclination to the horizontal.*
   This relation is illustrated in the figure above; it reduces to the law for free fall when the plane is vertical ($\theta = \pi/2$), and to the law of inertia when the plane is horizontal ($\theta = 0$). By using inclined planes with small inclinations, Galileo was able to test the hypothesis of uniform vertical acceleration.

   Galileo's law implies that the terminal speed of an object sliding down a frictionless inclined plane from an initial state of rest depends only on the distance through which it has fallen, not on the angle of inclination, because the terminal velocity v is related to the distance d by the formula

$$v^2 = 2(g \sin \theta)d;$$

but $d = h/\sin \theta$, where h is the height of the starting point (see figure above); hence $v^2 = 2gh$.

Galileo took great pride in this formula because it furnished a purely geometric definition of velocity. It reduced the measurement of velocity—a tricky business before the invention of an accurate clock—to the measurement of distance: to measure the speed of a moving object, let it slide up a frictionless plane and notice how far it goes; to impart a velocity v to an object, let it slide down a frictionless plane through a vertical distance $v^2/2g$.

4. *Galileo's principle of relativity and the motion of projectiles.*

Consider, with Galileo, the following thought experiment. A weight is dropped from the top of a sailboat's vertical mast. Where does it hit the deck? Some of Galileo's contemporaries said: "The answer depends on whether the boat is moving or at rest. If it is at rest, the weight will strike the deck vertically below its point of release, but if it is moving, the weight will strike the deck behind this point." This is certainly the right answer. But Galileo argued that the path of the falling weight deviates from the vertical only because of air resistance. In a vacuum it would always strike the deck directly below its point of release, provided the boat was moving with constant speed in a fixed direction.

This assumption enabled Galileo to deduce that the path of a weight dropped from the mast of a uniformly moving boat, as viewed by someone standing on the shore, is a parabola. The argument is simple. The weight's motion in the vertical direction is not affected by its horizontal motion. Hence in t seconds the weight falls a vertical distance $d = \frac{1}{2}gt^2$, and its height above the deck is given by the formula

$$y = y_0 - \tfrac{1}{2}gt^2, \tag{3.5}$$

where y_0 is the height at the moment of release, $t = 0$. The horizontal motion of the weight must be exactly the same as the boat's horizontal motion, which we have assumed to be uniform. Hence the weight's horizontal coordinate x is given by

$$x = V_x t, \tag{3.6}$$

where V_x is the speed of the boat and x is measured from the weight's initial position. By combining Equations 3.5 and 3.6 we obtain the equation of the path:

$$y = y_0 - \tfrac{1}{2}(g/V_x^2)x^2 \tag{3.7}$$

(see the figure on the next page).

In this problem the only function of the boat is to impart a certain initial velocity V_x to the weight. A projectile launched in any other manner with the same horizontal velocity will follow precisely the same path.

If the weight has an initial vertical velocity V_y, then according to Equation 3.2 its height at time t is given by

$$y = V_y t - \tfrac{1}{2}gt^2. \tag{3.8}$$

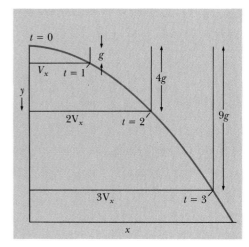

Path of a freely falling object in uniform forward motion.

As before, its horizontal coordinate $x = V_x t$. Hence its path, as viewed by a stationary observer, has the equation

$$y = y_0 + (V_y/V_x)x - \tfrac{1}{2}(g/V_x^2)x^2, \tag{3.9}$$

which is the equation of a parabola. This is the path of a projectile launched at height y_0 with horizontal velocity V_x and vertical velocity V_y.

The Rotation of the Earth and the Nature of Centrifugal Force

In *Dialogues Concerning the Two Chief World Systems*, Simplicio, the spokesman for Aristotelean physics, argues (p. 196):

And I allowed myself, simple-mindedly enough, to be convinced that stones would not be extruded by the whirling of the Earth! I take it back, then, and declare that if the Earth did move, then stones, elephants, towers, and cities would necessarily fly toward the heavens; and since that does not happen, I say that the Earth does not move.

Galileo's response to this Aristotelean argument reveals the precise limits of his understanding of centrifugal force and its connection with gravity. Galileo knew that a heavy object whirled at the end of a string tends to stretch the string. He also knew that when the string breaks, the object moves with constant speed along a tangent to its former trajectory. Starting from these premises, he argued substantially as follows.

Let the circle $AB'C'$ above represent the Earth's equator, A, B', and C' being successive positions of a particle carried along by the Earth's rotation. If gravity were suddenly abolished

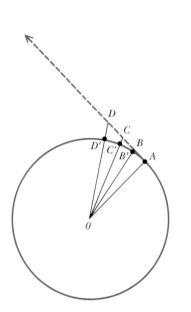

An object whirling at the end of a string. A, B', C', D' are its positions at equal time-intervals. $ABCD$ is its trajectory if the string is cut when the body is at A. If angle AOD is very small, $\overline{AB'} = \overline{AB'} = \overline{BC'} = \overline{CD'}$.

Christian Huygens (1629–1695), Dutch mathematician and physicist, who used symmetry arguments to derive conservation laws for momentum and energy. He also used what Einstein later called the principle of equivalence to derive a formula for centrifugal force; and he invented the pendulum clock.

when the particle was at A, it would travel with constant speed along the tangent ABC. If the angle AOB is sufficiently small, the distances $\overline{AB}$ and $\overline{AB'}$ in the figure are nearly equal, and $\overline{BB'}$ represents the distance through which a particle moving on the circular trajectory *falls away* from the straight-line trajectory AB in traveling from A to B'. The shorter the distance $\overline{AB}$, the smaller the distance $\overline{BB'}$; but the *ratio* $\overline{BB'}/\overline{AB}$ also gets smaller and smaller as B approaches A ($\overline{BB'}$ is in fact proportional to the *square* of $\overline{AB}$, so that $\overline{BB'}/\overline{AB} \propto \overline{AB}$). So far the argument is sound. But now Galileo makes a false step. He argues that, because the ratio $\overline{BB'}/\overline{AB'}$ approaches zero as $\overline{AB'}$ approaches zero, an arbitrarily weak gravitational force directed toward O will prevent a particle from leaving the surface of the Earth. But this is not true, because the ratio $\overline{BB'}/\overline{AB'}$ is not a valid measure of the relative importance of the centrifugal and gravitational forces.

The first correct account of centrifugal force was given by Christian Huygens (1629–1695) twenty years after Galileo's death. Huygens based his discussion on a physical principle that

The roller coaster exploits two principles clarified by Huygens: centrifugal force, which keeps the cars pressed to the rails as they travel around the loop; and the interchangeability of kinetic and potential energy.

Einstein, nearly three centuries later, called the "principle of equivalence": in any small region of space, a gravitational acceleration may be simulated (or canceled) by an appropriate motion of the frame of reference. Consider a heavy object suspended by a string; now imagine the same object whirling in a horizontal circle at the end of the same string. For some rate of whirling, the tension in the string is the same in both cases. Huygens *postulated* that the two physical situations are then *dynamically equivalent* and presented the following argument. In the first case—the hanging weight—the tension in the string results from the weight's tendency to fall at a uniformly accelerated rate; so in the second case—the whirling weight—the tension in the string must result from an identical tendency. *A whirling object tends to fall away from the center of its circular trajectory at a uniformly accelerated rate.*

Galileo's law of free fall, Equation 3.3, says that the distance traversed in free fall from an initial state of rest is proportional to the square of the elapsed time. According to the preceding argument, the distances $\overline{AB}$, $\overline{AC}$, and $\overline{AD}$ in the figure on p. 65 are equal to the corresponding distances $\overline{B'B}$, $\overline{C'C}$, and $\overline{D'D}$ shown here on the left. These distances are proportional to the squares of the corresponding elapsed times. But the elapsed times are proportional to the lengths of the circular arcs AB', AC', and AD' in the figure on p. 65. Thus

Uniform acceleration of a falling object; the trajectory *ABCD* of the lefthand figure on page 65 viewed in the rotating reference frame.

$$\frac{\overline{CC'}}{\overline{BB'}} = \frac{\overline{AC}^2}{\overline{AB}^2}.$$

This relation is approximately correct if the angle *AOC* is very small.

Using Galileo's law of free fall and his own principle of eqivalence, Huygens was now able to give a correct (though incomplete) account of centrifugal force. He used Galileo's formula $d = \frac{1}{2}gt^2$ to calculate the acceleration g of the body falling from A to D in terms of the distance $d = \overline{BB'}$ and the elapsed time $t = \overline{AB'}/v$:

$$g = \frac{2d}{t^2} = \frac{2\overline{BB'}}{(\overline{AB'}/v)^2},$$
(3.10)

where v is the velocity of the body in its circular orbit. Now $\overline{BB'}$ and $\overline{AB'}$ are both proportional to r, the radius of the circle. Hence, the quantity $\overline{BB'}/\overline{AB'}^2$ is inversely proportional to r, and $r\overline{BB'}/\overline{AB'}^2$ is independent of r. Equation 3.10 may therefore be written

$$g = \frac{Cv^2}{r}, \qquad C \equiv \frac{2r\overline{BB'}}{\overline{AB'}^2},$$
(3.11)

where C is a pure number whose value does not depend on the size of the circle and depends only weakly on the size of the small angle AOB. We will see later that C approaches unity as the angle AOB shrinks to zero.

Finally, Huygens argued that because the tension in a plumbline is proportional to the mass of the suspended object, the tension in the string restraining a whirling object must likewise be proportional to the object's mass. Thus the centrifugal force exerted by an object whirling with speed v at the end of a string of length r is given by the formula

$$F = \frac{mv^2}{r},$$
(3.12)

where m is the object's mass. The force F represents the object's effective weight; we could measure it with a spring balance as shown on the next page.

In the preceding argument Huygens uses, but does not actually mention, some basic concepts of the theory of motion that Newton formulated many years later: the proportionality between an object's weight (F) and its mass (m), and Newton's relation between force, mass, and acceleration ($F = mg$), and the equality of action (the object's weight) and reaction (the tension in the restraining string). In the history of science, use often precedes mention; new concepts usually make their appearance in concrete contexts long before their full generality is recognized.

Huygens was now in a position to answer Simplicio's objection to the hypothesis of a spinning Earth. At the equator the centrifugal force arising from the Earth's rotation acts exactly like gravity, but in the opposite direction. It reduces the weight of an object by an amount given by Equation 3.12, where r is the Earth's equatorial radius and v is the rotational speed at the

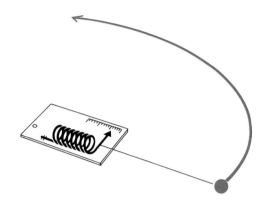

The effective weight of a whirling object as
measured by a spring balance.

equator. Because the reduction in weight is proportional to the mass (as measured by an Archimedean balance), the *fractional* reduction is the same for every object. It is just the ratio between the centrifugal acceleration given by Equation 3.11 with $C = 1$ and the gravitational acceleration $g = 980$ cm/sec^2. The Earth's rotational velocity at the equator is

$$v = \frac{2\pi r}{T} = \frac{2\pi \times 6.38 \times 10^8 \text{ cm}}{8.64 \times 10^4 \text{ sec}} = 4.64 \times 10^4 \text{ cm/sec},$$

where r = radius of the Earth and T = length of the day. Hence, the centrifugal acceleration at the equator is $v^2/r = 3.37$ cm/sec^2. The measured value of the gravitational acceleration g is 980 cm/sec^2; hence $(v^2/r)/g = 290 \simeq 17^2$. It follows that the Earth would need to spin 17 times as fast as it actually does—with a period of 24/17 hours—for the centrifugal acceleration at the equator to equal the gravitational acceleration. If it rotated still faster, then "stones, elephants, towers, and cities" would indeed "fly toward the heavens."

The Inverse-square Law

In 1663 it occurred to Isaac Newton that, just as a whirling stone is kept in a circular orbit by a pull directed toward the center of the circle, so the Moon must be kept in its nearly circular orbit by a pull directed toward the center of the Earth, and the planets must be kept in their nearly circular orbits by pulls directed toward the Sun (see the figure p. 69). This was neither the most original nor the most difficult step that Newton had to take to construct his theory of gravitation; yet it was less obvious than it may seem to a modern reader. Kepler, unaware of the law of inertia, believed that the planets, left to their own devices, would slow down and stop. He speculated that the planets are pushed along in their orbits by a transverse force transmitted from the rotating Sun to each planet by invisible flagella. Galileo went to the opposite extreme. He believed that no force at all was needed to keep the planets moving in their orbits. Huygens

Sir Isaac Newton (1642–1727) at age 46; portrait by Kneller.

"That by means of centripetal forces the planets may be retained in certain orbits, we may easily understand, if we consider the motions of projectiles . . . for a stone that is projected is by the pressure of its own weight forced out of the rectilinear path, which by the initial projection alone it should have pursued, and made to describe a curved line in the air; and through that crooked way is at last brought down to the ground; and the greater the velocity is with which it is projected, the farther it goes before it falls to the Earth. We may therefore suppose the velocity to be so increased, that it would describe an arc of 1, 2, 5, 10, 100, 1000 miles before it arrived at the Earth, till at last, exceeding the limits of the Earth, it should pass into space without touching it." (From Newton, *Principia*, Vol. II, p. 551.)

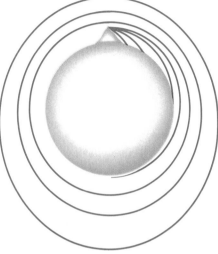

understood that if the planets and their moons were not constrained by forces, they would fly off in straight lines, but he refused to entertain the idea of a stringless pull, or *action at a distance*. Newton himself did not like the idea very much but wisely decided to work out its mathematical implications rather than worry about its metaphysical shortcomings.

Newton speculated that the pull exerted by the Earth on the Moon also causes the accelerations of falling bodies and projectiles. Now, Galileo had demonstrated experimentally that all bodies, regardless of their mass, composition, shape, or size, experience the same gravitational acceleration near the surface of the Earth. Newton forthwith assumed that the same rule applies to the Moon. The gravitational acceleration g_M at the Moon's distance r_M then depends only on the value of r_M and is given by Huygens' formula (with $C = 1$),

$$g_M = \frac{v^2_M}{r_M} = \frac{4\pi^2 r_M}{T^2_M}, \tag{3.13}$$

where T_M is the Moon's orbital period.

Let us compare this acceleration with the gravitational acceleration g_E at the Earth's surface. As we saw in the preceding section, an artificial satellite skimming the surface of the Earth has a period of 24/17 hours. By analogy with Equation 3.13, we can write

$$g_E = \frac{4\pi^2 r_E}{T^2_E}, \tag{3.14}$$

where r_E is the radius of the Earth and $T_E = 1/17$ day. Dividing Equation 3.14 by Equation 3.13, we obtain

$$\frac{g_E}{g_M} = \left(\frac{r_E}{r_M}\right)\left(\frac{T_M}{T_E}\right)^2. \tag{3.15}$$

The Moon's distance r_M is about 60 Earth radii r_E, and the Moon's orbital period is 27.3 days; hence

$$\frac{g_E}{g_M} = \frac{1}{60} \times (17 \times 27.3)^2 \simeq 3590 \simeq \left(\frac{r_M}{r_E}\right)^2. \tag{3.16}$$

This calculation shows that the gravitational acceleration at a distance of 60 Earth radii is about 60^2 times smaller than the gravitational acceleration at the surface of the Earth, which suggests the rule $g \propto 1/r^2$.

In exactly the same way, we can compare the gravitational accelerations of the planets toward the Sun. In Equation 3.15 we may interpret the letters E and M as labels for two of the planets—Earth and Mars, say. Kepler's third law of planetary motion (Chapter 2) states that $(T_M/T_E)^2 = (r_M/r_E)^3$. Inserting this formula in Equation 3.15, we obtain

$$\frac{g_E}{g_M} = \left(\frac{r_M}{r_E}\right)^2. \tag{3.17}$$

Since E and M may refer to any two planets, it follows that the gravitational accelerations of the planets toward the Sun are inversely proportional to the squares of their distances from the Sun.

The Moon as a Falling Body

Having postulated that each planet is pulled toward the Sun by a force whose strength varies inversely as the square of its distance, Newton now had to investigate whether a planet that was not moving in a circle would trace out an elliptical orbit at a rate determined by Kepler's law of areas. To solve this problem, he had to invent a new way of looking at accelerated motion and a new mathematical language to express it.

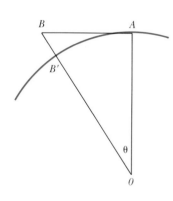

The figure to the left is a detail of the figure on p. 65, illustrating Huygens' theory of centrifugal force. As we saw, Huygens argued that if the string restraining a whirling stone were cut when the stone was at A, the stone would fly off along the tangent AB; he *assumed* that in the rotating reference frame, the stone would appear to drop at a uniformly accelerated rate. Newton reversed this argument: he asserted that the whirling stone (or a planet moving at constant speed in a circular orbit) is continually falling away from the straight trajectory it

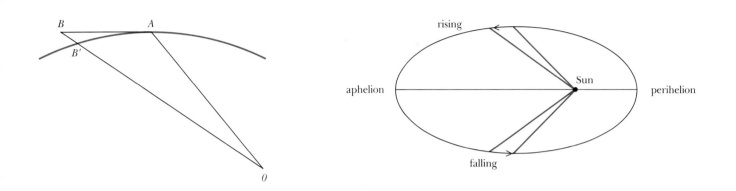

would follow if it were not continually pulled toward the center of the circle. Although the stone is continually falling toward the point O, it never gets any closer, because the direction in which it is falling is continually changing. The stone must continually fall just to stay at the same distance from the center of its orbit.

This may seem a very complicated way of looking at uniform circular motion, but it has a decisive advantage over Huygens' more intuitive description: it generalizes at once to nonuniform, noncircular motion. In the lefthand figure above, AB' is a short piece of a noncircular orbit, AB is tangent to the orbit, and the short segment BB' points toward a distant center of attraction at O. Because the tangent AB is not perpendicular to the radius OA, the motion is not purely "horizontal" (i.e., at right angles to the direction of the pull). The short segment BB' is not the total "vertical" displacement but only the part of the vertical displacement that is caused by the pull directed toward O. The curved segment AB' is exactly analogous to an oblique segment of the parabolic trajectory of a projectile moving near the surface of the Earth. A planet moving in a noncircular orbit "rises" (i.e., moves away from the Sun) as it travels from perihelion to aphelion, then "falls" back toward the Sun as it moves from aphelion to perihelion (see above, right). The vertical component of the planet's motion is accelerated toward the Sun, and the magnitude of the acceleration is inversely proportional to the square of the distance.

To express these ideas in precise terms, Newton invented a new kind of mathematics, which he called "the method of first and last ratios." (Nowadays it is called—less appropriately as well as less poetically—the differential calculus.) To illustrate the method, let us calculate the acceleration of a particle in uniform circular motion. Huygens *assumed* that the vertical displacement $\overline{BB'}$ in the figure on the facing page is proportional to the square of the elapsed time and hence to the square of the distance $\overline{AB'}$. That is, he assumed that the ratio $\overline{BB'}/\overline{AB'}^{2}$ does not depend on the angle AOB if that angle is very small. Newton *proved* that the ratio is close to ½ when the angle is small, and that it gets closer and closer to this value as the angle shrinks to zero. Newton expressed this relation between the ratio and its "ultimate" value by saying that the ratio *converges* to ½. The figure on the next page illustrates the convergence of the ratio to its limiting value.

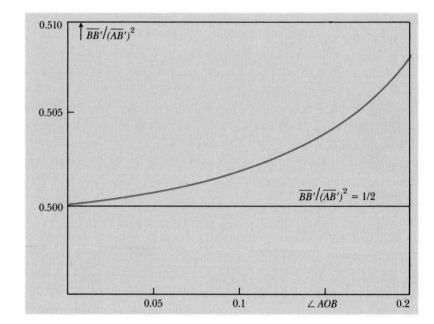

Newton used the method of first and last ratios to prove many deep and beautiful theorems. Nevertheless, some of Newton's contemporaries objected to the method on principle. They argued that the "last ratio" of two "evanescent" quantities (*i.e.,* quantities shrinking toward zero) or the "first ratio" of two "nascent" quantities (*i.e.,* quantities growing from zero) is a ratio of the form 0/0 and hence meaningless. Newton points out in *Principia* (pp. 38–39) that

> *this objection is founded on a false supposition. For those ultimate ratios with which quantities vanish are not truly the ratios of ultimate quantities, but limits toward which the ratios of quantities decreasing without limit do always converge; and to which they approach nearer than by any given difference, but never go beyond, nor in effect attain to, till the quantities are diminished in infinitum.*

This explanation did not satisfy all of Newton's critics. Bishop George Berkeley, who was born in the year it was published, argued that first and last ratios are incomprehensible: "He who can conceive the beginning of a beginning, or the end of an end . . . may be perhaps sharpsighted enough to conceive these things. But most men will, I believe, find it impossible to understand them in any sense whatever." He goes on to remark that anyone credulous enough to accept Newton's mathematics need not "be squeamish about any point in Divinity."

Newton's definition of "ultimate ratio" depends on an intuitive perception of motion, just as Euclid's definitions of "point" ("that which has no part") and "line" ("breadthless length") depend on an intuitive perception of space. An intuitive definition fulfills its purpose if it evokes an appropriate mental image. Euclid's and Newton's definitions satisfy this modest require-

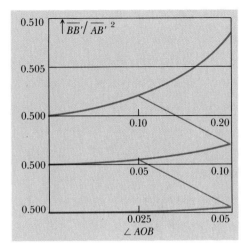

The ratio $\overline{BB'}/\overline{AB'}^2$ converges to the value ½ as the angle AOB shrinks to zero. That is, a plot of this ratio against angle AOB is indistinguishable from the horizontal line $\overline{BB'}/\overline{AB'}^2 = $ ½ in a sufficiently small neighborhood of the point $\angle\ AOB = 0$. This property serves to *define* the "ultimate ratio" of two "evanescent quantities."

ment; to ask more of them would be unreasonable. If we wish to know what a point or a line or an ultimate ratio "really is," we must study its *use*. It is true that Newton's notion of the first ratio of two nascent quantities or the last ratio of two evanescent quantities does not stand up under analysis. But the actual construction of such ratios is clear and unambiguous—provided certain conditions are met. The values of the quantity $\overline{BB'}/\overline{AB'}^2$, plotted against the angle AOB, lie on a smooth curve. If we halve the angle AOB and double the horizontal scale in the figure to the left, the curve remains smooth (see above). If we continue to halve the angle and double the scale, the plotted portion of the curve becomes indistinguishable from the straight line given by $\overline{BB'}/\overline{AB'}^2 = $ ½. It is this property of the *aggregate* of values of the ratio $\overline{BB'}/\overline{AB'}^2$ for small but finite values of the angle AOB that enables us to *assign* a definite value to the ultimate ratio.

Fluxions

Among the ultimate ratios that figure in Newton's theory, those that represent rates of change play especially important roles. Newton called them *fluxions*. Nowadays ultimate ratios are called *derivatives,* and fluxions are called *time derivatives*.

Let Q be any quantity whose value changes smoothly with time. Let ΔQ denote the change in Q during a time-interval Δt centered on time t. The ratio $\Delta Q/\Delta t$ represents the average rate of change of Q during this time-interval. Because we have assumed that Q changes smoothly with time, this ratio converges to a definite limit as the time-interval Δt shrinks to zero. Newton denoted this limit, the rate of change of Q at time t, by $\dot{Q}$, a notation still in use. (An alternative notation for the same limit, introduced by Leibniz, is dQ/dt. We may think of dQ and dt as values of ΔQ and Δt so small that their ratio does not differ significantly from its limiting value.)

Constructing $\dot{Q}$ at a point P on the curve $Q(t)$. Choose a nearby point R. In a first approximation $\dot{Q}$ is the slope of the chord PR, $i.e.$, the ratio RS/PS. Halve the time-interval PS to obtain the point S' and draw the chord PR'. Its slope $R'S'/PS'$ is a second approximation to $\dot{Q}$. Continue this process. (It is helpful to halve the scale of the relevant part of the diagram at each step, as indicated in the right side of the figure.) In the limit, the chord PR becomes the tangent PT, whose slope ST/PS is the derivative $\dot{Q}$ at P.

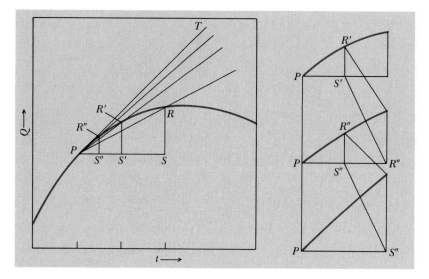

If Q represents the distance that a particle has traveled from a given starting point, $\Delta Q/\Delta t$ represents its average speed during the time-interval Δt, and $\dot{Q}$ represents its instantaneous speed. If the particle's motion is sufficiently smooth, its instantaneous speed $\dot{Q}$ also varies smoothly with time. The quantity $\Delta\dot{Q}$ represents the change in the speed $\dot{Q}$ during a time-interval Δt, and the ratio $\Delta\dot{Q}/\Delta t$ represents the average rate of change of $\dot{Q}$ during this time-interval. Finally, $\ddot{Q}$ (or $d\dot{Q}/dt$ or d^2Q/dt^2; all these notations are equivalent) represents the instantaneous rate of change of the instantaneous rate of change of Q. Newton called such quantities fluxions of fluxions; nowadays they are called second time derivatives.

Bishop Berkeley considered the fluxion of a fluxion to be an especially ridiculous concept, like the ghost of a ghost. Yet second—and higher—derivatives are easy enough to construct. We begin by plotting Q against time (see above). At any point on this curve we can carry through the construction illustrated above, which shows that the value of $\dot{Q}$ at any given value of t is the *slope* of the curve that represents Q at the point that corresponds to the given value of t.

Since we can in principle calculate the value of the slope $\dot{Q}$ at each value of t, we can construct a new curve that represents the variation of $\dot{Q}$ with t. We can now repeat the construction illustrated above, and calculate the rate of change of $\dot{Q}$ at any value of t. This rate of change is equal to the slope of the curve that represents $\dot{Q}$ at the value of t under consideration.

We can now go on to construct the third time derivative $\dddot{Q}$, the slope of the curve that represents $\ddot{Q}$. There is no end to this process. Any segment, no matter how small, of the curve $Q(t)$—Q plotted against t—contains all the information we need to construct all of the derivatives of Q at any point on this segment. Each derivative tells us something about how the curve is changing. The first derivative, the slope of the curve, specifies the rate of change at a given

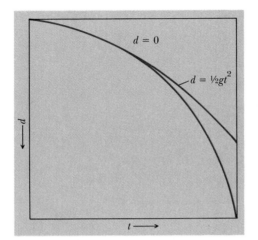

Fall of a particle, initially at rest, toward a fixed point mass. The line $d = 0$ and the curve $d = \frac{1}{2}gt^2$ represent successive approximations to the curve $d(t)$.

moment. The second derivative specifies the rate at which the rate of change is changing; the third derivative tells us how fast the rate of change of the rate of change is changing; and so on. The more derivatives we know at a given moment, the more accurately we can predict future values of Q and retrodict past values. If we knew the values of all the derivatives at a single moment, we could reconstruct the entire curve. The entire history of any smoothly varying quantity is implicit in its tendencies at a single moment of time.

Let us consider an example: a particle falling from rest along a straight line toward the center of the Earth, idealized as a point-mass. In the figure above, the distance d through which the particle has fallen is plotted against the time t. At the initial instant $t = 0$ the particle is at rest and the slope of the curve $d(t)$ vanishes: $\dot{d} = 0$. If we were to extrapolate the particle's behavior from this datum alone, we would represent $d(t)$ by the horizontal straight line $d = 0$ tangent to the curve at $t = 0$, representing a permanent state of rest. But because the particle is accelerated toward the center of the Earth, its speed is increasing at a finite rate, g, the gravitational acceleration appropriate to the particle's initial distance. Taking into account the initial rate of change of the speed, we would represent the distance traveled by Galileo's formula $d = \frac{1}{2}gt^2$ (see above).

This provides a much better representation of the motion, but it fails to take into account the fact that the acceleration itself increases as the particle's distance from the center of the Earth diminishes. Information about the rate of growth of the acceleration is implicit in the derivatives of the acceleration $\ddot{d}$ at the initial moment $t = 0$.

By taking such information into account, we could further improve our extrapolation. The effects of successive derivatives become noticeable at progressively later times. The further into the future we wish to extrapolate the behavior of $d(t)$, the more of its derivatives at $t = 0$ we need to know.

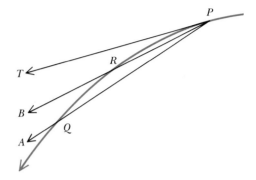

$\overrightarrow{PA}$ is the particle's average velocity for the time-interval during which it moves from P to Q; $\overrightarrow{PB}$ is its average velocity for the shorter time-interval corresponding to PR; and $\overrightarrow{PT}$, the tangent to the trajectory at P, is the instantaneous velocity.

Velocity, Acceleration, and Force

Newton's theory of motion, and the mathematical language he invented to express it, hinge on the law of inertia. As we saw earlier, Galileo formulated a law of inertia for horizontal motion at the surface of the Earth: if the Earth were a perfectly smooth sphere, a body sliding without friction on its surface would travel with constant speed along a great circle and eventually return to its starting point; no force would be needed to keep it moving. Newton recognized that this "law of horizontal inertia" is not as simple as it seemed to Galileo. A body sliding without friction on the surface of a smooth spherical Earth is not really moving freely. Two opposing forces act on it: the Earth's gravitational pull, directed toward the center of the Earth, and an equal push (reaction) exerted by the Earth's surface in the opposite direction. The horizontal motion is unaccelerated because no horizontal force acts on the body. But if the Earth's gravitational pull were suddenly abolished, the body would move off on a tangent to its circular path. And if it were not for the resistance offered by the Earth's solid surface, the body would fall toward the center of the Earth along a spiral path. The idealized motion is simple because the Earth's gravitational pull and the push exerted by the Earth's surface cooperate to keep the object moving in a perfectly circular orbit in which they exactly cancel each other. (If the object begins to sink it is pushed back by the surface; if it begins to rise it is pulled back by gravity.) *But if no forces at all were acting on a body it would continue to move at constant speed in a straight line.*

This is Newton's *first law of motion.* In more concise language, *the velocity of a free particle is constant in time.* A particle's velocity is a *vector:* it has both magnitude (the particle's speed) and direction. It can be represented by an arrow whose length equals the particle's instantaneous speed, measured in some convenient unit, and whose direction is parallel to the instantaneous direction in which the particle is moving. (For a review of vector arithmetic, see Appendix.) The figure above shows that at any moment a particle is moving in the direction of the tangent to its trajectory. If the trajectory is curved, the direction of motion changes continually. It follows from Newton's first law of motion that a particle moving in a curved trajectory cannot be moving freely; it must be acted upon by a force.

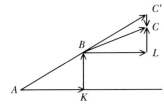

Displacement of a projectile during successive short time-intervals.

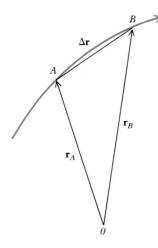

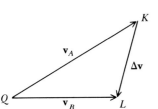

How does an applied force affect a particle's velocity? Galileo's theory of projectiles suggested the answer. Galileo argued that the vertical motion of a projectile is not affected by its horizontal motion. The vertical motion is uniformly accelerated; the horizontal motion is uniform. The top figure to the left shows a projectile's displacement during two short, equal time-intervals. During the first time-interval the projectile moves from A to B; its horizontal displacement is $\overrightarrow{AK}$, its vertical displacement $\overrightarrow{KB}$. During the following time-interval the projectile moves from B to C. Its horizontal displacement $\overrightarrow{BL}$ is equal to the horizontal displacement $\overrightarrow{AK}$ during the preceding time-interval; its vertical displacement $\overrightarrow{LC}$ is smaller than the preceding vertical displacement $\overrightarrow{KB}$, owing to the gravitational acceleration. In the absence of gravity, the projectile's displacement during the second time-interval would have been $\overrightarrow{BC'}$, equal in magnitude and direction to the displacement $\overrightarrow{AB}$. The displacement $\overrightarrow{C'C}$ results from, and represents, the vertical acceleration.

Newton recognized that not only displacement but also velocity, acceleration, and force are vectors, and that the laws of motion describe relations among vectors. The second figure down on the left shows that a particle's displacement $\overrightarrow{AB}$ during a time-interval Δt is equal to the change in its radius vector drawn from an arbitrary origin O: $\overrightarrow{OA}+\overrightarrow{AB}=\overrightarrow{OB}$, or $\overrightarrow{OB}-\overrightarrow{OA}=\overrightarrow{AB}=\Delta\mathbf{r}$, where $\Delta\mathbf{r}$ means the change in the radius vector $\mathbf{r}$ during the time-interval Δt. The average velocity during this time-interval is $\overrightarrow{AB}/\Delta t = \Delta\mathbf{r}/\Delta t$. The magnitude of this vector is $\overrightarrow{AB}/\Delta t$, the particle's average speed during the time-interval Δt. As this time-interval shrinks, the average speed converges to the instantaneous speed, and the direction of the average velocity (which is the same as the direction of the displacement $\overrightarrow{AB}$) converges to the direction of the tangent to the curve at the particle's instantaneous position. Thus the velocity $\mathbf{v}$ is the rate of change of the radius vector $\mathbf{r}$:

$$v = \dot{\mathbf{r}} \equiv d\mathbf{r}/dt. \qquad (3.18)$$

Analogously, the average acceleration during the time-interval Δt is the ratio $\Delta\mathbf{v}/\Delta t$, where $\Delta\mathbf{v}$ is the difference between the instantaneous velocities at points B and A separated by the time-interval Δt, as shown in the third figure down on the left. The analogy becomes clearer if the two velocity vectors are drawn from a common origin, as in the bottom figure on the left. The velocity vectors $\mathbf{v}_A$ and $\mathbf{v}_B$ in the third figure down on the left are analogous to the radius vectors $\mathbf{r}_A$ and $\mathbf{r}_B$ in the second figure down on the left, and the velocity displacement $\Delta\mathbf{v}$ is analogous to the position displacement $\Delta\mathbf{r}$. As the time-interval Δt shrinks, the ratio $\Delta\mathbf{v}/\Delta t$ converges to the instantaneous acceleration, which we denote by $\mathbf{a}$:

$$\mathbf{a} = \dot{\mathbf{v}} = d\mathbf{v}/dt. \qquad (3.19)$$

The magnitude a of the acceleration vector $\mathbf{a}$ equals the speed of the tip of the velocity vector $\mathbf{v}$ in the bottom figure on the left, and the direction of the acceleration coincides with the direction of the tangent to the curve swept out by the tip of the velocity vector.

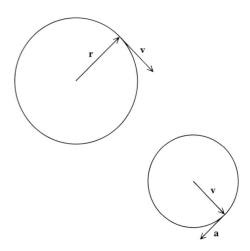

As the radius vector **r** sweeps out a circle, rotating at a constant rate, the velocity vector **v**, which is always perpendicular to the radius vector, sweeps out a second circle. The acceleration vector **a** is related to the velocity vector **v** in the same way as **v** is related to the position vector **r**.

The relations between the radius vector, the velocity, and the acceleration are especially simple for uniform circular motion (see above). As the tip of the radius vector traces out a circle of radius r, the tip of the velocity vector traces our a circle of radius $2\pi r/T$, where T is the period of the motion. Because the acceleration is related to the velocity in exactly the same way as the velocity is related to the radius vector, their magnitudes are in the same ratio: $a/v = v/r$ or $a = v^2/r$. And because the velocity is perpendicular to the radius vector, and the acceleration is perpendicular to the velocity, the acceleration and the radius vector are *antiparallel.* That is, the acceleration points to the center of the circular orbit. *A particle moving with constant speed v in a circular orbit of radius r experiences an acceleration of magnitude v^2/r directed toward the center of the circle.*

A stone whirling at the end of a string is kept in its circular orbit by the string's pull. Analogously, the Moon is kept in its nearly circular orbit by the pull of the Earth, and the planets are kept in their nearly circular orbits by the pull of the Sun. Each of these pulls acts in a definite direction and has a definite magnitude. The same is true of other forces. For example, the reaction of a surface to an object resting on it is perpendicular to the surface. Thus force, like displacement, velocity, and acceleration, is a directed quantity and so is represented by a vector.

In uniform circular motion the acceleration and the force that keeps the particle in its circular orbit are both directed toward the center of the circle. This suggests the following rule: *The acceleration of a particle is proportional to the applied force:* $\mathbf{a} \propto \mathbf{F}$.

Experience shows that the more massive an object is, the smaller the acceleration produced by a given applied force—for example, the force exerted by a spring that has been compressed by a fixed amount. The simplest rule that is consistent with this observation is Newton's *second law of motion: The acceleration $\mathbf{a}$ produced in an object of mass m by an applied force $\mathbf{F}$ is proportional to $\mathbf{F}$ and inversely proportional to m:* $\mathbf{F} = m\mathbf{a}$.

Newton's proof that a planet obeys Kepler's law of areas if and only if its acceleration is directed toward the Sun. If the planet's motion were unaccelerated, the planet would travel along the straight line ABC covering the equal distances $\overline{AB}$, $\overline{BC}$ in equal time-intervals Δt. The triangles SAB and SBC have equal areas because they have equal bases ($\overline{AB} = \overline{BC}$) and a common altitude $\overline{ST}$. Now, suppose the planet is accelerated toward S. During the time-interval Δt its velocity changes by $\Delta\mathbf{v} = \overrightarrow{BK} = \overrightarrow{CC'}$. Because BK and CC' are parallel, triangles SBC and SBC' have the same altitude (the distance between SB and CC') and hence the same area. Conversely, if they have the same area, CC' and SB are parallel, and $\Delta\mathbf{v}$ must point toward S (i.e., the point K must lie on SB).

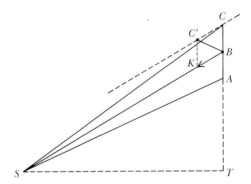

Kepler's Laws Interpreted

Is the acceleration of a planet moving in conformity with Kepler's laws directed toward the Sun? Propositions I and II of Newton's *Principia* answer this question: *A particle's acceleration is directed toward a fixed center if and only if the radius vector joining the center to the particle sweeps out equal areas in equal times* (see above). In other words, Kepler's law of areas is precisely equivalent to the statement that a planet's acceleration always points toward the Sun.

If a planet's acceleration is directed toward the Sun, the Sun must lie in its orbital plane. This is clear from the construction of the displacement $\overline{BC'}$ in the figure above.

The magnitude of the Sun's pull, F, depends only on the planet's distance r; the *direction* of the radius vector cannot affect the magnitude of the pull because, from the Sun's point of view, all directions are equivalent. Whatever the value of F at a given radius, we can construct a circular orbit with that radius. The planet's speed in this orbit is determined by the relation $F/m \equiv a = v^2/r$.

If the orbit is not a circle, however, the force exerted by the Sun varies from point to point. This variation determines the shape of the orbit. For most laws of variation, noncircular orbits are also nonperiodic; the orbiting object does not return to the same point with the same velocity. One of the few force laws for which all bound orbits are periodic is the inverse-square law, in which the Sun's pull diminishes as the inverse square of the distance: $F \propto 1/r^2$. Newton proved that for this law, and only for this law, the orbits are ellipses with the Sun at one focus. Newton's geometric proof is elegant and difficult. An easy algebraic proof, using elementary calculus, is given in the box on the next two pages.

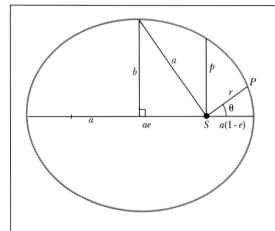

Kepler's First Law and the Inverse-Square Law of Gravitation

To prove: If the orbit is an ellipse with the Sun at one focus and the planet obeys Kepler's law of areas, the Sun's pull varies as the inverse square of the distance.

The radial and transverse velocity components are

$$v_r = \dot{r}, \qquad v_\theta = r\dot{\theta}. \tag{1}$$

Kepler's law of areas says that

$$rv_\theta = r^2\dot{\theta} = \text{constant} \equiv h \tag{2}$$

and implies that the acceleration is purely radial. The acceleration is

$$
\begin{aligned}
a_r &= \dot{v}_r - \frac{v_\theta^2}{r} \\
&= \ddot{r} - r\dot{\theta}^2
\end{aligned}
\tag{3}
$$

by Equation 1. The term v_θ^2/r arises from the changing direction of the transverse velocity, as in uniform circular motion. Thus

$$a_r = \ddot{r} - \frac{h^2}{r^3} \tag{4}$$

by Equations 2 and 3. The equation of an ellipse, as shown in the figure, is

$$\frac{1}{r} = \frac{1 + e \cos \theta}{p} \tag{5}$$

Differentiating,

$$\frac{d(1/r)}{dt} = \frac{-\dot{r}}{r^2} = \frac{-e\dot{\theta} \sin \theta}{p}$$

$$= \frac{-eh \sin \theta}{pr^2}$$

by Equation 2, whence

$$\dot{r} = \frac{eh \sin \theta}{p}.$$

Differentiating again,

$$\ddot{r} = \frac{eh\dot{\theta} \cos \theta}{p} = \frac{eh^2 \cos \theta}{pr^2}.$$

Substituting this formula in Equation 4, we obtain

$$a_r = \ddot{r} - \frac{h^2}{r^3} = \left(\frac{h^2}{r^2}\right) \frac{e \cos \theta}{p} - \frac{1}{r}$$

$$= \frac{-h^2}{pr^2} \tag{6}$$

(by Equation 5)

$$\equiv \frac{-M}{r^2}. \tag{7}$$

Q.E.D

The preceding argument is reversible. If we start with the inverse-square law (7), we can deduce that the orbit is described by Equation 5.

Are ellipses the *only* possible orbits for a particle whose attraction to a fixed center is inversely proportional to the square of its distance? The equation of the ellipse is

$$r = \frac{p}{1 + e \cos \theta}, \tag{3.20}$$

Kepler ellipses with a fixed perihelion distance *SP* and progressively more distance second
foci *F′*, *F″*. As the second focus
moves off to infinity, the ellipse
approaches a parabola.

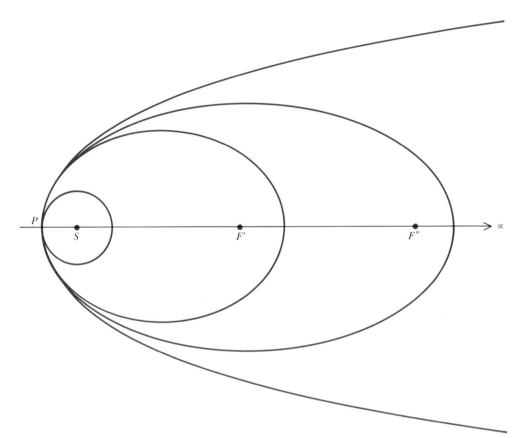

where θ is the angle between the radius vector and the major diameter of the ellipse, and p is a constant (the value of the distance r when $\theta = 90°$). The eccentricity e is the ratio between the Sun's distance from the center of the ellipse and the ellipse's major semidiameter; so it must lie between 0 and 1. When $e = 0$, the ellipse is the circle $r = p$. For progressively larger values of e, the ellipse is progressively more elongated; the second focus (the one not occupied by the Sun) recedes farther and farther into the distance (see above). When $e = 1$, the orbit is a parabola. But the formula makes sense even for values of e greater than 1: for such values it describes a hyperbola, as in the facing figure.

The inverse-square law links Kepler's first two laws to his third law. We may express the (constant) rate at which a planet's radius vector sweeps out area as the ratio between the area A of its elliptical orbit and its period T:

$$\frac{1}{2}h = \frac{A}{T} = \frac{\pi a b}{T}.$$

(3.21)

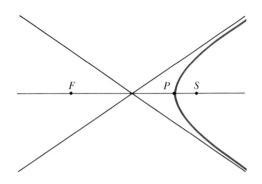

A particle coming from infinity executes a hyperbolic orbit with the Sun at one focus, S. A proton encountering a second proton executes a hyperbolic orbit with the second proton at the distant focus, F.

Here h is the constant defined by Equation 2 (p. 80). By Equations 6 and 7 (p. 81) $h^2 = pM$, where M is the constant of proportionality in the inverse-square law. From the figure on page 80 and Equation 3.20, the equation of the ellipse, we find that $p = a(1 - e^2)$, $b = a\sqrt{1 - e^2}$. Hence we may rewrite Equation 3.21 as

$$\frac{1}{2}h = \frac{1}{2}(pM)^{1/2} = \frac{1}{2}[Ma(1 - e^2)]^{1/2}$$
$$= \frac{\pi a^2(1 - e^2)^{1/2}}{T}$$

or

$$\left(\frac{2\pi}{T}\right)^2 a^3 = M, \tag{3.22}$$

which is Newton's form of Kepler's third law. It is surprising that the orbital eccentricity e does not figure in this relation, and even more surprising that Kepler was able to foresee this feature of the law.

Gravitation as a Universal Force

In discussing the motions of the planets, we treated the Sun as a fixed center of attraction. In discussing the Moon's motion, we treated the Earth as fixed. But if the Earth attracts the Moon, must it not also attract the Sun? And if the Sun is accelerated by the Earth's attraction, we cannot regard it as a fixed center of attraction for the planets. Such considerations led Newton to conclude that in a self-consistent theory of gravitation every particle in the universe must attract and be attracted by every other particle.

GRAVITATIONAL MASS AND INERTIAL MASS As a first step toward understanding the implications of this idea, let us consider a simplified model of the universe, one that contains only the Sun and the Earth. The Earth's acceleration a_E by the Sun is inversely proportional to the square of the distance r between the Earth and the Sun:

$$a_E = \frac{M_S}{r^2}. \tag{3.23}$$

The constant of proportionality M_S is called the *gravitational mass* of the Sun. Since we know the Earth's acceleration toward the Sun—by Huygens' formula, it is v^2/r, or $(2\pi/T)^2 r$, where T is the Earth's orbital period—we can assign M_S a definite numerical value.

Analogously, the Sun's gravitational acceleration by the Earth is given by the formula

$$a_S = \frac{M_E}{r^2}, \tag{3.24}$$

where M_E is the Earth's gravitational mass.

Gravitational mass is a measure of a body's attractive power. How is it related to its inertial mass, which is a measure of a body's resistance to acceleration by an applied force? The answer follows from the third of Newton's three laws of motion: *"To every action there is always opposed an equal reaction: or, the mutual actions of two bodies upon each other are always equal and oppositely directed."*

Thus the force exerted by the Sun on the Earth must be equal in magnitude and opposite in direction to the force exerted on the Sun by the Earth. By Newton's second law, the magnitudes of the two forces are $F_E = m_E M_S/r^2$ and $F_S = m_S M_E/r^2$. Hence the two forces (which act in opposite directions) will be equal in magnitude if $m_E M_S = m_S M_E$ or $m_E/M_E = m_S/M_S$. Because such a relation holds for every pair of particles in the universe, *inertial and gravitational mass are proportional to each other*.

The formula $a = M/r^2$ enables us to calculate a body's gravitational mass from the acceleration it produces in a test particle at any given distance. For example, consider a test particle moving in a circular orbit of radius r and period T. Its centripetal acceleration is given by $a = v^2/r = 4\pi^2 r/T^2$. The gravitational mass M required to produce this acceleration is $M = r^2 a = 4\pi^2 r^3/T^2$. Thus the gravitational mass of the Sun is $4\pi^2$(astronomical unit)^3/(year)^2. What is the Sun's inertial mass? We are free to assign any convenient value to the constant ratio M/m. The simplest course would be to set the ratio equal to unity, but convention has decreed that inertial mass be measured in a special and arbitrary unit, the gram (gm). This makes the ratio M/m a dimensional constant, denoted by G. Its value is

$$\frac{M}{m} \equiv G = 6.67 \times 10^{-8}\ \text{cm}^3 \cdot \text{sec}^{-2} \cdot \text{gm}^{-2}. \tag{3.25}$$

"THE STILL POINT OF THE TURNING WORLD" The central concept in Archimedes' theory of static equilibrium is that of *center of mass*. Two masses connected by a massless rod parallel to the *x*-axis balance (in the Earth's gravitational field) at a point whose *x*-coordinate X is given by

$$(m_1 + m_2)X = m_1 x_1 + m_2 x_2. \tag{3.26}$$

This condition is just a restatement of the rule that the masses lie on opposite sides of the balance point at distances $d_1 = |x_1 - X|$, $d_2 = |x_2 - X|$, which are inversely proportional to the masses: $d_1/d_2 = m_2/m_1$. If the rod connecting the masses is not parallel to the *x*-axis, the *y*-coordinate of the balance-point satisfies an analogous condition; and if the rod is not horizontal, the *z*-coordinate of the balance point also satisfies the same condition with z_1, z_2 in place of x_1, x_2 and Z in place of X. All three conditions are summed up in the vector equation

$$(m_1 + m_2)\mathbf{R} = m_1 \mathbf{r}_1 + m_2 \mathbf{r}_2, \tag{3.27}$$

where $\mathbf{r}_1$, $\mathbf{r}_2$, and $\mathbf{R}$ are radius vectors to mass 1, mass 2, and the center of mass, respectively, drawn from an arbitrary origin.

Returning to our idealized universe that consists of two point-masses representing the Sun and the Earth, let us calculate the velocity $\mathbf{V}$ and the acceleration $\mathbf{A}$ of the center of mass:

$$(m_1 + m_2)\mathbf{V} = (m_1 + m_2)\frac{d\mathbf{R}}{dt} = m_1 \frac{d\mathbf{r}_1}{dt} + m_2 \frac{d\mathbf{r}_2}{dt} \tag{3.28}$$
$$= m_1 \mathbf{v}_1 + m_2 \mathbf{v}_2,$$

$$(m_1 + m_2)\mathbf{A} = (m_1 + m_2)\frac{d\mathbf{V}}{dt} = m_1 \mathbf{a}_1 + m_2 \mathbf{a}_2 \tag{3.29}$$
$$= \mathbf{F}_1 + \mathbf{F}_2 = 0,$$

where in the last step, we have invoked Newton's law of action and reaction. Equation 3.29 says that the motion of the center of mass is unaccelerated. Hence its velocity $\mathbf{V}$ is constant, and its radius vector $\mathbf{R}$ is given by

$$\mathbf{R} = Vt + \mathbf{R}_0, \tag{3.30}$$

where $\mathbf{R}_0$ is the value of the radius vector at time $t = 0$.

We can simplify our description of how the two point-masses move by using a frame of reference that moves with the velocity $\mathbf{V}$ of the center of mass. Newton's three laws of motion hold in this new frame if they held in the original frame. The first law holds because a particle whose velocity $\mathbf{v}$ is constant in the old frame also has constant velocity $(\mathbf{v} - \mathbf{V})$ in the new frame.

Newton's second and third laws continue to hold because neither accelerations nor forces are altered when the same constant velocity is subtracted from the velocity of each particle. In the new frame the center of mass is at rest; and we might as well put it at the origin of our reference frame. Thus in the new frame $\mathbf{V} = \mathbf{R} = 0$.

The radius vector $\mathbf{R}$ of a system of n point-masses is defined by a formula analogous to Equation 3.27:

$$(m_1 + \ldots + m_n)\mathbf{R} = m_1\mathbf{r}_1 + \ldots + m_n\mathbf{r}_n. \tag{3.31}$$

The force acting on particle 1 is the sum of the forces exerted on this particle by the remaining particles:

$$\mathbf{F}_1 = \mathbf{F}_{12} + \mathbf{F}_{13} + \ldots + \mathbf{F}_{1n}. \tag{3.32}$$

Similarly,

$$\mathbf{F}_2 = \mathbf{F}_{21} + \mathbf{F}_{23} + \ldots + \mathbf{F}_{2n}, \tag{3.33}$$

and so on. The sum of the n forces $\mathbf{F}_i$ is a sum of contributions analogous to $(\mathbf{F}_{12} + \mathbf{F}_{21})$, one from each distinct pair of particles. Each such contribution vanishes, by Newton's law of action and reaction. Hence the sum of the forces vanishes. It follows from Equation 3.29, generalized to n particles, that the acceleration of the center of mass vanishes. Hence *the center of mass of an isolated system of interacting particles is unaccelerated.* As for two particles, we may conveniently choose a frame of reference in which the center of mass of the idealized system we are considering is permanently at rest at the origin.

MOMENTUM AND THE LAW OF CONSERVATION OF MOMENTUM   The quantity $m\mathbf{v}$ is called *momentum* and is denoted by $\mathbf{p}$. Newton's second law of motion, $m\mathbf{a} = \mathbf{F}$, may be expressed more compactly as

$$\frac{d\mathbf{p}}{dt} = \mathbf{F}. \tag{3.34}$$

That the velocity of the center of mass of an isolated system of interacting particles is constant implies that the system's total momentum is constant in time:

$$\mathbf{p}_1 + \ldots + \mathbf{p}_n = \mathbf{P} = (m_1 + \ldots + m_n)\mathbf{V} = \text{constant}. \tag{3.35}$$

Newton's law of action and reaction implies that interacting particles *exchange* momentum: an interaction between two particles does not change the sum of their momenta.

The product of mass and acceleration is equal to the rate of change of momentum if and only if the mass is constant in time:

$$\frac{d\mathbf{p}}{dt} = \frac{d(m\mathbf{v})}{dt} = \frac{m\,d\mathbf{v}}{dt} + \left(\frac{dm}{dt}\right)\mathbf{v}. \qquad (3.36)$$

For a system whose mass is changing—a rocket ejecting exhaust gases, a comet passing near the Sun, a star moving through a dense cloud of interstellar dust—the relation does not hold: $d\mathbf{p}/dt \neq m\mathbf{a}$. Which of the two forms of Newton's second law—$d\mathbf{p}/dt = \mathbf{F}$ or $m\mathbf{a} = \mathbf{F}$—holds when m is not constant in time? The first form, Equation 3.34, implies that the total momentum of an isolated system—the sum of the momenta of the rocket and its exhaust—is constant in time; the second does not. Now, as we will see in the next chapter, the momentum of an isolated system is constant for reasons that go even deeper than Newton's second and third laws of motion. This suggests that force should be equated to the rate of change of momentum, as Newton himself recognized and as experience confirms.

COMPACT SYSTEM OF INTERACTING PARTICLES IN AN EXTERNAL GRAVITATIONAL FIELD In discussing the Earth's motion around the Sun, we treated the Earth as if it were a point-mass instead of an extended system composed of very many interacting particles. We are now in a position to justify that idealization. We may think of the force acting on a particle of the Earth as the sum of two components: an internal component—the force exerted by other particles of the Earth—and an external component—the force exerted by the Sun and other distant masses. Thus we write Newton's second law for the ith particle in the form

$$m_i\mathbf{a}_i = \mathbf{F}_{i_{\text{int}}} + \mathbf{F}_{i_{\text{ext}}}. \qquad (3.37)$$

There is one such equation for every particle in the Earth. Let us sum these equations. The left side of the sum is $m\mathbf{A}$, where $m = m_1 + \cdots + m_n$, the combined mass of the particles—that is, the mass of the Earth. On the right, the sum of the internal forces vanishes, since, as we have seen, these forces cancel in pairs. So the result of summing Equation 3.37 is

$$m\mathbf{A} = \sum_i \mathbf{F}_{i_{\text{ext}}}. \qquad (3.38)$$

Because the Earth's diameter is very much smaller than the distance between the Earth and the Sun, the gravitational acceleration produced by the Sun is practically the same at every point in the Earth. Call this acceleration $\mathbf{g}$; then $\mathbf{F}_{i_{\text{ext}}} = m_i\mathbf{g}$ and when we sum the quantities $\mathbf{F}_{i_{\text{ext}}}$, we find

$$\sum_i \mathbf{F}_{i_{\text{ext}}} = m\mathbf{g}. \qquad (3.39)$$

From Equations 3.38 and 3.39 it follows that $\mathbf{A} = \mathbf{g}$: the center of mass moves as if the Earth's mass were concentrated at that point. A more detailed analysis would show that the difference between the acceleration $\mathbf{A}$ of the center of mass and the external gravitational acceleration $\mathbf{g}$ at that point is proportional to the *square* of the ratio between the diameter of the Earth and its distance from the Sun, as well as to a measure of the Earth's departure from sphericity. Thus the Earth's center of mass moves almost exactly as if its entire mass were concentrated in that point.

The acceleration of the ith particle relative to the center of mass of its parent system is $\mathbf{a}_i - \mathbf{A}$. From Equations 3.37 to 3.39, we see that

$$m_i(\mathbf{a}_i - \mathbf{A}) = \mathbf{F}_{i_{int}}. \tag{3.40}$$

Thus the accelerations of particles in or near the Earth, viewed in a frame of reference moving with the Earth's center of mass, are produced entirely by other particles of the system. The motion of the Earth's center of mass effectively cancels the Sun's gravitational attraction. Only the differences between the gravitational attractions at different points within the system have observable consequences. We will examine some of these consequences in Chapter 4.

The Gravitational Field

During the nineteenth century, physicists constructed a new way of looking at Newton's theory of gravitation. They shifted their attention away from the particles that give rise to gravitational accelerations, and onto the empty space between particles. The psychologist of art Rudolf Arnheim has characterized this shift as a perceptual development, analogous to the shift in the history of European painting from a "neat distinction between foreground and background" in pre-Renaissance painting to "an unbroken sequence of shape and color values" in post-Renaissance painting. In the designs of the contemporary Dutch artist M. C. Escher, for example, figure and ground assume equal importance.

The shift of attention from particles to the space between particles came first in the context of electromagnetism and was initiated by the experimental physicist Michael Faraday (1791–1867). James Clerk Maxwell (1831–1879) formulated Faraday's ideas in mathematical language. Here is how Maxwell described Faraday's way of thinking:

> *Faraday, in his mind's eye, saw lines of force traversing all space, where the mathematicians saw centres of force attracting at a distance; Faraday saw a medium where they saw nothing but distance; Faraday sought the seat of the phenomena in real actions going on in the medium, they were satisfied that they had found it in a power of action at a distance impressed on the electric fluids.*

Newton's theory, as we have considered it up until now, is a *particle theory*, a theory of particles and their interactions. Viewed in the new way, it becomes a theory in which particles and

Michael Faraday (1791–1867), discoverer of the law of electromagnetic induction, whose applications include electric motors and generators, here lecturing to the public at the Royal Institution in London.

their *gravitational field* play equal and complementary roles. Particles are the sources of the gravitational field, which in turn acts on the particles. Particles do not interact directly and at a distance, as in Newton's picture; instead, every particle is accelerated by the gravitational field at the place where it happens to be. The field picture abolishes action at a distance and replaces the Void by a Plenum.

Although the particle and field pictures are mathematically equivalent, we will see that the field picture is not only aesthetically pleasing, but also has certain practical advantages. Moreover, it is an indispensable stepping-stone to Einstein's theory of gravitation (Chapter 6). In Einstein's theory the field occupies the center of the stage; particles play supporting roles.

The figure on page 90 shows the gravitational field of a point-mass. Field lines radiate isotropically from the point-mass; their points of intersection with the surface of a sphere centered on the point-mass are evenly distributed over the surface. The field lines shown are, of course, merely representative; in reality, every point where the gravitational acceleration is not zero lies on a field line. It is convenient to stipulate that the number of field lines that terminate

The gravitational field of a
point-mass.

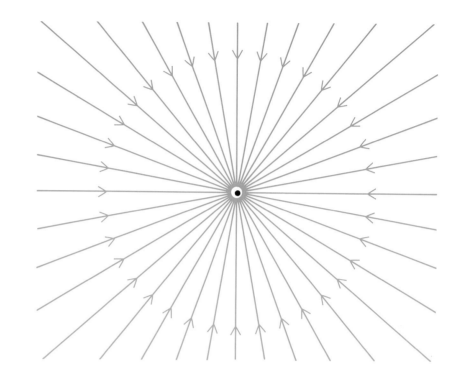

on a point-mass shall be a fixed multiple of that mass. Thus in the figure on the next page, which shows the gravitational field of two unequal point-masses, the ratio of the masses is the same as the ratio of the numbers of field lines that terminate on them.

The gravitational acceleration at any point is along the field line through that point. Thus in the figure on p. 91, the gravitational acceleration at each point is directed toward the point-mass that is the source of the gravitational field. In the figure above, the field lines are approximately radial and evenly spaced close to each of the two sources of the field.

The magnitude of the gravitational acceleration at a given point is proportional to the density of the field lines at that point. Where the field lines are close together, the field is strong (the acceleration is large); where they are far apart, the field is weak (the acceleration is small). Consider the field of a point-mass, as shown in the figure above. Let the total number of field lines that terminate on the point-mass be $4\pi N$. At every point on the surface of a sphere of radius r, the density of field lines is $4\pi N/4\pi r^2$, or N/r^2. Because we have stipulated that the number of field lines N is proportional to the mass M that is the source of the field, the figure above is a pictorial statement of Newton's law of gravitation as applied to a point-mass: the gravitational acceleration points toward the point-mass, and its magnitude is inversely proportional to the square of its distance from the point-mass.

The gravitational field of two unequal point-masses ($m_2 = 2\,m_1$). The figure shows field lines in a plane through the line connecting the two masses. Note that the number of field lines terminating in the mass m_2 is twice as great as the number terminating in the mass m_1; near each mass, the pattern of field lines resembles that of an isolated point mass; at distances much larger than the separation of the two masses, the field approximates that of a single point-mass $m_1 + m_2$ located at the center of mass O; and there is a single point P where the gravitational attraction vanishes and through which no field line passes.

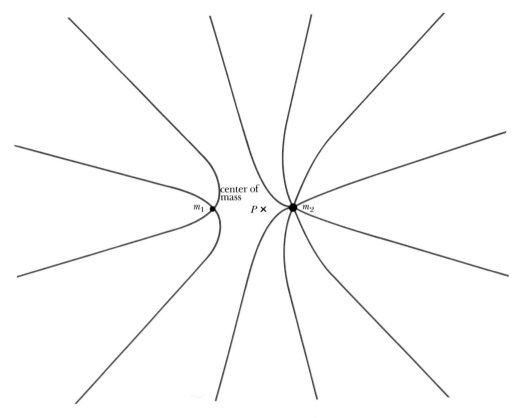

The gravitational field of a pair of unequal point-masses, shown on page 91, illustrates the three rules (all of which we have already encountered) for interpreting all such diagrams.

1. The gravitational acceleration at any point is along the field line through that point.

2. The magnitude of the gravitational acceleration at any point is proportional to the density of field lines at that point. (The constant of proportionality is the number of field lines per unit mass.)

3. Field lines terminate only in regions that contain mass; the number of field lines that terminate in a given volume of space is proportional to the mass within that volume.

Using the concept of the gravitational field, we can easily prove an important theorem that gave Newton much difficulty. In his discussion of the apple and the Moon, Newton assumed that the Earth attracts both objects as if its mass were concentrated in a point at its center. Is this assumption justified? Let us consider the *external* gravitational field of an idealized, perfectly spherical Earth, as shown on page 92. Because the mass distribution that produces the field has spherical symmetry (that is, it does not discriminate between points equally distant from the

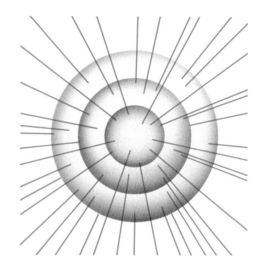

The gravitational field of a spherically symmetric distribution of mass.

center of symmetry), the field lines outside the Earth must be purely radial, and the points in which they pierce the surface of the idealized Earth must be distributed evenly on that surface. Moreover, by the third of the three rules just given, the number of field lines piercing the surface of the Earth is a fixed multiple of the mass within that surface. It follows that the external gravitational field is indistinguishable from the field that would be produced beyond the surface of the Earth if all of the Earth's mass were concentrated in a point at its center. This is the theorem that Newton wished to prove.

A companion theorem about spherically symmetric mass distributions is just as easy to prove: a spherical shell of uniform density exerts no gravitational force on a particle inside the shell. As in the preceding argument, we may infer from the symmetry of the mass distribution that the field inside the shell is purely radial. If it did not vanish, the field lines would have to meet at the center, which is impossible, because there is no mass at the center. Hence the field vanishes inside the shell.

This theorem has an important cosmological application. Astronomers detect mass in two main ways: by means of the light radiated by massive objects, especially stars; and by means of its gravitational effects. The first way is unreliable, because much of the mass in the universe may be invisible. For example, it may be bound in stars that have used up their available sources of nuclear energy and are now too dim to be recorded by the most sensitive light-detectors available. The second method is in general more reliable but has this limitation: the motions of (say) stars in a stellar system are not affected by invisible matter distributed in a spherical shell or halo around the system. As discussed briefly in Chapter 8, evidence accumulated during the

Maurits Cornelis Escher's "Day and Night" (1938).

past few years suggests that much, or even the bulk, of the mass in the universe resides in the nonluminous halos of galaxies. Earlier estimates of the masses of galaxies and of the average mass density of the universe have had to be increased by as much as a factor of ten to allow for this nonluminous contribution.

"Newtonian physics," said the French philosopher Henri Bergson, "descended from Heaven to Earth along the inclined plane of Galileo." In Chapters 2 and 3 we have traced this descent. In Chapter 4 we will see how Newton and his successors used his theory to reshape the heavens—to build a theoretical structure that for over two centuries supplied a framework for interpreting astronomical observations.

CHAPTER FOUR

NEWTON'S THEORY AND THE ASTRONOMICAL UNIVERSE

The universe, by its immeasurable greatness and the infinite variety and beauty that shine from it on all sides, fills us with silent wonder. If the presentation of all this perfection moves the imagination, the understanding is seized by another kind of rapture when, from another point of view, it considers how such magnificence and such greatness can flow from a single law, with an eternal and perfect order.

IMMANUEL KANT

Isaac Newton would have had no difficulty understanding a modern theoretical paper on the dynamics of star clusters or galaxies; he would have recognized it as an application of his own theory of gravitation. But he would not have recognized the astronomical universe to which that theory was being applied. Newton's *System of the World* (Book III of the *Principia*) dealt almost entirely with phenomena known to Hipparchus: the motions of the Moon and the planets, tides, the precession of the equinoxes, the orbits of comets. But between Newton's day and ours, telescopic observations have increased the radius of the observable universe by a factor of 100 million, and have revealed systems and structures undreamed of in the eighteenth century. In this chapter we will see how Newton's theory and its ramifications have succeeded in explaining the most conspicuous features of astronomical systems that range in size from light-seconds to tens of millions of light-years. We will also see, however, that Newton's theory does not provide an adequate theoretical framework for understanding the structure and dynamics of the universe as a whole.

Binaries

By postulating that the Earth is spinning, the Pythagoreans removed the need for a crystalline sphere to support the stars in their daily rotation. And once the stars had been set adrift, there was no longer any reason to suppose that they all lay at the same distance from the Earth. On the other hand, there was no compelling reason to believe that the stars do *not* occupy a thin shell centered on the Earth. Copernicus believed that the Sun is the "Lamp of the Universe" and sits enthroned at its center. Kepler held a similar view and cited the following experiment:

if . . . you pierce through a wall with only a pin, so that the sun can shine through the hole, a greater brightness is poured through from the beams than all the fixed stars shining together in a cloudless sky would give.

(Left) Replica of Isaac Newton's reflecting telescope (1671).

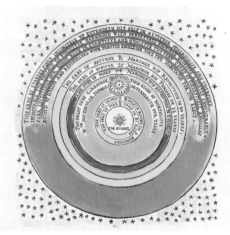

Thomas Digges' version of the Copernican universe, with the solar system embedded in an infinite sea of stars.

Scottish mathematician James Gregory (1638–1675).

The opposing view, that the stars are suns sprinkled more or less uniformly throughout an infinite space, was put forward by Greek philosophers in the fifth century B.C. and was revived in the sixteenth century by Thomas Digges and Giordano Bruno. In the following century Huygens carried out a refined version of Kepler's experiment to estimate the distance of the bright star Sirius, assuming it to be intrinsically as bright as the Sun. Like Kepler, Huygens was unable to make a pinhole small enough to look like a star when the Sun shone through it into a darkened room; so he devised a simple optical arrangement that enlarged the Sun's image until its diameter was 30,000 times that of the pinhole. The illuminated pinhole then "appeared of much the same clearness with Sirius." Huygens concluded that if Sirius is intrinsically as bright as the Sun, it is 30,000 times as distant as the Sun. (In fact, Sirius is much brighter than the Sun and therefore much more distant.)

In the *Principia* Newton describes an equally ingenious method, invented by James Gregory, for estimating the distance of Sirius. The method hinges on the observation that Sirius is about as bright as the planet Saturn. But Saturn shines by light reflected from the Sun. Its luminosity L_{Saturn} is thus a product of three factors: the Sun's luminosity L_{Sun}; the fraction f of the Sun's light intercepted by Saturn; and the fraction A (for *albedo*) of the incident sunlight reflected by the planet. Knowing Saturn's distance from the Earth, we can calculate its diameter from its angular size; knowing its diameter and its distance from the Sun, we can then calculate the fraction f. The albedo A is harder to estimate. Newton simply assumed that Saturn was made of some rocklike substance and guessed that its albedo was ¼ (the modern value is 0.43). Kepler's third (harmonic) law gave the relative dimensions of the planetary orbits, and the French expedition of 1672 (see Chapter 2) provided a reasonably accurate absolute scale. Newton could now calculate the factor f and thus estimate L_{Saturn}/L_{Sun}. The rest of the argument parallels Huygens'. Because Saturn and Sirius appear equally bright, the quantity Af must equal the square of

William Herschel and his reflecting telescope. The tube is 40 feet long.

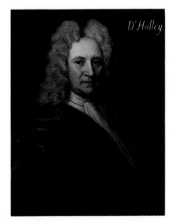

Edmond Halley (1656–1742) discovered that the "fixed stars" are not fixed.

the ratio of their distances from the Earth. From this argument Newton concluded that Sirius must be 100,000 times more distant than the Sun if the two stars are equally luminous.

Considering the difficulties inherent in both Huygens' and Newton's estimates, the agreement between them is surprisingly good. Huygens was comparing a bright pinhole in a darkened room with his *recollection* of Sirius on a clear night. And neither he nor Newton had any notion of what a "reasonable" estimate of Sirius' distance would be.

To decide between, on the one hand, Kepler's view that Sirius and other stars are much fainter than the Sun and, on the other, the view of Huygens and Newton that Sirius is as bright as the Sun, astronomers needed a direct estimate of Sirius' distance, an estimate that did not depend on an assumption about its intrinsic brightness. The only way of estimating the distance of an object whose intrinsic properties are completely unknown is to measure its parallactic shift (see Chapter 2). If Sirius were at the distance estimated by Newton, it would trace out annually on the celestial sphere an ellipse with a semidiameter of 2 arc-seconds (see the figure on p. 35). In Newton's day astronomers could not have measured such a small parallax (a star's parallax is the semidiameter of its parallactic ellipse), but they tried anyway. Stellar parallax became the Holy Grail of eighteenth-century astronomy, and its pursuit resulted in three major serendipitous discoveries: the aberration of starlight (by James Bradley in 1727; see Chapter 5); the motion of "fixed" stars on the celestial sphere (by Edmond Halley in 1718); and double stars (by William Herschel in 1803). A genuine parallactic shift was first accurately measured in 1836 by F. G. W. Struve at the Pulkova Observatory near St. Petersburg (now Leningrad).

William Herschel (1738–1822) was a talented composer and oboist who abandoned music for astronomy at the age of 35 after reading a popular textbook. A few years later he had an idea that, he hoped, would enable him to measure the first stellar parallax. Herschel had noticed that close to many bright stars there is a faint star. Like Huygens and Newton, Herschel assumed that all stars have the same intrinsic brightness. The bright star in a close pair would then be much less distant than the faint star; hence its parallax would be much larger. This reasoning led Herschel to expect that the angular separation between the members of such

If the members of a close pair
are at substantially different distances from the Earth, as
Herschel assumed, their parallactic ellipses will have substantially different diameters; hence
the angular separation of the
pair will vary periodically with a
period of one year.

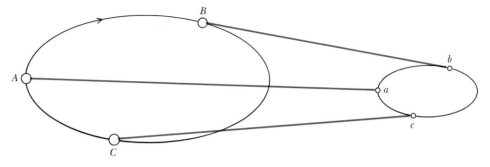

pairs would vary with a period of exactly one year, as shown above. If he could detect such a change, he could estimate the parallax of the brighter star.

Between 1782 and 1784, Herschel measured the separations of some 700 double stars. He then remeasured their separations and found several changes; none of these, however, could be interpreted as parallactic shifts. By 1803 Herschel was convinced that he had discovered something much more interesting: orbital motions. Most of the double stars on his list were not stars at greatly different distances that happened to lie in almost the same direction, but physical pairs held together by their mutual gravitational attraction and revolving about their center of mass in accordance with Newton's laws.

Herschel's discovery had several important implications. It provided the first direct evidence that Newton's law of gravitation holds outside the solar system. It showed that Huygens and Newton had been badly mistaken in their belief that all stars are intrinsically about as bright as the Sun, for the brighter member of many physical pairs is 100 times (or more) as luminous as its companion. It suggested (and subsequent observations confirmed) that double systems are common among stars. Finally, and most importantly, observations of double stars, interpreted in the light of Newton's theory, yield the most direct and reliable estimates we have of stellar masses.

The theory needed to interpret observations of double stars is remarkably simple and elegant. Consider two mutually gravitating stars in a frame of reference whose origin coincides permanently with their center of mass, as depicted in the top figure on the facing page. The radius vectors of the two stars then satisfy the relation $M_1\mathbf{r}_1 + M_2\mathbf{r}_2 = 0$, from which it follows that $\mathbf{r}_1$, $-\mathbf{r}_2$, and $\mathbf{r}_{12} = \mathbf{r}_1 - \mathbf{r}_2$ (the radius vector drawn from star 2 to star 1) are all parallel, and that

$$r_1 : r_2 : r_{12} = M_2 : M_1 : M_1 + M_2.$$

These relations enable us to recast the relation

$$\mathbf{a}_1 = -\frac{M_2\mathbf{r}_{12}}{r_{12}^{\,3}}, \tag{4.1}$$

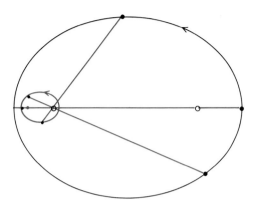

Each member of a double star traces out a Keplerian ellipse, with the pair's center of mass at a common focus of the two ellipses. In addition, the relative orbit (the orbit of one star relative to the other) is a Keplerian ellipse.

which equates the acceleration of star 1 to the gravitational attraction per unit mass exerted by star 2, into the following three forms:

$$\mathbf{a}_1 = -\frac{M_2' \mathbf{r}_1}{r_1^3},$$

$$\mathbf{a}_2 = -\frac{M_1' \mathbf{r}_2}{r_2^3}, \qquad (4.2)$$

$$\mathbf{a}_{12} = -\frac{(M_1 + M_2) \mathbf{r}_{12}}{r_{12}^3},$$

where $\mathbf{a}_{12} = \mathbf{a}_1 - \mathbf{a}_2$, the acceleration of star 1 relative to star 2; $M_2' = M_2^3/(M_1 + M_2)^2$; and $M_1' = M_1^3/(M_1 + M_2)^2$. Each of the three Equations 4.2 has the same form as the equation that governs the motion of a particle moving under the attraction of a *fixed* point-mass; hence, the motion governed by each equation is Keplerian. The first two equations tell us that each particle describes an ellipse with the center of mass at a common focus, and that the radius vectors drawn from the center of mass to each star sweep out equal areas in equal times. The third equation tells us that the *relative orbit*—the orbit of star 1 relative to 2, or of star 2 relative to star 1—is also a Keplerian ellipse, and that the radius vector drawn from either star to the other sweeps out equal areas in equal times. Of course, all three motions must have the same period, given by Newton's form of Kepler's third law (Equation 3.22):

$$\left(\frac{2\pi}{T}\right)^2 = \frac{M_2'}{a_1^3} = \frac{M_1'}{a_2^3} = \frac{M_1 + M_2}{a^3}, \qquad (4.3)$$

where a_1, a_2, and a denote the major semidiameters of the three ellipses (*not* the magnitudes of the three accelerations). The three semidiameters are related in the same way as the magnitudes of the three radius vectors:

$$a_1 : a_2 : a = M_2 : M_1 : M_1 + M_2.$$

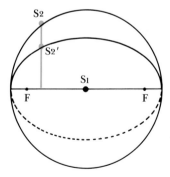

Star S_2 moves in a circular orbit relative to S_1. Viewed in a direction that is not perpendicular to the plane of the orbit, the orbit appears to be an ellipse, but S_1 is still at the center of the apparent orbit instead of at its focus. Thus a foreshortened circular orbit is distinguishable from an elliptical orbit.

We can use Equation 4.3 to estimate the combined mass $M_1 + M_2$ if we can measure the period T, which is easy, and the orbital semidiameter a, which is more difficult. To pass from a measurement of angular separation to an estimate of linear separation, we need to know the binary's distance. In addition, we must allow for the fact that, unless the plane of the orbit is perpendicular to the line of sight, the apparent orbit is foreshortened. The foreshortened orbit is also an ellipse, but its major diameter is in general shorter than it would be if the orbit were seen face on. Fortunately, the tilt of the orbital plane distorts the relative orbit in another way: it moves the reference star away from the focus of the ellipse. For example, if the relative orbit is a circle, the foreshortened orbit is an ellipse with the reference star at its center. By measuring this shift, we can deduce how the plane of the orbit is tilted relative to the plane of the sky, and so reconstruct the true relative orbit.

Triples

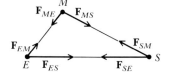

The equations that govern the motions of three mutually gravitating particles do not seem much more complicated than those that govern the motions of two mutually gravitating particles. The acceleration of each particle results from the gravitational accelerations produced by each of the other two particles, as shown in the figure to the left. But except in one special case, the motions themselves are exceedingly complicated. For two centuries after the publication of Newton's *Principia*, mathematicians tried to find explicit solutions to the equations that govern such motions. In the process, they developed powerful mathematical techniques that subsequently found important applications in celestial mechanics and quantum theory, but they failed to solve the three-body problem. Finally, near the end of the nineteenth century, H. Bruns and H. Poincaré proved that solutions of the kind that mathematicians had been seeking do not exist.

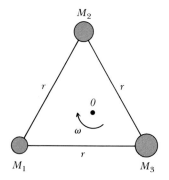

The only exact and stable solution to the three-body problem is illustrated in the lower figure on the left. The three particles, whose masses may have any values, lie at the vertices of an equilateral triangle that rotates about its center of mass with constant angular velocity ω ($= 2\pi/T$) given by a formula identical with Newton's form of Kepler's third law: $\omega^2 = M/r^3$, where M is the combined mass of the three particles. This solution to the three-body problem was discovered by the French astronomer, physicist, and mathematician Joseph Louis Lagrange (1736–1813). Examples of Lagrange's rotating equilateral triangle actually exist in the solar system: the Sun, Jupiter, and certain minor planets called Trojans lie at the vertices of rotating equilateral triangles, as shown in the middle figure on the facing page.

The impossibility of solving the three-body problem exactly (except in the one special case) leaves the astronomer with two choices: numerical solution of the governing equations, and approximation. The two approaches complement one another. With the help of a sufficiently powerful computer, we can predict the motions of three mutually gravitating particles over long periods of time. Such calculations, however, do not yield insight into the character of the motions; they enable us to *describe* the motions—an essential first step—but not to *understand* them.

Jules Henri Poincare (1854–1912), perhaps the most profound and versatile mathematician of his day.

The Sun, Jupiter, and Trojan satellites represent points on Lagrange's rotating equilateral triangles.

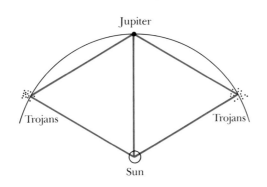

Jupiter

Trojans

Trojans

Sun

Joseph Louis Lagrange (1736–1813) cast Newton's theory in a new and more powerful form that made explicit the parallelism between, on the one hand, translational motion, linear momentum, and force and, on the other hand, rotational motion, angular momentum, and torque.

The method of approximation works best when (as the English theoretical physicist Paul Dirac remarked in a related context) the motions we wish to understand can be split up into two parts, of which one is simple, the other small. Consider the triple system consisting of the Sun, the Earth, and the Moon. The *simple* part of the motion has several distinct components: the Earth's Keplerian motion around the Sun; the Moon's Keplerian motion around the Earth; and the Earth's rotation about an axis whose direction is fixed. The *small* part of the motion also has several components, including the following.

1. The Earth's seas, oceans, and atmosphere rise and fall with a period of half a lunar day.

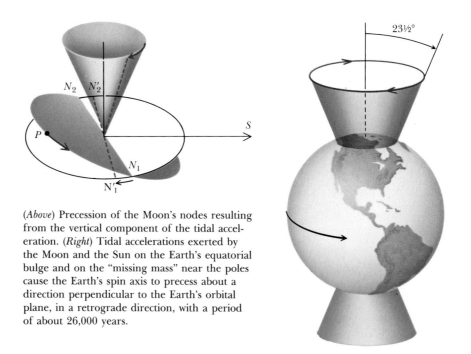

(*Above*) Precession of the Moon's nodes resulting from the vertical component of the tidal acceleration. (*Right*) Tidal accelerations exerted by the Moon and the Sun on the Earth's equatorial bulge and on the "missing mass" near the poles cause the Earth's spin axis to precess about a direction perpendicular to the Earth's orbital plane, in a retrograde direction, with a period of about 26,000 years.

2. The month is about an hour longer than the value predicted by the formula $(2\pi/T)^2 = M/a^3$, where M is the mass of the Earth, and a is the major semidiameter of the Moon's orbit.

3. The points at which the Moon's orbit intersects the Earth's orbital plane (*nodes of the lunar orbit*) regress (move backward along the orbit) with a period of 18.6 years (above left).

4. The major diameter of the Moon's orbit rotates in the forward direction (*i.e.*, in the direction of the Moon's orbital motion) with a period of slightly less than 9 years.

5. The Earth's axis is not absolutely fixed; it precesses about the perpendicular to the Earth's orbital plane, in the retrograde direction, with a period of 26,000 years (see above, right).

These effects are all small. Each of them results from *differential* gravitational attraction. As we saw in Chapter 3, the acceleration of the Earth's center of mass by the combined gravitational pull of the Sun and the Moon cancels this pull for an observer who shares the motion of the center of mass. An observer in free fall or on board a nonrotating spacecraft whose engines have been shut off experiences no gravitational acceleration. But the cancellation is exact only at the center of mass. Everywhere else there is a slight residual acceleration, and it is this residual acceleration that raises tides in the Earth's oceans and atmosphere, that adds an hour to the month, that causes the major diameter of the Moon's orbit to advance and its line of nodes to regress, and that causes the Earth's axis to sweep out a slightly wavy cone once every 26,000 years. The residual acceleration is also the reason that the innumerable tiny particles that make up the ring systems of Saturn, Jupiter, and Uranus cannot coalesce into solid moons. It is in the explanation of these varied phenomena that the scope and power of Newton's theory manifest themselves most impressively. Kepler's laws provided a strongly overdetermined but not completely accurate description of the motions of the Moon and planets. Newton's theory showed not only that Kepler's laws, in slightly more general form, can be deduced from a simple mathematical rule for calculating gravitational acceleration, but also that a variety of other phenomena that no one had imagined were related could be explained—quantitatively as well as qualitatively—by applications of the same simple rule.

The Tide-Raising Acceleration

The tidal acceleration $\overrightarrow{PT}$ experienced by a drop of water at P is the vector difference between the solar gravitational accelerations at P and O, the center of the Earth. The tidal acceleration is symmetric about the line of centers OS; that is, the pattern of accelerations shown in the figure on page 106 is the same on every plane through OS. The tidal acceleration also has mirror symmetry about the plane through O perpendicular to OS; that is, the accelerations on one side of this plane are the mirror images of those on the other side. The radial and tangential components of the tidal acceleration have different symmetries and are shown separately, along with their resultant. The thin curve, whose radial distance from the circle is proportional to the radial component of the tidal acceleration, outlines the figure of the *equilibrium tide*, the tide that would be raised in a deep, frictionless sea of uniform depth. As shown in the figure on the next page, the tide-raising acceleration at a point P whose radius vector from the center of mass O is $\mathbf{r}$ is the difference between the gravitational accelerations at P and O due to the mass M at S (radius vector $\mathbf{R}$):

$$\mathbf{a} = M\left(\frac{\mathbf{R} - \mathbf{r}}{|\mathbf{R} - \mathbf{r}|^3} - \frac{\mathbf{R}}{R^3}\right)$$

$$= M\mathbf{R}\left(\frac{1}{|\mathbf{R} - \mathbf{r}|^3} - \frac{1}{R^3}\right) - \frac{M\mathbf{r}}{|\mathbf{R} - \mathbf{r}|^3}.$$

We assume that $r \ll R$. Then

$$|\mathbf{R} - \mathbf{r}|^2 = R^2 - 2rR\cos\theta + r^2 \simeq R^2 - 2rR\cos\theta,$$

and

$$|\mathbf{R} - \mathbf{r}|^{-3} = R^{-3}\left(1 + \frac{3r}{R}\cos\theta\right).$$

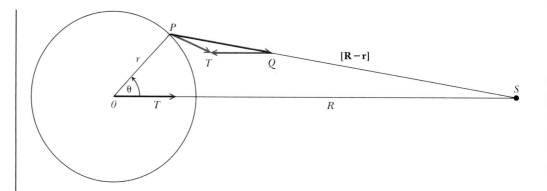

A point-mass at S produces gravitational accelerations $\overrightarrow{OT}$ at O and $\overrightarrow{PQ}$ at P. These accelerations differ both in magnitude and (except when P lies on the OS) in direction. The vector $\overrightarrow{QT}$ is equal in length and parallel to $\overrightarrow{OT}$, but points in the opposite direction. That is, $\overrightarrow{QT} = -\overrightarrow{OT}$. Hence the differential or tidal acceleration experienced by a particle at P—its acceleration relative to the point O (which represents the center of mass of the system of which the particle is a member)—is represented by the vector $\overrightarrow{PQ} - \overrightarrow{OT} = \overrightarrow{PQ} + \overrightarrow{QT} = \overrightarrow{PT}$. The tidal acceleration is symmetric about the line of centers OS, even when the distance ratio $\overline{PO/PS}$ is not small. If this ratio is much smaller than unity, the tidal acceleration is also symmetric with respect to reflection in the plane through O at right angles to the line OS.

In the applications discussed in the text, the points O, P, and S, have the following interpretations: (a) O = center of the Earth, P = a particle on the Earth's surface, S = the Moon or the Sun; (b) O = center of mass of the Earth and the Moon, P = a particle circling this point at the Moon's distance, S = the Sun; (c) O = the center of a hypothetical satellite of Saturn, P = a particle on the satellite's surface, S = the center of Saturn.

Inserting this formula into the preceding formula for the tidal acceleration and keeping only terms proportional to r/R, we obtain

$$\mathbf{a} = \left(\frac{Mr}{R^3}\right)(3\hat{\mathbf{R}}\cos\theta - \hat{\mathbf{r}}),$$

where $\hat{\mathbf{R}}$ and $\hat{\mathbf{r}}$ are vectors of unit length in the directions $\mathbf{R}$ and $\mathbf{r}$, respectively. Let $\hat{\boldsymbol{\theta}}$ denote a vector of unit length perpendicular to $\mathbf{R}$, i.e., a unit vector in the tangential direction. We can resolve $\hat{\mathbf{R}}$ into components along the radial and tangential direction:

$$\hat{\mathbf{R}} = \hat{\mathbf{r}}\cos\theta - \hat{\boldsymbol{\theta}}\sin\theta,$$

whence

$$a = \left(\frac{Mr}{R^3}\right)[(3 \cos^2 \theta - 1)\hat{\mathbf{r}} - 3 \sin \theta \cos \theta \cdot \hat{\boldsymbol{\theta}}].$$

Let us examine some properties of the tidal acceleration. In the figure on the facing page, O represents the center of mass of the Earth and the Moon; P, the Moon's center of mass; and S, the Sun's center of mass. Our frame of reference, whose origin is fixed at O, is falling freely in the Sun's gravitational field. In this frame of reference the gravitational acceleration at a point P on the Moon's orbit is the (vector) difference between the solar gravitational accelerations at P and O. This residual or *tidal* acceleration is calculated in the box on page 103. The figure on page 106 illustrates the two following properties of the tidal acceleration.

Property 1 *The tidal acceleration is symmetric about the line OS, and nearly symmetric about the plane through O at right angles to OS.*

Thus the tidal acceleration points directly away from the center of mass when P lies directly between the center of mass and the Sun, and also when it lies on the opposite side of the center of mass from the Sun; and its magnitude is nearly the same in both cases.

Property 2 *Averaged over a sphere centered on the origin, the tidal acceleration vanishes; on a circle whose plane contains the Earth–Sun line OS, it is predominantly radially outward; along a circle whose plane is perpendicular to the Earth–Sun line, it is directed toward O.*

Two additional properties of the tidal acceleration that are not obvious from the figure follow from the definition of the tidal acceleration (see the box):

Property 3 *The magnitude of the tidal acceleration at any point P is directly proportional to the distance $\overline{OP}$.*

Property 4 *The tidal acceleration is inversely proportional to the cube of the Sun's distance and directly proportional to its mass.*

Newton was especially proud of his theory's success in explaining the "ebb and flow of the sea," a problem that had engaged and baffled both Kepler and Galileo. In *Dialogue Concerning the Two Chief World Systems*, Galileo set forth a wholly erroneous theory of ocean tides, which he attributed to the Earth's combined daily and annual motions. Kepler correctly attributed the tides to the Moon's gravitational pull, but could not explain how the Moon can cause the sea to rise when it is directly underfoot as well as when it is directly overhead. Newton's theory accounted for the three principal regularities displayed by the tides:

1. The rhythm of the tides is regulated primarily by the Moon, not the Sun. At any given place, high and low tides occur at fixed lunar times.

2. High tide and low tide occur twice each lunar day, although the heights of successive high tides may differ greatly.

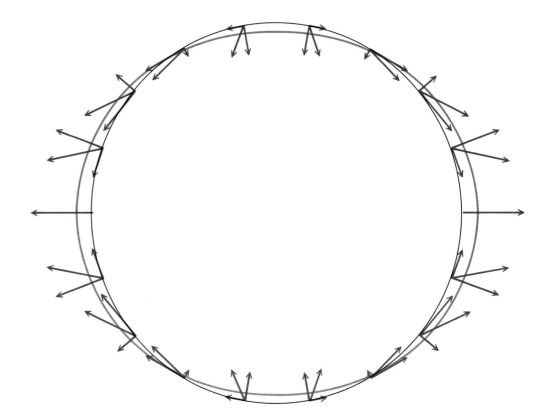

The tidal force and its radial and tangential components at representative points on the surface of an idealized spherical body. The line joining the center of the sphere to the center of the distant point-mass that produces the tidal acceleration is horizontal and lies in the plane of the figure. Any plane through this line cuts the sphere in a circle, here shown in black. We can construct the tidal acceleration at any point on this circle in the manner illustrated in the figure on page 104 and explained in the legend to that figure. The field of tidal acceleration illustrated above is represented symbolically by the equation at the top of page 105. The radial component of the tidal acceleration (represented by arrows that point directly toward or away from the center of the black circle) is pro-

portional to the expression $(3 \cos^2 \theta - 1)$; the tangential component (represented by arrows tangent to the circle) is proportional to the expression $-3 \sin \theta \cos \theta$. Thus the tangential component vanishes at the points $\theta = 0$ and $\theta = \pi$ where the radial component is largest, and also at the points $\theta = \pm \pi/2$ where the radial component has its largest negative (inward) value. The red curve is a section of the tidally distorted surface of an idealized liquid sphere, or of a deep ocean of uniform depth. The radial distortion (that is, the radial distance between the red and black curves) is a fixed fraction of the radial component of the tidal acceleration. Thus the tidally distorted surface is an ellipsoid generated by rotating the red curve about its horizontal axis.

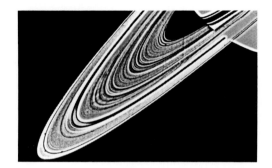

Part of Saturn's ring system. A two-image mosaic made from photographs taken by Voyager 1 on 6 November 1980 from a distance of eight million kilometers. Roche's theory of tidal disruption predicts that the particles that make up the rings cannot coalesce into large bodies held together by their mutual gravitation.

3. The difference in height between high tide and low tide is greatest at the new and full phases of the Moon and least at the quarter phases.

To see how regularities are related to the preceding four properties of the tidal acceleration, let us return to the figure on p. 104, where we now interpret O as the center of the Earth, P as a drop of water in an idealized ocean of uniform depth, and S as the Moon (or the Sun).

Why does the Moon, rather than the Sun, regulate the tides? The Sun is about 27 million times as massive as the Moon, but it is also 390 times more distant. Hence, by Property 4 (p. 105), the tidal acceleration caused by the Moon is $(390)^3/27{,}000{,}000 \simeq 2\frac{1}{4}$ times that caused by the Sun.

Why is the length of a tidal cycle *half* a lunar day? And why do successive high tides differ in height? The tidal acceleration is greatest at points on the line OS in the figure and least at points that lie on circles centered on O at right angles to this line. If the Earth's axis of rotation were perpendicular to the plane of the diagram, a rotating particle P would experience two equal maxima and two equal minima of the tidal acceleration each lunar day. Because the Earth's axis is not perpendicular to the line of centers, successive maxima and minima are usually unequal, but the interval between them is always half a lunar day.

Finally, why does the height of high tide vary with the phase of the Moon? Although the Moon dominates the tides, the Sun's contribution to the tidal acceleration is nearly half that of the Moon. The height of the tide should therefore depend noticeably on the relative positions of the Sun and the Moon. When the centers of the Sun, the Earth, and the Moon all lie on a straight line (that is, when the Moon is new or full), the solar and lunar tidal accelerations reinforce one another and the resulting tides (spring tides) have their greatest amplitude. At the quarter phases of the Moon, the solar and lunar tidal accelerations act at right angles, and the resulting tides (neap tides) have their smallest amplitude.

The Sun and the Moon excite tidal oscillations in the Earth's atmosphere as well as in its seas and oceans. In the atmosphere the solar tide, with a period of half a solar day, has a much greater amplitude than the lunar tide, with a period of half a lunar day, because the atmosphere happens to have a natural period of oscillation close to 12 solar hours. The atmosphere behaves like a resonator, greatly amplifying the periodic motions produced by the solar tidal acceleration.

The Moon and the Sun also produce tides in the solid Earth. The magnitude of the observed effect depends on and can be used to estimate the Earth's rigidity.

The Sun's tidal acceleration affects the length of the month. The radial component of the tidal acceleration is predominantly outward along the Moon's orbit. Thus it tends to reduce the Earth's gravitational attraction. Because the Moon's orbit is nearly circular, reducing the gravitational attraction at the Moon's distance by a fraction f has nearly the same effect as reducing the effective mass of the Earth by the same fraction. Now, the Moon's orbital period is inversely proportional to the square root of the combined mass of the Earth and the Moon (Equation 4.3). If the Earth's mass were multiplied by the factor $1 - f$ (with $f \ll 1$), the Moon's orbital period would be multiplied by the factor $1 + \frac{1}{2}f$. The fraction f is the ratio between the solar tidal acceleration, averaged around the Moon's orbit, and the gravitational acceleration by the Earth:

$$f = \frac{1}{2}\frac{M_S r/R^3}{M_E/r^2} = \frac{1}{2}\left(\frac{M_S}{M_E}\right)\left(\frac{r}{R}\right)^3 \tag{4.4}$$

$$= \frac{1}{2} \times \frac{3 \times 10^5}{(390)^3} = 2.5 \times 10^{-3}.$$

The fractional increase in the length of the month, according to this crude estimate, is $\frac{1}{2}f$ or about 1 part in 800. The observed fractional increase is about an hour, or 1 part in 650. More accurate calculations reproduce the observed value.

Tidal forces tend to disrupt satellites. In the figure on p. 104, let S represent the center of Saturn, O the center of a Saturnian satellite of negligibly small mass, and P a particle at the surface of the satellite. Saturn will raise a tide in the solid body of the satellite, just as the Moon raises a tide in the Earth's solid body. Since the tide-raising acceleration increases rapidly with decreasing separation R between the centers of mass of Saturn and the satellite, the satellite will be unable to cohere under its own gravitation if the separation R is less than a certain limiting value. Let us try to estimate this value.

The tidal acceleration alters the satellite's shape, lengthening the diameter that lies along the line of centers and shortening diameters at right angles to this line. In an equilibrium configuration the satellite keeps the same face turned toward the planet; otherwise it would experience a changing tidal acceleration. Hence the satellite rotates with an angular velocity equal to the angular velocity of its orbital revolution. This angular velocity, ω, is given by the now familiar formula $\omega^2 \equiv (2\pi/T)^2 = M/R^3$, where T is the orbital period, M the mass of Saturn, and R the separation between the centers of Saturn and the satellite. Thus three forces act on a particle P at the surface of the satellite: the tidal force of Saturn, the gravitational attraction of the satellite, and the centrifugal force arising from the satellite's rotation. A *necessary* condition for equilibrium is that the radial component of the resultant of these three forces should be directed inward (toward the center of the satellite) at every point of the surface. This condition is most severe at the two points where the line of centers OP pierces the surface, where it takes the form

$$\frac{M'}{r^2} - \frac{2Mr}{R^3} - \omega^2 r > 0, \tag{4.5}$$

where M' denotes the mass of the satellite. The three terms on the left side of this inequality represent, respectively, the acceleration due to the gravitational pull of the satellite, the tidal acceleration, and the centrifugal acceleration (in a frame of reference rotating with the satellite). Setting $\omega^2 = M/R^3$ and simplifying, we obtain

$$\frac{M'}{r^3} > \frac{3M}{R^3}. \tag{4.6}$$

This condition says that the satellite's mean density must exceed three times the mean density of a sphere of radius R centered on Saturn. When the separation R is less than the value allowed by Inequality 4.6, the satellite certainly cannot be in equilibrium.

It does not follow, however, that an equilibrium configuration exists if Inequality 4.6 is satisfied. In an equilibrium configuration, the resultant of the gravitational forces and the centrifugal force must have both an inwardly directed radial component at every point on the satellite's surface and a vanishing tangential component, because there is nothing that could balance a tangential force acting on a particle at the surface. (The inwardly directed radial force acting on a small piece of the surface is balanced by the pressure exerted by the underlying material.) This is a much stronger condition than the one expressed by Inequality 4.6. Its implications were worked out in 1850 by the French mathematician Edouard Roche. Roche proved that a liquid satellite has one and only one equilibrium configuration whenever the separation R exceeds a certain critical value. In this configuration the satellite is an ellipsoid with three unequal axes, the longest of which lies along the line of centers. The critical separation, below which no equilibrium configuration is possible, is defined by an inequality of the same form as Inequality 4.6, but with the coefficient 15 in place of 3 on the right side. This change increases the critical separation by the factor $(15/3)^{1/3} = 1.7$, a substantial change. Saturn's diameter is about 60,500 km. For a satellite whose density is equal to Saturn's mean density, Roche's critical radius is 2.455 times the radius of the planet, or slightly less than 150,000 km. The diameter of Saturn's ring system is about 136,000 km. Thus the rings are in a region where material of Saturn's density cannot hold together under its own gravitation for an extended period. Roche's theory correctly predicts that the rings are composed of particles small enough to be held together by chemical cohesion—that is, by the forces that hold together ordinary solids. The weaker Inequality 4.6 predicts a critical radius of only 88,000 km, well

Thomas Wright (1711–1786). English amateur astronomer and theologian whose speculations about the Milky Way inspired Immanuel Kant.

Immanuel Kant (1724–1804) published an important cosmological treatise in 1755 before turning to philosophy; an engraving based on a 1791 painting by Döbler.

within Saturn's ring system. Roche's sophisticated theory is really needed to explain why the particles that make up Saturn's rings have not coalesced into larger bodies.

The Hierarchy of Self-Gravitating Systems: Kant's Model of the Stellar Universe

Newton and Huygens postulated that the stars are sprinkled more or less uniformly throughout an infinite space. Yet bright stars, faint stars, and unresolved starlight are all concentrated in the bright irregular band that the Romans called *Via Lactea*, the Milky Way. In 1750 Thomas Wright published *An Original Theory or New Hypothesis of the Universe, Founded upon the Laws of Nature, and Solving by Mathematical Principles the General Phenomena of the Visible Creation; and Particularly the Via Lactea.* Wright argued that the appearance of the Milky Way can be explained by the supposition that the stars are "all moving the same way [in circular orbits around a distant center] and not much deviating from the same plane, as the planets in their heliocentric motion do round the solar body." Immanuel Kant, then a private tutor in natural science and mathematics, read an account of Wright's book in a Hamburg newspaper and was inspired by it to construct a cosmological theory of his own. This was printed in 1755 under the title *Universal Natural History and Theory of the Heavens.* (A title that Kant had considered earlier is more descriptive: *Cosmogony, or an Attempt to Deduce the Origin of the Universe, the Formation of the Heavenly Bodies, and the Causes of Their Motion, from the Universal Laws of the Motion of Matter, in Conformity with the Theory of Newton.*)

Kant begins by presenting an improved version of Wright's argument. He points out, as Wright failed to do, that the band of the Milky Way defines a great circle on the celestial sphere. Since the stars are at varying distances from the Sun, this implies that their spatial distribution is strongly concentrated toward a plane that passes very close to the Sun (see facing page). The observation that the Milky Way runs along a great circle rules out an alternative model suggested by Wright in which the stellar system has the form of a hollow spherical shell instead of a disc.

Kant next tries "to discover the cause that has made the positions of the fixed stars come to be in relation to a common plane." Here he makes a decisive step beyond Wright. Wright had suggested an analogy between the planets revolving around the Sun and the visible stars revolving about a distant center. Kant suggested that the solar system and the stellar system are not only analogous but *homologous*—i.e., that they are similar in structure because the motions of the bodies that compose them are governed by the same physical laws (Newton's laws of motion and gravitation) and because they were formed by the same process. Kant conjectured that the stars that make up the Milky Way revolve in the same direction in nearly circular, nearly coplanar orbits about a distant center, like the planets around the Sun, whereas stars that lie far from the Milky Way are moving in elongated and highly inclined orbits around the same center, like comets around the Sun. In short, the Milky Way is to the stellar system what the Zodiac is to the solar system; and the stars that lie outside the Milky Way are the comets of the stellar system.

A 360° photomontage of the night sky, showing the Milky Way. The horizontal line that bisects the figure is a great circle on the celestial sphere.

Up to this point Kant's hypothesis could be described as a logical development of Wright's seminal insight in the light of Newtonian physics. But now Kant makes an imaginative leap. It would be difficult to improve his own account of it:

I come now to that part of my theory which gives it its greatest charm, by the sublime idea which it presents of the plan of the creation. The train of thought which has led me to it is short and natural; it consists of the following ideas. If a system of fixed stars which are related in their positions to a common plane, as we have delineated the Milky Way to be, be so far removed from us that the individual stars of which it consists are no longer sensibly distinguishable even by the telescope; if its distance has the same ratio to the distance of the stars of the Milky Way as that of the latter has to the distance of the Sun; in short, if such a world of fixed stars is beheld at such an immense distance from the eye of the spectator situated outside of it, then this world will appear under a small angle as a patch of space whose figure will be circular if its plane is presented directly to the eye, and elliptical if it is seen from the side or obliquely. The feebleness of its light, its figure, and the apparent size of its diameter will clearly distinguish such a phenomenon, when it is presented, from all the stars that are seen single.

We do not need to look long for this phenomenon among the observations of the astronomers. It has been distinctly perceived by different observers. They have been astonished at its strangeness; and it has given occasion for conjectures, sometimes to strange hypotheses, and at other times to probable conceptions which, however, were just as groundless as the former. It is the "nebulous" stars which we refer to, or rather a species of them, which M. de Maupertuis [Discours sur la Figure des Astres; Paris, 1742] thus describes: "They are," he says, "small luminous patches, only a little more brilliant than the dark background of the heavens; they are presented in all quarters; they present the figure of ellipses more or less open; and their light is much feebler than that of any other object we can perceive in the heavens."

Kant argued that if the "luminous patches" were enormously distended stars, their surface brightnesses would be much higher than they actually were. But a stellar *system* like the Milky Way, viewed from a sufficiently great distance, would look just like one of the luminous patches:

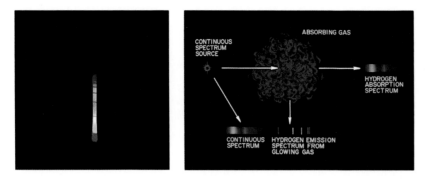

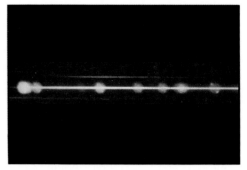

The spectra of stars and gas clouds. Light from most natural sources is a mixture of monochromatic components. Newton discovered that a prism separates these components. The spectrum of an incandescent solid is continuous, most of the energy being concentrated in a broad band centered on a color whose wavelength is inversely proportional to the absolute temperature of the radiating surface. An incandescent gas emits light in a few very narrow bands, called lines, whose wavelengths are characteristic of the emitting substance. Light from an incandescent solid that has passed through an intervening tenuous gas is depleted or augmented in the set of very narrow bands characteristic of the gas. If the gas is hotter than the radiating surface, the radiation is augmented at these wavelengths, and the continuous spectrum is crossed by a series of fine bright lines; if the intervening gas is cooler than the radiating surface, the spectrum is crossed by a series of fine dark lines at the same wavelengths. The Sun and most other stars have continuous spectra crossed by dark lines produced in their outer layers (left). Glowing gas clouds in interstellar space have discrete spectra (right).

an ellipse of low, uniform surface brightness. Again, Kant's argument is technically sound and his conclusion is correct. Slightly over a hundred years later William Huggins, an English amateur astronomer, built the first astronomical spectroscope and used it to make systematic observations of stars and nebulae. He found that the spectra of spiral nebulae, which Kant had identified with distant stellar systems, had continuous spectra similar to those of ordinary stars, exactly as Kant's hypothesis predicted (see above). In contrast, the spectra of nebulae that appeared to be smooth and structureless even when viewed through a powerful telescope consisted entirely of narrow, bright lines. These were evidently true nebulae, tenuous clouds of hot gas in our own stellar system.

Kant recognized a basic structural property of the astronomical universe that Newton had not imagined and that most professional astronomers failed to appreciate until well into the twentieth century: the astronomical universe is a hierarchy of self-gravitating systems. The Sun's satellites are themselves miniature planetary systems; most stars have companions; single and double stars congregate in small groups and larger clusters, which in turn congregate in galaxies; and galaxies themselves are units in an extended hierarchy of self-gravitating systems.

THE HIERARCHY OF SELF-GRAVITATING SYSTEMS

(Below) Self-gravitating systems: (*Top row*) A Galactic star cluster (the Pleiades). A globular star cluster (M5 in the constellation Serpens). The Andromeda Galaxy and its two satellites NGC 205 and NGC 22. (*Bottom row*) Stephan's quintet, a small group of galaxies. A cluster of galaxies in Virgo. A rich cluster of galaxies in the constellation Coma Berenices (the Coma cluster).

System	Mass in solar masses	Diameter in light-years	System	Mass in solar masses	Diameter in light-years
Jupiter	10^{-3}	3×10^{-6}			
Solar system Double star	1	10^{-3}	Galaxy + satellites Double galaxy	10^{12}	3×10^{5}
Multiple-star system	10	5×10^{-3}	Multiple-galaxy system	10^{13}	1.5×10^{6}
Rich star cluster	10^{6}	30	Rich galaxy cluster	10^{15}	3×10^{7}

Data on the masses and diameters of the self-gravitating systems that make up the hierarchy are collected in the table above. Photographs of representative systems are also shown above.

Kant, of course, did not have such data to work with. The first estimate of a stellar mass (other than the Sun's) was still 50 years in the future; the first estimate of a stellar distance, 80 years in the future. Nevertheless, Kant conjectured that dense clumps of stars like the Pleiades (see photo on preceding page) would turn out to be clusters of stars held together by their mutual gravitation, as indeed they are. He recognized that self-gravitating systems of vastly

Hypothetical section of the stellar system.

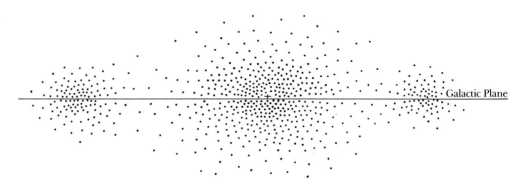

Galactic Plane

Arthur Stanley Eddington (1882–1944), a pioneer in the theory of stellar structure and the dynamics of stellar systems; a chalk drawing done in 1928–1929 by Sir William Rothenstein.

different size and mass can have identical structure. The Moon and the Earth, Jupiter and the Sun, the twin stars Castor and Pollux, the Galaxy* and its companion, the Andromeda Galaxy, may all be described, in a first approximation, as pairs of mutually gravitating point-masses. Table 4.1 has been arranged to exhibit the structural and dynamical analogies between systems of stars and systems of galaxies.

Like most highly creative scientists, Kant thought deeply about the relation between theory and observation. He recognized that a scientific cosmology must be based on observation, but he also recognized a less obvious and less widely appreciated truth—that only observation guided by theoretical insight is likely to uncover the deep regularities underlying phenomena:

The Nebulous Stars, properly so called, and those about which there is still dispute as to whether they should be so designated, must be examined and tested under the guidance of this theory. When the parts of nature are considered according to their design and a discovered plan, there emerge certain properties in it which are otherwise overlooked and which remain concealed when observation is scattered without guidance over all sorts of objects.

The last sentence was prophetic. Kant's picture of the universe could have guided the work of succeeding generations of observational astronomers. In fact, it did not. Throughout the nineteenth century and well into the twentieth, astronomers used measurements of stellar positions, parallaxes, colors, brightnesses, spectra, proper motions, and radial velocities to elaborate models of the Galaxy that Kant would have dismissed as naive and unphysical. In 1914 A. S. Eddington, the leading theoretical astronomer of his day, summarized current views of the structure of our stellar system in the figure reproduced above. In this picture the Sun, marked by a cross, lies close to the center of a lens-shaped system surrounded by "a series of irregular agglomerations of stars of wonderful richness, diverse in form and grouping, but keeping close to the fundamental plane." Not until the penultimate page of his book does Eddington mention

*Just as "Sun" denotes our own sun, "Moon" our own moon, so "Galaxy" refers to our own galaxy, the Milky Way.

An opaque dust cloud obscuring part of a rich star field: the Horsehead Nebula.

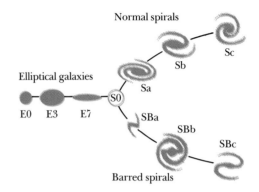

Hubble's original tuning-fork diagram, redrawn from *The Realm of the Nebulae* (1936).

the possibility that the system may be rotating. As late as 1920 many astronomers believed that the Galaxy is larger than any other stellar system, and that the spiral nebulae are its satellites. Astronomers who held such views had far more data at their disposal than Kant. But, as Kant understood so well, data do not speak for themselves. To interpret the data, astronomers had to make various assumptions. For example, Shapley assumed that interstellar space is transparent outside the black clouds we see projected against the bright background of the Milky Way (above, left). This assumption eventually proved to be false. Once the effects of interstellar dimming and reddening were understood, data that had spoken for a heliocentric Galaxy and a Galactocentric universe told a different story, the story that Kant had constructed almost two centuries earlier.

Stereotyped Structures: Stellar Populations

Whereas two mutually gravitating particles revolve about one another in Keplerian ellipses, three mutually gravitating particles can have very many qualitatively distinct possible configurations. The variety of possible configurations of 10^{11} mutually gravitating particles boggles the imagination; yet galaxies exhibit little structural variety. Between 1920 and 1923 Edwin Hubble devised a simple scheme for classifying galaxies (above, right). That his scheme is as serviceable today as it was when Hubble published it testifies not only to Hubble's skill as a taxonomist but also to the severely limited structural variety of galaxies.

Consider elliptical galaxies. In Hubble's classification they are distinguished by a single parameter, the axial ratio of the projected elliptical image. When images with the same axial ratio are reduced (or enlarged) to a common scale, they are virtually indistinguishable. What is even

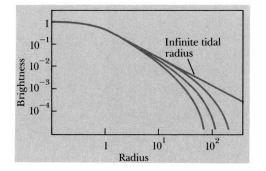

Brightness profiles in rich, spherical, fully "relaxed" stellar systems for different values of the tidal radius, as calculated by Ivan King. The curve labeled "infinite radius" represents the brightness profile of a hypothetical system of infinite mass and radius. The other curves correspond to systems with finite tidal radius. The tidal radius is the distance at which the tidal forces excited by matter outside the system are comparable to the gravitational attraction of the system itself.

more remarkable, the light profiles of elliptical galaxies closely resemble those of globular star clusters and of rich spheroidal galaxy clusters like the one illustrated on p. 113 (the Coma cluster). How can we interpret this finding? If a system composed of a great many interacting particles is left to itself for a long time, it "relaxes" into a simple, stereotyped state. The air in a sealed room, for example, relaxes into a state of uniform temperature and uniform density. Analogously, a rich self-gravitating system relaxes toward a state in which its temperature (measured by the random motion of its members) is uniform and its density decreases outward in accordance with a law first derived in 1882 by the German mathematician A. Ritter. In the figure above, the curve labeled "infinite tidal radius" represents this law. The central regions of rich, nearly spherical systems conform closely to the law, but at great distances from the center, the observed brightness profiles fall off faster than the law predicts.

The reason is that at a certain easily calculable distance from the center of any actual system, the tidal forces exerted on a member of the system by external systems become comparable in magnitude to the attraction exerted by the system itself. This distance is called the tidal radius. Ivan King of the University of California at Berkeley has calculated the brightness profiles of quasirelaxed self-gravitating systems with various tidal radii, and compared these theoretical profiles (shown above) with observed brightness profiles of star clusters and elliptic galaxies. The predicted and observed profiles agree almost perfectly, showing that rich, spheroidal, self-gravitating systems are indeed quasirelaxed.

The structure of a quasirelaxed system depends on only three parameters, besides the tidal radius: the mass density at its center, the radius at which the mass density equals some definite fraction (one-half, say) of the central density, and the degree of flattening. These three parameters are themselves determined by the values of three dynamical quantities: mass, energy, and angular momentum. In an isolated dynamical system, the values of these three quantities (and, in general, of these quantities only) remain constant in time, however much the system's inter-

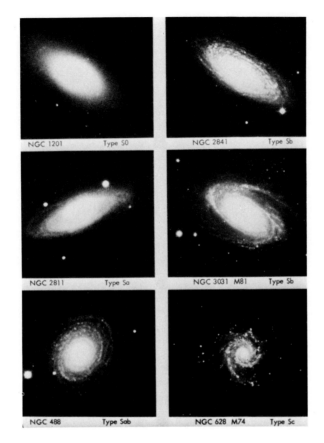

A montage of six galaxies with NGC numbers and classifications.

nal structure may change. This suggests that rich spheroidal systems owe their stereotyped structure to evolutionary processes that tend to destroy all structural information except that contained in the values of the three conserved quantities, mass, energy, and angular momentum. We will see that the mass and the energy jointly determine a system's central density and diameter, and that all three quantities jointly determine its degree of flattening.

In his cosmogonic treatise of 1755, Kant pointed out that the solar system is made up of two distinct populations: a *disc population*, consisting of planets moving in orbits that are nearly circular and nearly coplanar; and a *halo population*, consisting of comets moving in elongated orbits that make large angles with the plane of the planets. He suggested that the Galaxy likewise is composed of a disc population and a halo population with properties analogous to those of the two populations that make up the solar system. In 1944 Walter Baade, an astronomer at the Mount Wilson Observatory, put forward a more general and more detailed version of Kant's hypothesis. He proposed that the membership of stellar systems is drawn from two qualitatively distinct stellar populations. *Population I* inhabits the discs of spiral galaxies. In our own Galaxy it includes the Sun and its neighbors. Its most distinctive constituents are hot blue

Left: The Andromeda Galaxy, photographed in blue light shows bright blue stars of Population I in the spiral arms. Right: NGC 205, companion of the Andromeda Galaxy, photographed in yellow light, shows stars of Population II, the brightest of which are 100 times fainter than the bright blue stars of Population I. The inset shows Andromeda and NGC 205.

stars (supergiants), Cepheids (a kind of variable star, which we will discuss presently) with periods longer than a day, star clusters like the Hyades and the Pleiades, and interstellar dust and gas. These constituents have the same statistical properties in all spirals.

The stars that inhabit elliptical galaxies and the spheroidal component of spiral galaxies are drawn from *Population II*. Elliptical galaxies are nearly pure Population II, whereas spirals are mixtures of the two populations, with Population I dominant in the disc and Population II dominant in the central bulge and in the halo.

Baade's bold conjecture has been amply confirmed by subsequent observations. It implies that there are two basic stereotyped structures among stellar systems: spheroids and discs. In the following sections we will consider some of the most basic physical principles needed to understand this state of affairs.

Energy

One of Archimedes' lost books, *On Levers*, presumably gave a unified account of his theory of simple machines: levers, arrangements of pulleys, screws, winches, and the like. Each of Archimedes' surviving books develops the logical consequences of a single idea. What was the idea on which Archimedes based his theory of simple machines? It may well have been the earliest version of what we now call the principle of conservation of energy.

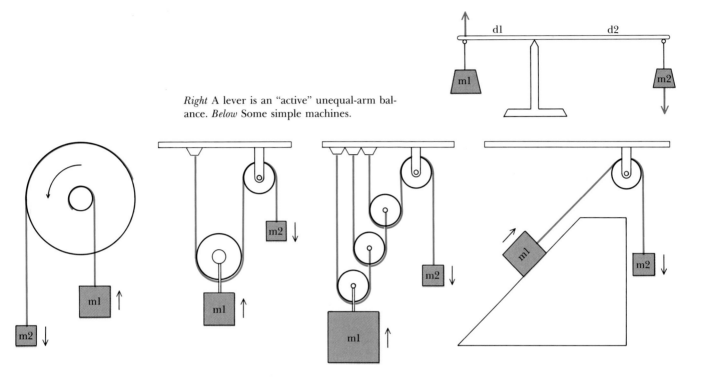

Right A lever is an "active" unequal-arm balance. *Below* Some simple machines.

Consider Archimedes' principle of the lever. A lever is an unequal-arm balance functioning in an active rather than a passive mode (see above). Suppose that a 10-pound weight and a 100-pound weight are connected by a weightless rod balanced on a knife edge. If a grain of sand is added to the 10-pound weight, it will slowly sink, causing the 100-pound weight to rise. From the upper figure above, we see that Archimedes' principle of static equilibrium, which requires the knife edge to divide the connecting rod in inverse ratio to the weights, implies Archimedes' principle of the lever, which states that a mass m_2 at distance d_2 from the fulcrum can be raised by applying a weight m_1 at a distance d_1 on the other side of the fulcrum, provided that $m_1 d_1 = m_2 d_2$. This conclusion does not require the rod to be horizontal. It implies that the upward displacement Δz_2 of the mass m_2 and the downward displacement $-\Delta z_1$ of the mass m_1 satisfy the relation $m_1 \Delta z_1 + m_2 \Delta z_2 = 0$ or

$$m_1 z_1 + m_2 z_2 = \text{constant.} \tag{4.6}$$

The same rule governs the equilibrium configurations of the simple machines illustrated in the lower figure above, in each of which the slow descent of a mass m_1 is accompanied by the slow rise of a mass m_2.

Simon Stevin (1548–1620), Dutch mathematician; postulated the impossibility of perpetual motion, from which Huygens later deduced the first version of the principle of conservation of energy.

When we raise a weight, we do work (in the everyday sense). Let us *define* the work done as the product of the applied force and the resulting displacement. In the simple machines we have been considering, the two weights are initially in equilibrium and begin to move, very slowly, when a grain of sand is added to one of them. Hence the applied force exactly balances the weight mg, and the work done on one of the weights when its height changes by an amount Δz is $mg\Delta z$. When this quantity is negative, that is, when the weight sinks, the weight does work on its environment. Equation 4.6 implies that, in a simple machine working close to equilibrium, the work done *by* one of the weights is equal to the work done *on* the other.

With an eye to the future, let us give the name *potential energy* to the quantity mgz and label it U; that is, $U = mgz$. Then Equation 4.6, multiplied through by the gravitational acceleration g, says that the sum of the potential energies is constant in a simple machine working close to equilibrium:

$$m_1 g z_1 + m_2 g z_2 \equiv U_1 + U_2 = \text{constant.} \tag{4.7}$$

This may well have been the unifying principle underlying Archimedes' theory of simple machines. Its generalization to any number of weights in a near-equilibrium configuration is obvious and immediate: the combined potential energy does not change as the individual weights rise and sink.

We saw in Chapter 3 how Galileo arrived at the formula $v^2 = 2gh$ for the terminal speed v of a body falling from rest with constant acceleration g through a vertical distance h. We may write this law in the slightly more general form

$$\tfrac{1}{2}mv^2 + mgz \equiv K + U \equiv E = \text{constant,} \tag{4.8}$$

which also holds for bodies that are not initially at rest (a body can be imagined to acquire its initial speed by falling from rest through an appropriate vertical distance). The quantity $K = \tfrac{1}{2}mv^2$ is called the *kinetic energy*. Equation 4.8 says that a body can exchange kinetic energy for potential energy or vice versa; the total energy $K + U = E$ of a body sliding without friction on a roller coaster does not change as the body rises and falls, slows down and speeds up.

Consider with Huygens the following experiment (see the top facing figure). Two blocks, initially at rest at heights h_1, h_2, slide down frictionless planes onto a horizontal table where they collide. After the collision each block slides up an inclined plane until it comes to rest. Call the new heights h_1', h_2'. Suppose we had a film of this experiment. Would we be able to tell which was the right way to run it through the projector? In other words, could we use such an experiment to distinguish between the forward and backward directions of time? Certainly the parts of the record in which the blocks slide up and down their inclined planes are completely reversible, according to Galileo's theory of motion: If a block attains a speed v in falling from rest through a vertical distance h, it will rise an equal distance if it starts with speed v. Huygens postulated that in the absence of friction, a collision between two blocks sliding on a horizontal

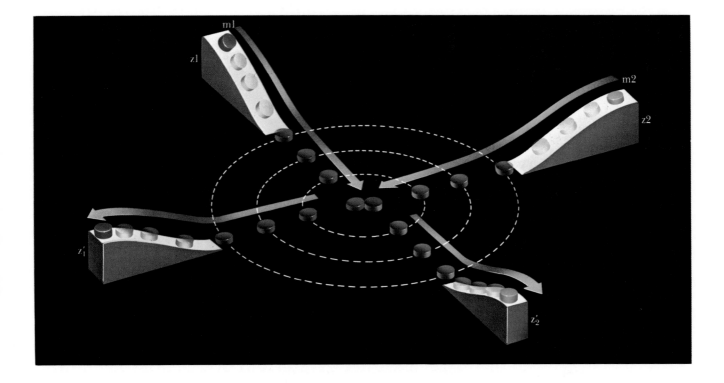

Huygens used the thought experiment illustrated here to deduce the law of conservation of kinetic energy in elastic collisions from the law of conservation of potential energy that governs levers, pulleys, and other simple machines.

plane is also reversible. He then constructed an argument to show that the center of mass of the two blocks must be at the same height in the initial and final states. Since the height Z of the center of mass is given by Archimedes' formula $(m_1 + m_2)Z = m_1z_2 + m_2z_2$, this conclusion is formally identical with Archimedes' principle of simple machines, expressed by Equation 4.6 or 4.7. Huygens' argument runs as follows.

Suppose that the center of mass were not at the same height in the initial and final states. Because every possible sequence of events is reversible, we would then be able to find initial conditions that would make the center of mass higher in the final state. We could then construct a machine that would work in a cycle, and whose only effect on the outside world would be to raise a weight by a fixed amount during each cycle. To construct such a machine, connect the two blocks in their final state by a massless rod, connect the balance point of this rod to one end of a lever, and use this lever to raise a weight while lowering the center of mass of the two blocks until it is at the same height as the center of mass in the initial state. The blocks can now be moved, with the expenditure of an infinitesimal quantity of work, to their initial positions, thereby completing the cycle (see figure on next page).

Simon Stevin (1548–1620), Galileo's contemporary and a countryman of Huygens, had earlier put forward the remarkably bold and prescient hypothesis that it is impossible to construct

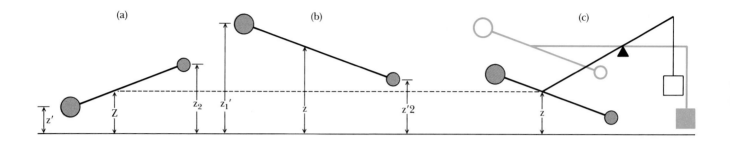

If, in the experiment illustrated on page 121, the center of mass of the two blocks was not at the same height in the initial and final states, one could use the experimental arrangement to construct a machine that would operate in a cycle and whose only effect on the outside world would be to raise a weight a little higher after each cycle. Following Stevin, Huygens postulated that such a device—a perpetual motion machine—cannot exist, and deduced that the center of the mass must be at the same height in the initial and final states.

such a machine—a machine that spontaneously generates and indefinitely maintains motion without otherwise affecting its environment. Huygens concluded that the center of mass of the two weights must be at the same height in the initial and final configurations:

$$m_1 z_1 + m_2 z_2 = m_1 z_1' + m_2 z_2'. \tag{4.9}$$

This equation is formally identical with Archimedes' principle, as expressed by Equation 4.6, but it applies to a much wider class of physical processes, namely, those illustrated in the figure on p. 121. Finally, we may use Galileo's formula $v^2 = 2gh$ to turn Equation 4.9 into the *law of conservation of kinetic energy for elastic collisions:*

$$
\begin{aligned}
K \equiv K_1 + K_2 &= \tfrac{1}{2}m_1 v_1{}^2 + \tfrac{1}{2}m_2 v_2{}^2 \\
&= \tfrac{1}{2}m_1(v_1')^2 + \tfrac{1}{2}m_2(v_2')^2 \\
&= K_1' + K_2' \equiv K',
\end{aligned}
\tag{4.10}
$$

which says that elastically colliding bodies *exchange* kinetic energy; one's loss equals the other's gain.

From Equations 4.8 and 4.9 we can actually deduce the more general law

$$E \equiv E_1 + E_2 = \text{constant}, \tag{4.11}$$

where E, the (total) energy, is given by Equation 4.8. This law states that the sum of the two energies stays constant during the entire process, as the blocks slide down their inclined planes, collide, and slide back up again.

Potential Energy of a Particle in the Field of a Point-Mass.

Consider, a body falling directly toward the center of the Earth, so that its velocity $v = dr/dt$, and its acceleration $a = dv/dt$. The acceleration is given by Newton's law of gravitation:

$$\frac{dv}{dt} = -\frac{M}{r^2}. \tag{1}$$

Multiply Equation 1 on the left by v and on the right by dr/dt $(=v)$. The left side is then

$$v\left(\frac{dv}{dt}\right) = \frac{d}{dt}(\tfrac{1}{2}v^2),$$

and the right side is

$$-\left(\frac{M}{r^2}\right)\left(\frac{dr}{dt}\right) = \frac{d}{dt.}\left(\frac{-M}{r}\right).$$

If the rates of change of two quantities are equal, the quantities themselves differ by a constant; so

$$\tfrac{1}{2}v^2 - \frac{M}{r} = \text{constant} \equiv \frac{E}{m}. \tag{2}$$

Next consider a projectile or a satellite moving in an arbitrary direction. In place of Equation 1 we have the vector form of Newton's law of gravitational acceleration,

$$\frac{d\mathbf{v}}{dt} = -\frac{M\mathbf{r}}{r^3}, \tag{3}$$

or

$$\frac{dv_x}{dt} = -\frac{Mx}{r^3},$$

$$\frac{dv_y}{dt} = -\frac{My}{r^3}, \tag{4}$$

$$\frac{dv_z}{dt} = -\frac{Mz}{r^3}.$$

Guided by the preceding derivation of Equation 2, we multiply the first of Equations 4 by v_x on the left and by dx/dt on the right, the second by v_y on the left and dy/dt on the right, and the third by v_z on the left and dz/dt on the right, and add the three resulting equations. As before, we can express the left and right sides of the equation constructed in this way as rates of change of quantities whose difference must accordingly be constant in time. In this way we obtain an equation identical with Equation 2, but valid for a body moving in any trajectory whatever under the attraction of a fixed point-mass at the origin. The constant in Equation 2 represents the body's energy per unit mass: if it is negative, the trajectory is a Keplerian ellipse; if zero, the trajectory is a parabola; if positive, a hyperbola.

Galileo's law of energy conservation (Equation 4.8) for projectiles and falling bodies near the surface of the Earth is a limiting case of a law that holds for projectiles or satellites whose distance from the center of the Earth may vary by any amount. The general law has the same form as Equation 4.8, but the potential energy U is given by the formula $U = -mM/r$, where M is the mass of the Earth, and r is the projectile's distance from the center of the Earth (see the preceding box). Now consider a pair of mutually gravitating particles. Their accelerations are given by

$$m_1 \frac{d\mathbf{v}_1}{dt} = -m_1 \frac{M_2 \mathbf{r}_{12}}{r_{12}{}^3},$$
$$m_2 \frac{d\mathbf{v}_2}{dt} = m_2 \frac{M_1 \mathbf{r}_{21}}{r_{12}{}^3}, \tag{4.12}$$

in which $m_1 M_2 = m_2 M_1$ and $\mathbf{r}_{12} = \mathbf{r}_1 - \mathbf{r}_2 = -\mathbf{r}_{21}$. During a short time-interval Δt, particle 1 undergoes a displacement $\mathbf{v}_1 \Delta t$, particle 2 a displacement $\mathbf{v}_2 \Delta t$. Each of these displacements contributes to the change Δr_{12} in the separation r_{12}. Keeping this fact in mind, we can construct an argument like that in the preceding box in order to arrive at a new conservation law:

$$\tfrac{1}{2}m_1 v_1{}^2 + \tfrac{1}{2}m_2 v_2{}^2 - \frac{Gm_1 m_2}{r_{12}} \equiv K_1 + K_2 + U_{12} \equiv E = \text{constant}, \tag{4.13}$$

where, for the sake of symmetry, the product $m_1 M_2$ ($= m_2 M_1$) has been written as $Gm_1 m_2$. The new feature of this equation is that the potential energy, $U_{12} = -Gm_1 m_2/r_{12}$, is not a sum of independent, individual contributions from the two particles, but a true interaction energy.

From the conservation law for a pair of gravitationally interacting particles (Equation 4.13), it is a short and easy step to the corresponding law for an isolated system containing any number of mutually gravitating particles:

$$\sum_i \tfrac{1}{2}m_i v_i^2 - \sum_{i<j} \frac{Gm_i m_j}{r_{ij}} \equiv \sum_i K_i + \sum_{i<j} U_{ij} \tag{4.14}$$

$$\equiv K + U = E = \text{constant},$$

where the sums run over all particles and over all distinct pairs of particles.

What does the law of energy conservation mean? As we have just seen, the law of conservation of energy for a system of mutually gravitating particles is a straightforward mathematical consequence of Newton's second law of motion and his law of gravitational force. Nongravitational forces, such as those produced by an extended or compressed spring, are also associated with potential energy, and a discussion like that which led to Equation 4.14 leads to a more general law of the form

$$K + U_{grav} + U_{elastic} + \cdots = E = \text{constant}, \tag{4.15}$$

where the dots stand for contributions to the potential energy by whatever interactions may be present in the system.

But the law of conservation of energy also holds in contexts more general than those governed by Newton's theory. For example, it holds in contexts where mechanical energy is transformed into heat or into electromagnetic radiation. Energy is also conserved in isolated systems governed by the laws of quantum mechanics and by Einstein's special and general theories of relativity. The law of conservation of energy must evidently have its roots in principles common to all these theories. Huygens' proof that kinetic energy is conserved in elastic collisions suggests the nature of these principles. Huygens did not specify the law of interaction between the colliding bodies in his thought experiment, but he did stipulate that it should not discriminate between the two directions of time. This stipulation exemplifies what have come to be called *symmetry* (or *invariance) principles*. The deepest answer that physics has been able to give to the question "Why is energy conserved?" is "Because the laws governing elementary processes discriminate neither between different moments in time nor between the two directions of time."

Huygens used another symmetry principle, that of the relativity of uniform motion, to derive the earliest version of the law of conservation of momentum. His derivation, slightly generalized, runs as follows. Suppose that the experiment illustrated in the figure on p. 121 takes place on the bank of a river, where it is observed by a passenger in a passing boat. Galileo had previously argued that uniform motion is relative: no experiment can tell us whether the bank or the boat is "really" at rest, provided their relative motion is uniform. The passenger on the boat must therefore agree with the experimenter on the bank that kinetic energy is conserved during an elastic collision. But if the boat has velocity $\mathbf{V}$, the colliding blocks, as seen from the boat, will have velocities $\mathbf{v}_1 - \mathbf{V}$, $\mathbf{v}_2 - \mathbf{V}$ before the collision and $\mathbf{v}_1' - \mathbf{V}$, $\mathbf{v}_2' - \mathbf{V}$ after the collision. If we write down an equation analogous to Equation 4.10 expressing the conservation of kinetic energy in the boat's frame of reference, we can easily find that this new equation and Equation 4.10 are both satisfied, whatever the relative velocity **V**, if and only if

$$m_1\mathbf{v}_1 + m_2\mathbf{v}_2 = m_1\mathbf{v}_1' + m_2\mathbf{v}_2'. \tag{4.16}$$

This equation says that the combined momentum of the two bodies is the same before and after an elastic collision. In the next chapter we will use Huygens' method to derive the relativistic generalization of Newton's formula $\mathbf{p} = m\mathbf{v}$ connecting velocity and momentum.

Gravitational Potential

In Chapter 3 we saw how the pattern of field lines determines the magnitude and direction of the gravitational acceleration at every point in space. The preceding discussion of energy enables us to construct a simpler and deeper representation of the gravitational field.

Consider a test particle of mass m in a static gravitational field. When the particle undergoes a displacement $\Delta\mathbf{r}$, its potential energy *decreases* by an amount ΔU equal to the work done by the field on the particle. The force **F**, the displacement $\Delta\mathbf{r}$, and the change in potential energy ΔU are connected by the relation

$$\Delta U = -F \,\Delta r \cos\theta, \tag{4.17}$$

where θ is the angle between the force and the displacement. Suppose first that the displacement is perpendicular to the force ($\theta = 90°$). Then $\cos\theta = 0 = \Delta U$. That is, the displacement connects points at which U has the same value. In other words, *the field lines (which are everywhere parallel to the gravitational force) are perpendicular to the surfaces on which U is constant.* Now consider a displacement *along* a field line ($\theta = 0$). For such a displacement Equation 4.17 shows that $\Delta U = -F \,\Delta r$, or $\Delta r = -\Delta U/F$. Because Δr is measured along a field line and the field lines are perpendicular to the surfaces on which U is constant, the relation $\Delta r = -\Delta U/F$ means that the distance between neighboring surfaces U = constant is proportional to the difference ΔU between the values of U on the two surfaces and inversely proportional to F, the strength of the field at the point **r**. These relations between the system of surfaces U = constant and the system of field lines enable us to construct either system if we are given the other.

It is easy to construct the surfaces U = constant. As we have seen, the potential energy per unit mass of a particle in the field of a point-mass fixed at the origin is $U/m = -M/r$, so the surfaces U = constant are spheres. For two fixed point-masses M_1, M_2 at $\mathbf{r}_1$, $\mathbf{r}_2$,

$$U/m = -M_1/|\mathbf{r} - \mathbf{r}_1| - M_2/|\mathbf{r} - \mathbf{r}_2|.$$

The corresponding surfaces U = constant are shown on page 127. More generally, let

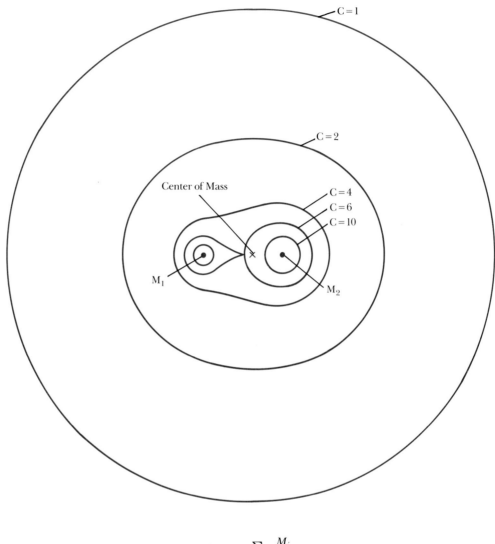

$$\phi(\mathbf{r}) = -\sum_{i} \frac{M_i}{|\mathbf{r} - \mathbf{r}_i|}$$

and rewrite equation (4.17) as

$$\phi(\mathbf{r} + \Delta\mathbf{r}) - \phi(\mathbf{r}) \equiv \Delta\phi = -(F/m)\,\Delta r \cos\theta.$$

This relation holds even if the point-masses that give rise to the field are not fixed, provided $\phi(\mathbf{r} + \Delta\mathbf{r})$ and $\phi(\mathbf{r})$ refer to the same moment of time. The relation implies that the surfaces $\phi = $ constant are everywhere perpendicular to field lines, and that their spacing is everywhere inversely proportional to the strength of the field.

The quantity ϕ is called the *gravitational potential*, and the surfaces $\phi = $ constant are called *potential* or *level surfaces*. The level surfaces are the surfaces we would map out with a carpenter's level. A block sliding without friction on a level surface moves with constant speed along a locally straight line. Galileo's principle of "circular inertia" illustrates this: a block sliding without friction on the surface of a spherical planet moves along a great circle at constant speed.

Angular Momentum

The law of conservation of momentum implies that the velocity of the center of mass of an isolated system is constant in magnitude and direction. Analogously, the law of conservation of angular momentum, which we will derive here, implies that the *spin* of an isolated system is constant in magnitude and direction.

Newton's second law implies that the total momentum of a system of interacting particles changes at a rate equal to the external force acting on the system. Analogously, the total spin of a system of interacting particles changes at a rate equal to the external *torque*.

The momentum of a system of particles is proportional to the velocity of its center of mass, and the factor of proportionality is the system's mass. Analogously, the spin of a rigid body rotating about an axis of symmetry is proportional to its angular velocity, and the constant of proportionality is the *moment of inertia* about this axis. Mass is a measure of a body's resistance to linear acceleration; moment of inertia (about a given axis) is a measure of its resistance to changes in its spin (about that axis).

These analogies are summarized (and expressed symbolically) in the table below. (See the Appendix for an explanation of the vector product denoted by ×.)

MOTION OF CENTER OF MASS

Translational Motion		Rotational Motion	
Quantity	Definition	Quantity	Definition
Mass	$m = \Sigma m_i$	Moment of inertia about an axis of symmetry	$I = \Sigma m_i R_i^2$
Momentum	$\mathbf{P} = \Sigma m_i \mathbf{v}_i$	Spin	$\mathbf{S} = \Sigma m_i (\mathbf{r}_i \times \mathbf{v}_i)$
Velocity	$\mathbf{V} = \dfrac{\mathbf{P}}{m}$	Angular velocity	$\boldsymbol{\omega} = \dfrac{\mathbf{S}}{I}$
Applied force	$\mathbf{F} = \Sigma \mathbf{F}_i$	Applied torque	$\mathbf{K} = \Sigma \mathbf{r}_i \times \mathbf{F}_i$
Relation between force and momentum	$\dot{\mathbf{P}} = \mathbf{F}$	Relation between torque and angular momentum	$\dot{\mathbf{S}} = \mathbf{K}$

Newton's derivation of Archimedes' law of the unequal-arm balance leads in a natural way to the definitions of torque and spin. Although Newton himself did not formulate these concepts explicitly, he realized that Archimedes' equilibrium condition guarantees that the dumbbell formed by the two masses and the massless rod connecting them will not begin to rotate about its point of support (see facing figure). Newton imagined that the point of support was the center of a wheel, and that the weights were suspended from horizontal spokes. By an argument given in the caption on p. 129, he proved that if Archimedes' formula is satisfied, the

The masses m_1, m_2 are suspended from points A, B at distances d_1, d_2 from the point of support O of a massless horizontal rod AB. Suppose that m_2 is smaller than m_1. Draw a circle of radius d_2 about the center O, and suppose that the points A, B lie on horizontal spokes. From the point A drop a perpendicular to the horizontal line AB, meeting the circle in P. We may imagine that the mass m_1 is suspended from the point P instead of from A because if the string holding the mass is pinned at the points A and P and the section of string AP is removed, nothing will change. Resolve the force $m_1\mathbf{g}$, represented in the figure by $\overrightarrow{PS}$, into a radial component $\overrightarrow{PR}$ and a tangential component $\overrightarrow{PT}$. The radial component simply exerts a pull on the axle of the wheel and stresses the spoke OP. The tangential component, whose magnitude is $m_1 g \sin \theta$, where θ is the angle between the radial direction OP and the vertical direction PS, tends to turn the disk counterclockwise. The force m_2 exerted on the wheel by the mass m_2 is also tangential and tends to turn the wheel clockwise. Since the turning effect of a tangential force is the same at all points on a given circle, the wheel will be in equilibrium under the two forces if $m_2 = m_1 \sin \theta$ or $m_2 = m_1(d_1/d_2)$ or $m_1 d_1 = m_2 d_2$, which is Archimedes' criterion.

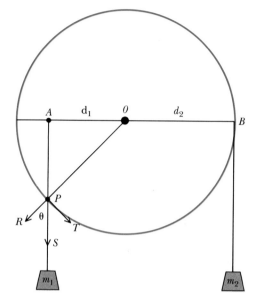

forces exerted by the weights are equivalent to equal tangential forces applied to the rim of the wheel and tending to turn it in opposite directions.

Let us give the name *torque* to the turning effect of an applied force, and denote it by $\mathbf{K}$. As explained in Appendix 1,

$$\mathbf{K} = \mathbf{r} \times \mathbf{F} = \frac{d\mathbf{J}}{dt}, \tag{4.18}$$

where $\mathbf{J}$ is the *angular momentum*. That is, torque is the rate of change of angular momentum, just as force is the rate of change of momentum.

The torque on a particle vanishes when the applied force points directly toward or away from the fixed point O, because then the angle θ between $\mathbf{r}$ and $\mathbf{F}$ is zero, and $\mathbf{r} \times \mathbf{F} = rF \sin \theta = 0$. Since the force acting on each member of a pair of mutually gravitating particles is directed toward their center of mass, the torques about this point vanish, and each particle's angular momentum is constant in time. This is a familiar conclusion, somewhat disguised:

$$\mathbf{r} \times \mathbf{p} = m(\mathbf{r} \times \mathbf{v}) = m(r \times d\mathbf{r})/dt.$$

The vector $\mathbf{r} \times d\mathbf{r}$ is perpendicular to the plane defined by two adjacent radius vectors, and its magnitude is twice the area of the triangle formed by these two vectors and the short line connecting their tips (see above, right). Thus, the constancy of $\mathbf{J} = \mathbf{r} \times \mathbf{p}$ implies that each orbit lies in a plane (the same plane, since the line connecting the particles passes through the center of mass), and that the radius vector of each particle sweeps out equal areas in equal times.

In a system of three mutually gravitating particles, the resultant of the gravitational forces acting on a given particle is not in general directed toward or away from the system's center of mass; hence, the individual angular momenta are continually changing. Their sum, however, remains constant in time:

$$\mathbf{K}_1 + \mathbf{K}_2 + \mathbf{K}_3 = \frac{d}{dt}(\mathbf{J}_1 + \mathbf{J}_2 + \mathbf{J}_3) = \mathbf{r}_1 \times \mathbf{F}_1 + \mathbf{r}_2 \times \mathbf{F}_2 + \mathbf{r}_3 \times \mathbf{F}_3$$
$$= \mathbf{r}_1 \times (\mathbf{F}_{12} + \mathbf{F}_{13}) + \mathbf{r}_2 \times (\mathbf{F}_{21} + \mathbf{F}_{23}) + \mathbf{r}_3 \times (\mathbf{F}_{31} + \mathbf{F}_{32})$$
$$= (\mathbf{r}_1 - \mathbf{r}_2) \times \mathbf{F}_{12} + (\mathbf{r}_1 - \mathbf{r}_3) \times \mathbf{F}_{13} + (\mathbf{r}_2 - \mathbf{r}_3) \times \mathbf{F}_{23} = 0,$$

where $\mathbf{F}_{12}$ is the force exerted by particle 2 on particle 1, which is parallel to the vector $\mathbf{r}_1 - \mathbf{r}_2$ joining particle 2 to particle 1. This result, a consequence of Newton's law of action and reaction, obviously holds for any isolated system of interacting particles, provided the interaction between any two particles is along their line of centers. (Such interactions are called *central*.) *The total angular momentum of any isolated system of centrally interacting particles is constant in time.* The particles in such a system exchange angular momentum, just as they exchange ordinary momentum and energy.

The total angular momentum or *spin* of a spinning planet is related in a simple way to its rate of rotation, which we denote by the vector $\boldsymbol{\omega}$. This vector points in the direction that the rotation would drive a right-handed screw and has magnitude $\omega = 2\pi/T$, where T is the period of rotation. If the planet rotates about its symmetry axis, its spin, which we denote by $\mathbf{S}$, is proportional to $\boldsymbol{\omega}$:

$$\mathbf{J} = \mathbf{S} = I\boldsymbol{\omega}, \qquad I = \sum_i m_i R_i^2, \tag{4.20}$$

where R_i is the distance of the ith particle from the axis of rotation, and the sum extends over all particles in the system. For a uniform sphere of radius a, $I = \frac{2}{5}ma^2$. For the Earth, $I \simeq \frac{1}{3}ma^2$, because the Earth's density increases toward the center.

It follows from the law of conservation of angular momentum that, in the absence of applied torques, the Earth's spin axis must point in a fixed direction, and that the length of the sidereal day must be constant. This conclusion is not as obvious as it may seem at first sight. The angular velocity vector of a rigid body that is not spinning about a symmetry axis does not keep a fixed direction but precesses about the fixed direction of its angular momentum $\mathbf{J}$ (which in this case does not coincide with $\boldsymbol{\omega}$). The symmetry axes of spinning planets and stars coincide with their axes of symmetry, because the departures from sphericity in these objects are caused by the centrifugal force, which is symmetric about the spin axis. An external gravitational field exerts no net torque on a spherically symmetric body (see figure on facing page). Even a nonspherical

If *OS* is an axis of symmetry of a body whose center of mass is at *O*, the contributions to the torque exerted by a mass at *S* from symmetrically placed points *P*, *P'* cancel.

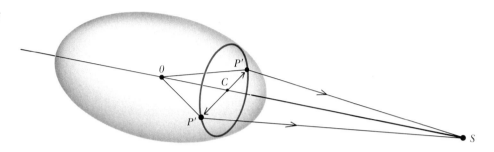

body experiences no net torque from a distant point mass if its mass distribution is symmetric about the line joining the point mass to the system's center of mass (see above).

We are now in a position to understand the precession of the equinoxes and the regression of the Moon's line of nodes. If the Earth's equatorial plane, the plane of the Moon's orbit, and the plane of the Earth's orbit all coincided, neither the Moon nor the Sun would exert a torque on the spheroidal Earth, by the argument explained in the caption to the above figure. Because the Earth's equatorial plane is inclined to the orbital plane of the Moon and the Earth, the torques exerted by the Moon and the Sun do not vanish. We will see in a moment that the residual lunar torque, averaged over a month, tends to twist the Earth's spin axis into the "vertical" direction, *i.e.*, the direction perpendicular to the Moon's orbital plane (see the upper figure on p. 132). However, the change in angular momentum is in the same direction as the torque, which is perpendicular to the spin itself and to the vertical direction; hence the torque drives the tip of the spin vector around a circle centered on the vertical direction, causing the spin vector to sweep out a cone, as shown in the upper figure on p. 132. Thus the lunar torque, averaged over a month, changes neither the magnitude of the spin nor its inclination to the vertical. Moreover, if we define the direction of the spin as *direct*, then the upper figure on p. 132 shows that a torque that tends to twist the spin axis into the vertical direction causes the spin to precess in the *retrograde* direction.

The same argument applies to the solar torque, averaged over a year. Because the directions perpendicular to the Moon's and the Earth's orbital planes nearly coincide, the averaged torques are nearly parallel. The two torques have nearly the same ratio as the tidal accelerations that produce them, about 2¼.

These qualitative considerations make it plausible to attribute the precession of the Earth's spin axis to tidal torques exerted by the Moon and the Sun, but the argument remains incomplete because we have not yet ascertained whether the *magnitude* of the observed effect—the precession period of 26,000 years—is consistent with its presumed cause. Many an ingenious and plausible scientific hypothesis has foundered on that rock. As explained in the legend to the upper figure on p. 132, the angular rate of precession is $K/(S \sin \alpha)$, where K is the torque, S

The vertical component **F** of the gravitational pull of the Moon M exerts a net torque **K** on the equatorial ring QQ' and the negative-mass polar caps P, P' (see figure on the facing page). This torque is parallel to the line QQ' in which the equatorial plane meets the Moon's orbital plane, and "seeks" to twist the spin **S** into the vertical direction. It causes the tip of the spin vector to travel around a circle of radius $2\pi S \sin \alpha$ at the rate K. Hence the precession period is $(2\pi S \sin \alpha)/K$ and its angular rate is $K/(S \sin \alpha)$.

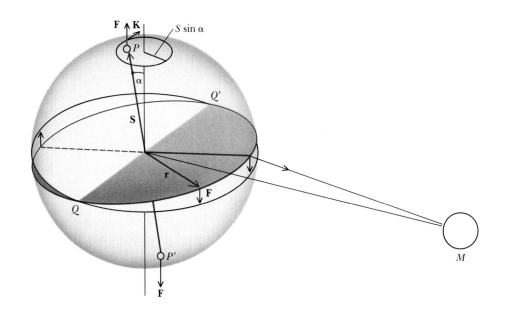

The vertical component of the Sun's tidal pull on the Moon, whose mass we may consider to be distributed evenly around its orbit, exerts a torque **K** parallel to the line NN' in which the Moon's orbit intersects the plane of the ecliptic, causing the Moon's orbital angular momentum **L** to precess about the vertical direction.

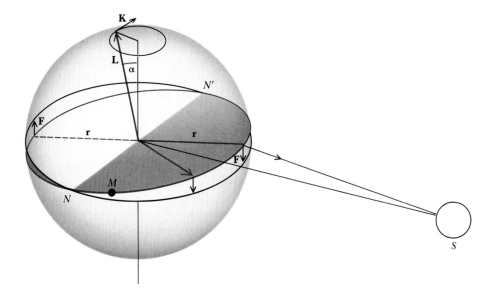

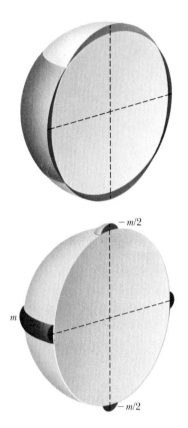

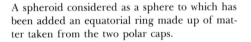

A spheroid considered as a sphere to which has been added an equatorial ring made up of matter taken from the two polar caps.

the spin, and α the angle between the spin vector and the vertical direction about which it precesses. So to estimate the rate of precession we need to estimate the torque; to estimate the torque we need to know the tidal forces exerted by the Moon and the Sun on the Earth, which are described in the box on page 103, and the way in which the distribution of mass in the Earth departs from sphericity.

Apart from local irregularities represented by mountains and valleys, the Earth is a spheroid whose polar diameter is smaller than its equatorial diameter by about 1 part in 300. In estimating the torque, we may think of the Earth as a sphere (which experiences no net torque) to which has been added an equatorial ring composed of matter removed from the two polar caps (above). It is a straightforward matter to calculate the torques exerted by the Moon and the Sun on a ring equal in mass to the Earth's equatorial bulge. This calculation shows, however, that these torques account for only about one third of the observed rate of precession. The source of the discrepancy is obvious from the above figure: in estimating the torque, we must allow for the missing mass at the poles as well as the extra mass at the equator. The "negative-mass" polar caps turn out to contribute twice as much to the torque as the equatorial bulge.

There is a check. The Moon's orbital angular-momentum vector precesses in a retrograde sense about a direction perpendicular to the plane of the Earth's orbit (see the lower figure on p. 132). Since the period of precession (18.6 years) is much greater than the Moon's orbital period (27.3 days), the effect is much the same as it would be if the Moon's mass were spread out along its orbit in a thin ring. When I estimate the rate of precession due to the Sun's tidal torque on such a ring, using the formula used earlier to estimate the torque exerted by the Moon and the Sun on the Earth's equatorial bulge, I get 17.8 years—close enough to the observed value, considering the crudeness of the estimate.

Evolutionary Processes

Astronomical systems owe their distinctive features partly to the processes by which they were formed and partly to processes that have operated since they were formed. In this section we will consider briefly some processes of the second kind; we will discuss the formation of astronomical systems in Chapter 8.

   Large-scale radial motions tend to decay. Left to itself, tea that has been vigorously stirred will come to rest. The kinetic energy initially present in organized large-scale motions flows into motions at progressively smaller scales, and eventually into the random thermal motions of water molecules. Analogously, and for much the same reasons, organized large-scale motions in a rich self-gravitating system composed of atoms, stars, or galaxies tend to decay into less-organized motions on smaller scales, and eventually into the random thermal motions of the system's "particles." Owing to the law of conservation of angular momentum, rotational motions about the spin axis are protected; they cannot decay. But organized motions that do not contribute to the total angular momentum are not protected, and eventually they do decay.

   Systematic radial motions (which contribute nothing to a system's angular momentum) are in fact present only in special kinds of self-gravitating systems. Among stars, for example, pulsation, collapse, and explosions are triggered by specific instabilities. An expanding star cluster— an interesting but very rare phenomenon—is the site of recent star formation, a process that releases large quantities of energy during short periods of time.

   In a rich system without systematic radial motions, the total kinetic energy K and the potential energy U satisfy the following simple relation, known as the *virial theorem:*

$$2K + U = 0. \tag{4.21}$$

This equation is the analog of a relation between the kinetic energy, the potential energy, the volume, and the pressure of an imperfect gas, derived by Rudolf Clausius in 1870. In Chapter 8 we will use a cosmological analog of Clausius' relation. Since $K + U = E$, it follows that

$$K = -\tfrac{1}{2}U = -E. \tag{4.22}$$

The quantity $-E$ is called the *binding energy*. It is the energy needed to disperse the system's constituent particles. According to Equation 4.22, the binding energy and the kinetic energy are equal in a self-gravitating system that has no systematic radial motions.

Equation 4.22 connects the three most basic physical properties of a self-gravitating system of particles: its mass M, its effective diameter D, and its velocity dispersion $\langle v^2 \rangle$. The last two quantities are defined by the formulas

$$U \equiv -\frac{mM}{D,} \qquad K \equiv \tfrac{1}{2}m\langle v^2 \rangle, \tag{4.23}$$

Substituting these values for U and K in Equation 4.22, we obtain

$$\langle v^2 \rangle = \frac{M}{D}; \tag{4.24}$$

that is, the velocity dispersion of the particles that make up a self-gravitating system is the quotient of the system's gravitational mass and its effective diameter.

We can use this relation to deduce the average temperature of the Sun (which is proportional to the velocity dispersion of its particles) from its measured mass and diameter. In this way we find that the Sun's average temperature is a few million degrees on the Kelvin scale. At such temperatures, hydrogen and helium, which account for at least 98 percent of the Sun's mass, are fully ionized into their constituent nuclei and electrons, and so are such light atoms as carbon, nitrogen, and oxygen, which account for most of the remaining mass. Because nuclei and electrons have very small radii—about 10^{-12} cm—the material at the center of the Sun is a nearly ideal gas despite the fact that its density is 100 times the density of water.

In applications to systems whose "particles" are stars and galaxies, we can use Equation 4.24 to deduce the mass M from estimates of the diameter D and the mean squared velocity $\langle v^2 \rangle$. Such estimates yield the gravitating mass within the radius of a system of "markers" whose diameter and dispersion velocity have been estimated. Such estimates have led to the important finding that 90 percent or more of the mass bound in galaxies and galaxy clusters is nonluminous. We will discuss these estimates and their implications further in Chapter 8.

Gaseous systems lose energy by radiation. The energy of an isolated system of gravitating particles is rigorously conserved. But if astronomical systems were made of simple gravitating particles, we would not be able to see them. Starlight is a form of energy, and its existence is a sign that mechanical energy—kinetic energy plus potential energy—is not always conserved. In gaseous systems, collisions between gas molecules readily transform kinetic energy of molecular motions (heat) into radiation, which escapes. Thus, the total mechanical energy of a radiating system continually decreases.

This conclusion has a curious consequence, first pointed out in 1870 by the American mathematician Homer Lane. Consider a spherical self-gravitating gas cloud. In equilibrium the

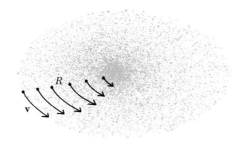

In an ideal disc the gravitational acceleration is directed toward the center of the disc by symmetry. Every particle moves in a circular orbit at a speed v that makes its centripetal acceleration equal to the gravitational acceleration.

cloud's kinetic energy is given by Equation 4.22: $K = -E$. Hence as the cloud radiates away energy, and E, which is negative, becomes more negative, K increases. Hence the average temperature of the cloud, which is a fixed multiple of the kinetic energy per molecule, increases as the cloud radiates away energy. A self-gravitating gas cloud gets hotter as it cools! Such a cloud has *negative* heat capacity: its temperature decreases when energy is added to it, increases when energy is removed.

Since Lane was a mathematician and not an astronomer, he did not appreciate the absurdity of this conclusion. Astronomers, unable to find a flaw in Lane's mathematical argument, sensibly concluded that a real star must solidify when the density at its core becomes comparable to the density of water, after which the star will cool in the "normal" way—i.e., like a hot poker. This theory of stellar evolution was widely held well into the 1920s and is discussed in leading textbooks of the 1930s. In fact, there is nothing whatever wrong with Lane's argument. A self-gravitating gas cloud *does* have negative heat capacity. Although it is close to mechanical equilibrium, it is not close to thermodynamic equilibrium, the state of maximum entropy. A self-gravitating gas cloud has no state of maximum entropy.

Gravitational interactions do not readily transfer angular momentum. An isolated system can radiate away energy but not angular momentum. Moreover, external forces exert small torques on nearly spherical systems. These facts suggest that dissipative self-gravitating systems—systems that have a way to transform internal kinetic energy into heat and radiation—tend to evolve into cold, rotating discs in which gravitational attraction is everywhere balanced by centrifugal force (see above). Discs are systems that have radiated away any kinetic energy that was not bound up in the rotational motions that contribute to the system's angular momentum. In an ideal disc, the orbits of all the particles are concentric, coplanar circles. The planetary system and the satellite systems of Jupiter and Saturn approximate ideal discs. The discs of spiral galaxies have presumably evolved from less flattened systems by dissipating internal kinetic energy while conserving angular momentum.

"The Realm of the Nebulae"

In 1923 Edwin Hubble set out to chart the spatial distribution of galaxies using observations with the 100-inch telescope on Mount Wilson. The 100-inch telescope was the most powerful in the world, and no other astronomer combined Hubble's breadth of scientific vision with his insight into the technical problems that he needed to solve to transform that vision into a scientific picture of the world. It was perhaps the happiest conjunction of a powerful telescope and a powerful mind since the time of Galileo.

What did Hubble expect to find? The distribution of bright galaxies on the celestial sphere is conspicuously clumpy; so their distribution in space must also be clumpy. It was reasonable to anticipate that galaxies, like stars, would be hierarchically clustered. The lowest levels of the clustering hierarchy were clearly discernible in the spatial distribution of nearby galaxies. The Small and Large Magellanic Clouds were evidently satellites of the Galaxy; the Galaxy and Andromeda seemed to be the most massive members of a larger aggregate, the Local Group; and still larger aggregates, some of them with hundreds of member galaxies, showed up clearly in long-exposure photographs made with the 100-inch telescope. Given that galaxies, like stars, belong to a hierarchy of self-gravitating clusters, what can we say about their spatial distribution on the very largest scales? There are three qualitatively distinct possibilities.

1. *There is a largest self-gravitating system of galaxies, the Metagalaxy, which stands in the same relation to individual galaxies as the Milky Way does to individual stars.*

This view was advocated as late as 1965 by the Swedish theoretical physicists Oskar Klein and Hannes Alfvén.

2. *The clustering hierarchy never terminates; there are systems whose masses and diameters exceed any given limits and whose average densities are smaller than any given limit.*

This model of the universe was suggested in 1908 by the Swedish astronomer C. V. L. Charlier.

3. *There is a largest scale of clustering. Averaged over regions whose dimensions greatly exceed this scale, the spatial distribution of galaxies is uniform.*

This was the hypothesis that Hubble set out to test.

The distribution of bright galaxies is clumpy over very large angular scales. Hubble realized that the only way he might detect an underlying uniformity would be to photograph galaxies so distant that the largest clumps would subtend very small angles. The 60-inch and 100-inch reflectors enabled Hubble to photograph very faint and distant galaxies, though each photograph covered only a small area of the sky. The 100-inch telescope, for example, produces sharp images in an area equal to that covered by the full moon. So Hubble was forced to sacrifice breadth of sky coverage for depth. During a period of several years, using both large reflectors, he collected "1,283 separate samples rather uniformly scattered over 75 per cent of the sky."

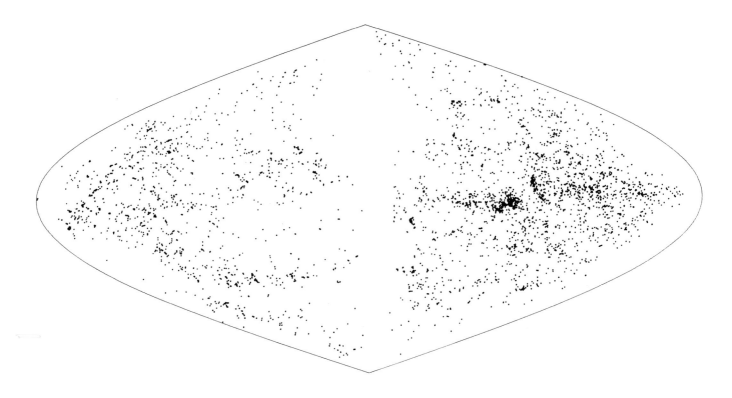

Equal areas on this plot of 4,364 bright galaxies, by Gérard De Vaucouleurs of the University of Texas, represent equal areas of the sky. Galaxies are absent from a central vertical strip and along the edges of the plot. These regions correspond to an irregular band centered on the Galactic equator, the "zone of avoidance," produced by absorbing matter near the plane of the Galaxy. The northern Galactic hemisphere (right side) is noticeably richer in galaxies than the southern Galactic hemisphere (left side). This imbalance is caused by the great cluster of galaxies in Virgo, which lies near the north galactic pole.

At high Galactic latitudes—in the two Galactic polar caps—the distribution was approximately uniform. At lower Galactic latitudes the number of galaxies per square degree brighter than a given limit decreased systematically toward the Galactic equator. And very close to the Galactic equator no galaxies at all were to be seen (see the figure above).

The region in which no galaxies at all are seen varies in width from 10° to 40°. Its boundaries are ragged and unsymmetric. Hubble concluded, as had others before him, that this "zone of avoidance" is produced by a relatively small number of dense, dark clouds. He then went on to study the distribution on the sky of galaxies outside the zone of avoidance. He found that in areas of the sky free from obvious obscuration the number of galaxies per square degree decreases systematically toward the Galactic equator. He could account for this decrease by postulating a uniform light-absorbing layer centered on the Galactic plane (see figure on facing page).

Hubble turned next to the distribution of galaxies in depth. The uniform distribution of faint galaxies on the celestial sphere in regions unobscured by Galactic dust clouds implies either that the universe is spherical (and we happen to be close to the center) or that the spatial distribution of galaxies is uniform apart from local irregularities. We can distinguish observationally between these possibilities by noticing how rapidly the number of galaxy images per square degree increases as the limiting brightness decreases. Suppose that all galaxies have the

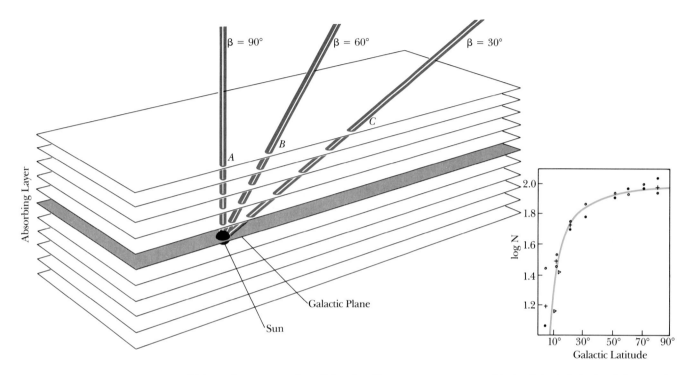

"Schematic representation of the absorbing layer. As observed from the vicinity of the Sun, extragalactic objects are obscured according to the lengths of the light paths through the absorbing layer. The obscuration is least for objects seen in the direction of the Galactic poles and increases as the latitudes diminish.

The average number of nebulae per unit area (brighter than a given limiting magnitude) increases with the Galactic latitude of the area, in a manner which closely follows the cosecant law,

$$\log N = \text{constant} - 0.15 \operatorname{cosec} \beta.$$

Circles and discs represent data from the northern and the southern Galactic hemispheres, respectively; crosses are means of the two. The pair of triangles represent supplementary data in low latitudes."

HUBBLE, *The Realm of the Nebulae*

same intrinsic brightness and that they are sprinkled uniformly in space. Because the apparent brightness is inversely proportional to the square of the distance, and the volume of a sphere is proportional to the cube of its radius, the number of galaxies brighter than a given apparent brightness l is proportional to $l^{-3/2}$. The same proportionality holds for a homogeneous mixture of galaxies with different intrinsic brightnesses. Thus Hubble could test the hypothesis of statistical uniformity by counting galaxy images to successively fainter limits. He found that the counted numbers did indeed satisfy the predicted relation (provided he allowed for dimming caused by the cosmic expansion, which we are about to discuss). In 1936 he summarized his findings in these words:

There is no evidence of a thinning-out, no trace of a physical boundary. There is not the slightest suggestion of a supersystem of nebulae isolated in a larger world, Thus, for purposes of speculation, we may apply the principle of uniformity, and suppose that any other equal portion of the universe, selected at random, is much the same as the observable region. We may assume that the realm of the nebulae is the universe and that the observable region is a fair sample.

The Cosmic Expansion and the Cosmic Distance Scale

The spectrum of a galaxy is a composite of the spectra of the stars that contribute most of its light. The absorption lines in the spectrum of each star are shifted in wavelength by an amount proportional to the radial component of the star's velocity relative to the Earth. The fractional change in wavelength is independent of the wavelength and is given, for small fractional shifts, by the formula

$$\frac{\Delta\lambda}{\lambda} = \frac{v_r}{c},$$

where v_r is the radial component of the star's relative velocity (positive for recession, negative for approach) and c is the speed of light. In the composite spectrum, each line is broadened by the wavelength shifts caused by the motions of individual stars relative to the galaxy's center of mass, and its central wavelength is shifted by an amount proportional to the radial velocity of the galaxy's center of mass relative to the Earth (see figure on facing page). Thus we can deduce a galaxy's speed of approach or recession from measurements of the fractional shifts in the wavelengths of its spectral lines.

The spectra of galaxies are faint and difficult to measure. Galaxies have low surface brightness to begin with, and the spectrograph spreads out their images, further reducing their surface brightness. In 1912 V. M. Slipher succeeded in measuring Doppler shifts in the spectrum of the brightest external galaxy, the Andromeda Galaxy, with a spectrograph attached to the 24-inch telescope at the Lowell Observatory in Arizona. Two years later he published a list of radial velocities for 13 spirals.

Most of Slipher's measured velocities were positive, indicating motion away from the Sun, and several were much larger than any previously measured astronomical velocities. The spirals were evidently receding from the Galaxy in all directions and with enormous velocities.

By 1925 Slipher had measured radial velocities for 45 spirals. A pattern was now beginning to emerge. All but a few of the brightest, and presumably nearest, galaxies were receding from the Sun, and the speed of recession tended to increase with decreasing apparent brightness. The largest velocity of recession in Slipher's sample was 1,800 kilometers per second. It seemed as if the galaxies were rushing away from the Milky Way in all directions, their speeds increasing with increasing distance. Did such a pattern of motion make sense?

Hubble's inference that the galaxies are distributed uniformly in space, apart from local clustering, imposes a powerful constraint on the kinds of large-scale motion we can reasonably

Formation of a galactic spectral line. (a) A single atom emits light at a series of discrete wavelengths, including the wavelength designated λ_0. The received wavelength of the light is shifted by an amount $\Delta\lambda$ proportional to the relative velocity of the atom and the receiver. The shift is toward the red (longer wavelengths) if the source is receding from the observer, toward the violet if the source is approaching. (b) The light emitted by a collection of atoms in a hot gas forms a spectral "line" whose width is proportional to the spread in the atoms' line-of-sight velocities. (c) Each small parcel of gas in the disc of a spiral galaxy viewed edge-on has its own line-of-sight velocity. The resulting spectral "line" is a superposition of contributions from individual parcels, each red- or blue-shifted by an amount proportional to its line-of-sight velocity. (d) Finally, the whose broad line is shifted in wavelength by an amount proportional to the relative velocity of the galaxy's center of mass and the observer. This figure illustrates the formation of a galactic emission line. Galactic absorption lines are shifted and broadened by analogous motions.

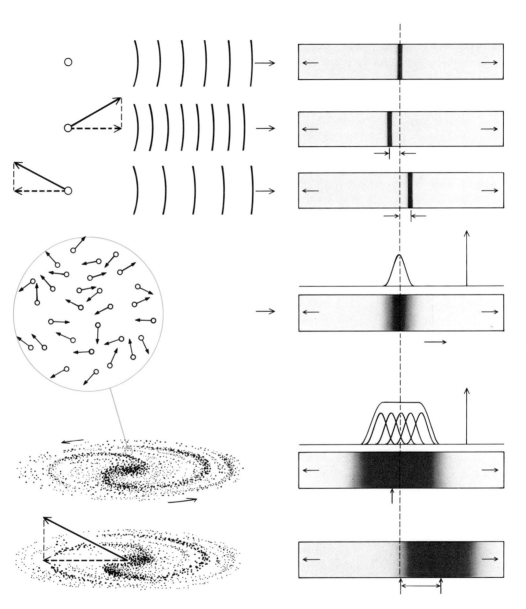

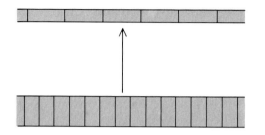

Uniform expansion in one dimension.

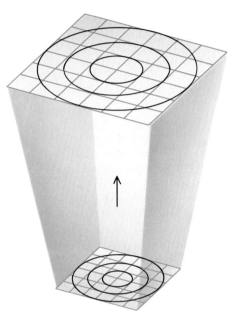

Uniform expansion in two dimensions: change
of scale with no change in shape.

expect to observe. In a uniform distribution, there are no preferred positions or directions (if
we ignore local irregularities). Large-scale motions would destroy this symmetry unless they
themselves shared it. What kinds of large-scale motions preserve large-scale spatial uniformity?

Let us first consider the problem in one dimension. Imagine an infinitely long, straight
rubber band marked off in inches. The only large-scale stretching motion that would not cause
the rubber band to become thinner in some places and thicker in others is one in which the
distance between every pair of successive marks increases at the same rate (see the lefthand
figure above). Suppose that the rubber band is stretching at such a rate that the distance between successive marks doubles in one minute. Then the distance between *any* two marks doubles in one minute. It follows that the relative velocity of any pair of marks is proportional to
their separation.

A stretching motion of the kind just described does not confer preferred status on any mark.
An observer stationed on any mark sees every other mark receding from him at a rate proportional to its distance; and the view is exactly the same from each mark. The center of this
one-dimensional universe is everywhere; its edge, nowhere.

Next, imagine a flat rubber sheet stamped with a square grid, like a sheet of graph paper.
The only large-scale stretching motion that does not cause the sheet to become thinner in some
places and thicker in others is one in which every square grows at the same rate, changing its
size but not its shape (above, right). This kind of stretching alters the size but not the shape of
any figure drawn on the sheet. As before, the relative velocity of any two marks is proportional

Uniform expansion in three dimensions.

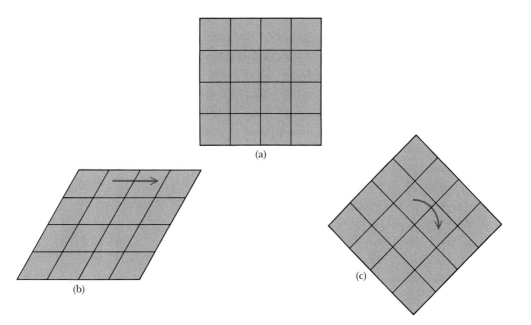

to their separation, and any point may be considered to be at the center of this two-dimensional expanding universe.

The situation is exactly the same in three dimensions. The only large-scale pattern of motion that preserves a uniform distribution of grid points is one in which every spherical surface expands at a rate proportional to its radius. Any point may be considered to be at the center of such an expanding medium, and the view is exactly the same from every point.

The kind of large-scale motion we have been considering is called *uniform expansion*. Uniform contraction is the reverse motion, and of course it too preserves spatial uniformity. Uniform expansion and contraction are the only large-scale motions that do not define a preferred position *or direction* in space. Uniform shear and uniform rotation (see figue on facing page) are examples of large-scale motions that do not define a preferred position but do define a preferred direction.

Slipher's discovery that faint galaxies are receding from the Milky Way at enormous speeds therefore suggests that the entire system of galaxies is undergoing a uniform expansion. No other interpretation of the data would be consistent with Hubble's inference that the spatial distribution of galaxies is uniform and isotropic apart from local irregularities. But is the system of galaxies really expanding uniformly? If so, the radial velocities of faint galaxies should be proportional to their distances. Galaxies ten times as distant (and 100 times as faint) as the most distant galaxies on Slipher's list would then have radial velocities of 18,000 kilometers per second; galaxies ten times more distant still, radial velocities of 180,000 kilometers per second— more than half the speed of light.

Henrietta Swan Leavitt (1868–1921), American astronomer; her discovery of a relation between the period of light variation and the luminosity of Cepheids in the Magellanic Clouds launched modern observational cosmology.

Ejnar Hertzsprung (1873–1967), Danish astronomer; discovered that yellow, orange, and red stars belong to two distinct classes, which he dubbed "giants" and "dwarfs"; calibrated Leavitt's period-liminosity relation by estimating the distances of 13 Galactic Cepheids.

To test the idea of uniform expansion, Hubble needed to estimate the distances of the spirals on Slipher's list. Estimating the distances of external galaxies and galaxy clusters is the central problem of observational cosmology. The underlying theory is elementary. Nearly all cosmological distance estimates are based on one of the two following laws.

The luminostity L and apparent brightness l of a distant object are related to the object's distance by the inverse-square law $l = L/r^2$.

An object's linear diameter D, angular diameter θ, and distance r are connected by the relation $\theta = D/r$ (if $\theta \ll 1$).

These formulas are valid for relatively nearby objects. Their relativistic generalizations are derived in Chapter 7. The deep and difficult problems concern their *application*. How do we find the luminosity L of a very faint star of unknown distance or the luminosity or linear diameter of a distant galaxy or galaxy cluster?

There is a standard strategy for tackling these problems. Cosmologists refer to the *cosmic distance ladder*. It is really more like a pyramid: at each level we gain access to materials that we use to construct the next level.

Stars close enough to have measured parallaxes—i.e., stars whose distances can be estimated by Thales' method of triangulation—provide the base of the pyramid. Some stars with parallaxes large enough to measure accurately are much brighter than the Sun and have distinctive characteristics, such as color, that allow other stars of the same type to be identified at distances well beyond the limits of triangulation. However, no star with measurable parallax is bright enough to be identifiable in an external galaxy, or even outside a relatively small neighborhood of the Sun in our own Galaxy.

The breakthrough that marked the beginning of modern observational cosmology came in 1912. Henrietta Swan Leavitt, a research assistant at the Harvard College Observatory, discovered a relation between the periods and the apparent brightnesses of Cepheids (a class of stars that vary in brightness) in the Small Magellanic Cloud. The Magellanic Clouds are a pair of small satellite galaxies—"moons" of the Milky Way—visible from the southern hemisphere. Cepheids vary regularly in brightness, with periods ranging from 1 to 100 days. The top figure on the facing page shows the light curve of the eponymous star δ Cephei, whose variability was reported in 1784 by John Goodricke. Leavitt's original plot of period against brightness is shown on page 145, along with a modern version of the same plot.

The Danish astronomer Ejnar Hertzsprung immediately recognized the revolutionary implications of Leavitt's discovery. Cepheids are exceptionally bright stars, among the brightest in the Magellanic Clouds. Their periodic light variation makes them easy to find, and the period of the light variation is easy to measure. Because the Small Magellanic Cloud subtends a small angle in the sky, its linear diameter and hence, presumably, its depth are small compared with its distance. Hence the Cepheids in the Cloud are all at nearly the same distance, and Leavitt's relation between period and apparent brightness implies a similar relation between period and intrinsic brightness or luminosity. Thus Cepheids would make ideal distance indicators if some

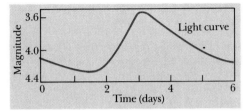

(Left) Light curve of the star δ Cephei.

(Right) Leavitt's original plot of brightness *versus* period for Cepheids in the Small Magellanic Cloud (inset) and a modern plot for Cepheids in the Large and Small Magellanic Clouds (LMC and SMC, respectively). **W** is a linear combination of blue and visual apparent magnitudes that is insensitive to reddening by interstellar matter.

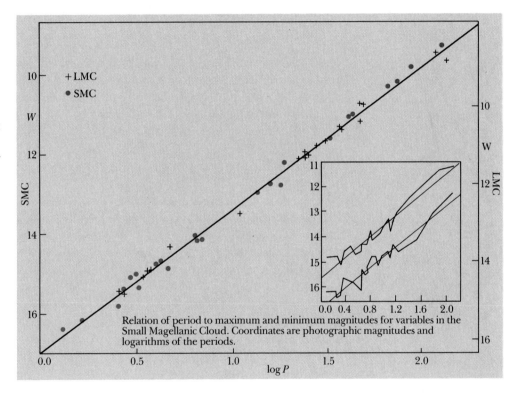

Relation of period to maximum and minimum magnitudes for variables in the Small Magellanic Cloud. Coordinates are photographic magnitudes and logarithms of the periods.

way could be found to calibrate Leavitt's period–luminosity relation, that is, to establish the intrinsic brightness of one or more Cepheids. Cepheids are also found in our own Galaxy; in 1913 there were 13 whose positions on the celestial sphere had been accurately recorded for many years. Hertzsprung was able to estimate the distances of these 13 stars by the method of *secular parallax* explained in the caption to the figure on page 146. From these estimated distances and measured apparent brightnesses, Hertzsprung could calculate the intrinsic brightness of the Galactic Cepheids and thus transform Leavitt's relation between period and apparent brightness into a relation between period and intrinsic brightness. Hertzsprung could now estimate the distance of the Small Magellanic Cloud, and although his estimate was only about a seventh of the modern value (about 200,000 light-years), it supported his conclusion that the

$Vt\overline{/ES}$

E

Vt

S

Sun's Velocity

The method of secular parallax. Suppose that a distant star E is at rest relative to a much more distant background (for example, of external galaxies). Suppose further that the Sun's motion has a component V perpendicular to the line of sight SE. During a time-interval t the direction of the line of sight will change by an amount $Vt\overline{/ES}$, and this shift will be measurable if the time-interval t is long enough. Unfortunately, we cannot assume that E is at rest relative to the distant background. Its motion is likely to have a transverse component comparable in magnitude to the transverse component of the Sun's motion. However, if we observe a group of similar stars at similar distances, the transverse components of their motions are likely to cancel out. So by constructing an appropriately weighted average of the angular displacements of stars belonging to such a homogeneous group, we can estimate the average parallactic displacement and thus the group's distance.

Magellanic Clouds are independent stellar systems, cleanly separated from the Milky Way. This was the first direct corroboration of Kant's hypothesis that there are galaxies beyond our own.

With the 100-inch telescope Hubble was able to find and study Cepheids in several nearby galaxies, whose distances he was therefore able to estimate. With these distance estimates in hand, he could now calibrate new distance indicators even brighter than Cepheids: novae, blue supergiants, and bright irregular variables. These stars all have easily identifiable characteristics, independent of apparent brightness; so they can be used as distance indicators in the same way as Cepheids.

Using these distance indicators, Hubble could now check out an especially attractive candidate for the role of distance indicator: the brightest star in a spiral. Unfortunately, the range of intrinsic brightnesses of brightest stars proved to be quite large. The average brightness of the brightest stars in spirals belonging to the same cluster of galaxies turned out to be much less variable, however, and with the help of this criterion Hubble was able to estimate the distance of the great cluster of galaxies in Virgo. Finally, having estimated the distance of the Virgo Cluster, Hubble could estimate the intrinsic brightnesses of its member galaxies. He could now use whole galaxies as distance indicators, and with their help he could estimate the distances of galaxies too distant for any of their stars to be resolved by the 100-inch telescope.

A preliminary compilation of distances for 22 galaxies on Slipher's list resulted in the plot shown in the upper figure on the facing page, which was published in 1929. The evidence for a linear relation between distance and velocity is suggestive but not overwhelming. To extend the plot to greater distances and higher velocities, Hubble enlisted the aid of Milton Humason. Using the 100-inch reflector, equipped with a new camera of very low focal ratio, Humason was able to record the spectra of galaxies so faint that they could not be seen at the telescope's Cassegrain focus (a focus just outside the tube of the telescope, used by observers to guide the telescope during long exposures). From Humason's measured redshifts and his own estimated distances, Hubble constructed the plot shown in the bottom figure on page 147. The proportionality between distance and velocity was now unmistakable.

Observational cosmology has made great strides since Hubble's day. Modern telescopes record the brightnesses, colors, and redshifts of galaxies in a volume of space ten million times larger than Hubble's sample. New kinds of detectors carried by balloons, rockets, and space-

"The formulation of the velocity-distance relation. The radial velocities (in km/sec), corrected for solar motion, are plotted against distances (in parsecs) estimated from involved stars and, in the case of the Virgo cluster (represented by the four most distant nebulae), from the mean luminosity of all nebulae in the cluster. The black discs and full line represent a solution for the solar motion using the nebulae individually; the circles and dashed line, a solution combining the nebulae into groups."

HUBBLE, *The Realm of the Nebulae*

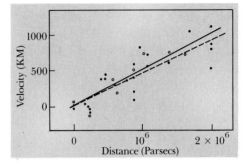

"Logarithms of velocities (in km/sec and corrected for solar motions) are plotted against apparent magnitudes of the fifth brightest nebulae in clusters (corrected for Galactic obscuration.) Each cluster velocity is the mean of the various individual velocities observed in the cluster, the number being indicated by the figure in brackets."

HUBBLE, *The Realm of the Nebulae*

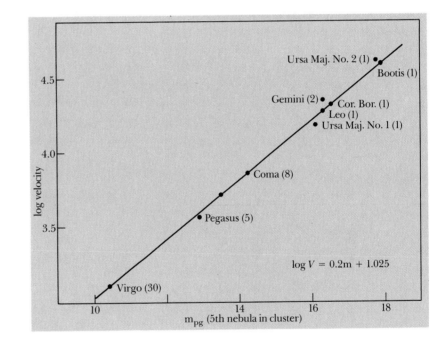

craft have widened the astronomer's window of the electromagnetic spectrum from a single octave (a factor of 2 in the frequency of the light) to more than 12 powers of ten. Recent observations have revealed aspects of the universe unsuspected by Hubble and his contemporaries. But they have also strongly confirmed the hypothesis that Hubble took as his starting point: that the spatial distribution, motions, and "demographic" characteristics of galaxies are the same everywhere and in all directions, apart from local irregularities. The figure on page 148 shows a modern version of Hubble's plot of redshift against apparent brightness.

Not all scientists accepted Hubble's picture of an expanding universe. Even today there are some who reject it, usually arguing along the following lines: "A scientific hypothesis, no matter how plausible, can never be conclusively proved, but it can be shown to be false by a single piece

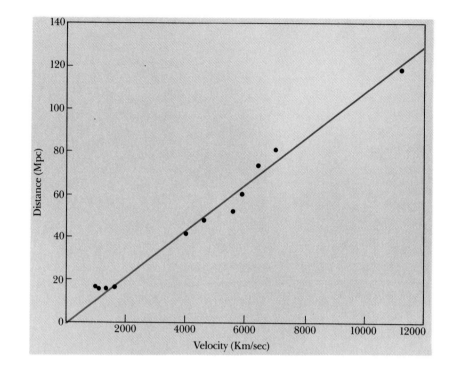

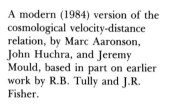

A modern (1984) version of the cosmological velocity-distance relation, by Marc Aaronson, John Huchra, and Jeremy Mould, based in part on earlier work by R.B. Tully and J.R. Fisher.

of contradictory evidence. Exhibit *A* (a photograph, a statistical analysis, or the like) contradicts the expansion hypothesis. Hence that hypothesis must be false." But none of the Exhibit *A*s so far placed in evidence unequivocally contradicts the expansion hypothesis. In every case the evidence is either very weak (for example, photographic evidence that certain objects with large redshifts are physically linked to bright nearby galaxies) or susceptible to an interpretation that does not contradict the expansion hypothesis.

We will see later that the expansion hypothesis draws support from observational evidence of a kind that did not exist when Hubble proposed the hypothesis in 1929, evidence relating to a uniform sea of radiation (the microwave background) that pervades all space (Chapter 7). The expansion hypothesis also has powerful theoretical support. Uniform cosmic expansion is a consequence of Einstein's theory of gravitation and the hypothesis of spatial uniformity (Chapter 7). It was, in fact, predicted by the Russian mathematician Alexander Friedmann in 1922, seven years before Hubble announced the velocity–distance relation.

Newton's Theory Breaks Down

Theories based on Newton's laws of motion and his law of universal gravitation successfully describe the structure and dynamics of self-gravitating systems, from planets and their satellites to galaxies, galaxy clusters, and clusters of galaxy clusters. When Einstein published his theory of gravitation in 1915, there existed in the whole realm of dynamical astronomy not a single well-established discrepancy between observational evidence and predictions based on Newton's theory. (Einstein's theory is sometimes said to have predicted a previously unexplained discrepancy between Newton's theory and observations of the motion of the planet Mercury; but as we will see in Chapter 6, in 1915 most astronomers believed that this discrepancy had already been explained by a Newtonian hypothesis.)

But if we try to apply Newton's theory to a uniformly expanding fluid of uniform density—the simplest model of the universe consistent with astronomical observations—we run into difficulties.

What is the gravitational acceleration at a given point? If the universe had a center O, we could argue that the distribution of mass must be symmetric about O and hence that the gravitational acceleration at a point at a distance r from O must be directed toward O and have magnitude $M(r)/r^2 \propto r$. But if the mass distribution is uniform, all points in space have equal status, and so, by symmetry, the gravitational acceleration must vanish everywhere. Hence the expansion must be unaccelerated; the relative velocity of any pair of particles must be constant in time.

So far there is no difficulty. But let us calculate the kinetic energy associated with the expansion. The relative velocity of two particles is proportional to their separation: $V = Hr$. A simple calculation shows that the kinetic energy K of an expanding sphere of radius r and mass m is $\frac{3}{10}mH^2r^2$. Thus the kinetic energy per unit mass, K/m, increases without limit with increasing radius r. If the universe is infinite, the kinetic energy per unit mass is also infinite. But if every finite volume contains an infinite quantity of energy it is hard to understand how gravitationally bound systems, whose energies are finite and negative, could have come into being.

One might suppose that the kinetic energy associated with uniform expansion is exactly canceled by negative potential energy. The potential energy U of a sphere of radius r is $-\frac{3}{5}mr^2(4\pi\rho/3)$. We could make the sum $K + U$ vanish by setting $H^2 = 8\pi\rho/3$. But this condition cannot be consistently imposed. Because the relative velocity of any pair of particles is constant in time, their separation must be proportional to the time, and the velocity-distance ratio must be inversely proportional to the time. Because the mass of a sphere bounded by material particles is constant in time, the density of such a sphere is inversely proportional to the cube of its radius and hence inversely proportional to the cube of the time: $\rho \propto t^{-3}$. But we have just seen that $H \propto t^{-1}$. Thus the condition $H^2 = 8\pi\rho/3$ cannot be met at more than a single moment of time, and we cannot avoid the conclusion that the energy per unit mass is infinite. In Chapter 7 we will see how Friedmann's idea that space itself is expanding resolves this difficulty.

CHAPTER FIVE

SPECIAL RELATIVITY: EINSTEIN'S THEORY OF UNACCELERATED MOTION

Henceforth space by itself and time by itself are doomed to fade away into mere shadows, and only a kind of union of the two will preserve an independent reality.

HERMANN MINKOWSKI

We saw at the end of Chapter 4 that Newton's theory of gravitation breaks down when we try to apply it to the simplest model of the universe that is consistent with astronomical observations, a uniform, infinite, expanding distribution of mass. We will see in Chapter 7 that Einstein's theory of gravitation makes possible a self-consistent, overdetermined theory of the universe. But Einstein did not set out to solve the cosmological problem. He began to rebuild Newtonian physics during the opening years of the twentieth century, before there was any evidence for large-scale cosmic uniformity, and long before anyone suspected that the universe was expanding. He was motivated not by any lack of agreement between existing theories and experimental or observational evidence, but by what he perceived to be a lack of symmetry in existing theories.

In his use of symmetry arguments to derive new physical laws, Einstein harks back to Huygens. We saw in Chapter 4 how Huygens derived the law of conservation of energy by postulating that collisions between ideal billiard balls are reversible in time, and how he derived the law of conservation of momentum by postulating that there is no preferred state of uniform motion for the description of such collisions. He derived the law of centrifugal force from an assumption that, suitably generalized, became the cornerstone of Einstein's theory of gravitation: the force acting on a weight whirling at the end of a string is indistinguishable from the force acting on the same weight when it is suspended in an appropriate gravitational field. Newton's theory of motion made Huygens' methods obsolete in the eyes of most physicists. Only in this century, through the work of Einstein and his successors, have we come to understand that symmetry principles are deeper than specific physical laws, even laws as general as those of Newton and Maxwell.

Einstein's theories of space, time, and energy—special relativity, the theory of gravitation, and relativistic cosmology—are really aspects of a single, unified theory. Cosmology, for example, is not merely an application of Einstein's theory of gravitation; it is an essential part of the theory itself. In this and the following two chapters, we examine the key steps in Einstein's construction of this theory.

(Left) Edwin P. Hubble at the 48-inch Schmidt telescope on Mt. Palomar.

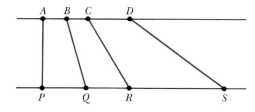

Aristotle's distance-time diagram. Corresponding positions (A, B, C, D) and "nows" (P, Q, R, S) are connected by dashed lines.

The Classical Theory of Unaccelerated Motion

By the time of Aristotle, Greek mathematicians had developed a sophisticated description of uniform or unaccelerated motion—that is, motion at constant speed in a straight line. This description entails two key ideas: first, the notion of an *event*; second, the notion of a *time line*. Although the terms *event* and *time line* are modern, the ideas they refer to seem, nowadays, almost self-evident. Yet they do not belong to the great stock of ideas that are common to all human cultures; they were invented by the Greeks, and never, so far as is known, independently reinvented. Without these ideas Galileo could not have begun to develop his theory of motion.

Let us consider a particle moving along a straight line. Greek mathematicians asserted that such a particle is in a definite *place* on the line at each *moment* of time. Thus they thought of a moving particle's *history* as being made up of *events*, each occurring at a definite place and a definite moment. They represented moments of time by points on a line and identified the *lengths* of segments of this line with the *durations* of time-intervals. In Aristotle's *Physics* we find diagrams like that above, in which the place and the moment that define each event are connected by a dashed line. Using this diagram, Aristotle could define a uniform motion as one in which

$$\overline{AB} : \overline{BC} : \overline{CD} : \cdot \ \cdot \ \cdot \ = \overline{PQ} : \overline{QR} : \overline{RS} : \cdot \ \cdot \ \cdot \ .$$

That is, a particle moving with constant speed covers equal distances in equal times.

Although this representation of uniform motion enabled Aristotle to compare one speed with another, Greek mathematicians apparently stopped short of representing speed itself as a quantity with a continuous range of possible values. This important step was taken by the French mathematician Nicole Oresme around the middle of the fourteenth century. Oresme drew the diagram shown on the facing page, top left. Here, for the first time, we find speed represented as a quantity that can have a continuous range of values, graphed against time. The horizontal line CD represents uniform motion at a speed represented by the distance $\overline{BC}$; uniformly accelerated motion is represented by the line AE. The speed at an arbitrary moment K is represented by a vertical line (ordinate) such as $\overline{KL}$ or $\overline{KM}$. Oresme argued that the distance traveled during the time $\overline{AB}$ by a particle whose uniformly accelerated motion is represented by

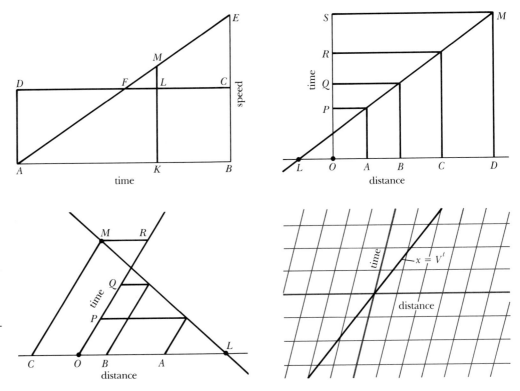

(*Top left*) Oresme's plots of speed *versus* time for uniform motion (line DC) and uniformly accelerated motion (line AE). Distance covered is represented by an area in such plots. (*Top right*) A space-time diagram representing uniform motion. (*Bottom left*) A space-time diagram with oblique axes. (*Bottom right*) An oblique space-time grid.

the line $\overline{AE}$ is equal to the distance traveled during the same time by a particle whose speed is constant, and is equal to the average of the accelerated particle's initial and final speeds:

$$\overline{BC} = \tfrac{1}{2}(0 + \overline{BE}) = \tfrac{1}{2}\,\overline{BE}.$$

A century and a half later, Galileo drew the same diagram and reached the same conclusion.

Galileo or one of his contemporaries probably drew the first genuine spacetime diagram, shown above, top right. Here the history of a particle moving with constant speed along a straight line is represented by a straight line, whose points, such as L and M, represent events in this history. The particle's speed is represented by the distance–time ratio $\overline{LD}/\overline{DM}$, which is the reciprocal of the slope of the line LM. Every unaccelerated motion can be represented by a straight line in this diagram, and every straight line represents an unaccelerated motion.

The distance and time axes in a spacetime diagram need not be perpendicular. In the following discussion we will often find it convenient to use oblique axes, as shown above, bottom left. The distance and time coordinates of events are measured along lines parallel to the coordinate axes. Thus the distance and time coordinates of the event M are $\overline{MR}$ and $\overline{MC}$. The speed of the particle whose history is represented by the line LM is $\overline{LC}/\overline{CM}$.

In the physics of Galileo and Newton, no place or time has preferred status. Thus the event O at the origin of the space time diagrams on p. 153 may be any event at all, and the distance and time axes may be any lines parallel to the distance and time axes shown in these figures. To emphasize the democracy of events, we may think of space and time coordinates as defined, not by a specific pair of axes, but by a *coordinate grid*, of which any point may serve as the origin, and any pair of intersecting grid-lines as coordinate axes. The bottom right figure on p. 153 shows a piece of an oblique coordinate grid.

Up to this point, the description of unaccelerated motion is the same in the theories of Galileo and Einstein. The two theories differ about how two observers in uniform relative motion will describe the same set of events, such as events in the history of a uniformly moving particle or of a photon. A clear understanding of the classical theory and its underlying assumptions will help us to understand the relativistic theory.

Photons

A beam of light exerts a measurable pressure on a piece of metal foil. It also heats the foil. Hence light carries both momentum and energy. In this respect a beam of light resembles a stream of material particles. But light also has a property that distinguishes it from beams of material particles. If we split a beam of monochromatic light into two parts, which we then allow to come together again after they have traversed slightly different distances, we will—if we have done the experiment with sufficient care—observe *interference fringes*. Experiments of this kind show that monochromatic light has a wavelike character. Two parts of the same beam can be made to interfere constructively or destructively, as shown in the figure. Bright fringes appear where the wavecrests of the two parts of the beam coincide; dark fringes, where the crests of one part of the beam coincide with the troughs of the other. Material particles cannot interfere in this way; hence, light cannot consist of tiny particles, as Newton believed.

In 1905 Einstein suggested that exchanges of energy and momentum between light and matter always occur in discrete packets whose size is proportional to the frequency of the light. More specifically, if an atomic nucleus, an atom, a molecule, or a crystalline solid absorbs or emits monochromatic light of frequency ν, its energy must increase or decrease by a multiple of the energy quantum $\Delta E = h\nu$, and its momentum must increase or decrease by a multiple of the momentum quantum $\Delta p = h\nu/c$, where c is the speed of light ($\simeq 3 \times 10^{10}$ cm/sec) and h ($\simeq 6.6 \times 10^{-27}$ gm·cm²/sec) is a universal constant known as Planck's constant. In an elementary interaction, one quantum of energy and one quantum of momentum are exchanged. Thus monochromatic light behaves, in its interaction with matter, as if it consisted of discrete packets, each carrying a quantity of energy $h\nu$ and a quantity of momentum $h\nu/c$. Moreover, these packets travel with the constant speed c. Unlike material particles, however, they cannot be localized in space. When we speak of a photon's x-coordinate or represent its history by a line in the x, t-plane, we are referring to the x-coordinate and the history of a plane wavefront of infinite extent, perpendicular to its direction of motion, the x-axis. If such a photon is absorbed by an atom, it will be absorbed in a definite place, the place where the atom happens to be. But the probability that it will be absorbed in a given region of its wavefront depends only on the area of the region, not on its location.

(*Above*) A plane, linearly polarized wave. **E** and **B** the electric and magnetic fields, are constant on planes perpendicular to the direction of propagation. (*Below*) A plane, circularly polarized wave. While **E** or **B** advances (moves to the right) one wavelength it moves through 360°.

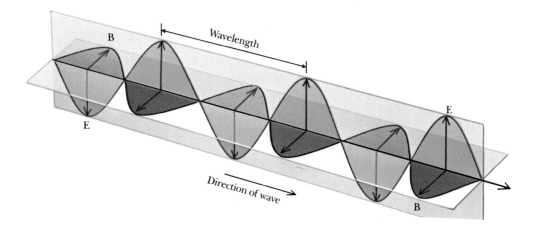

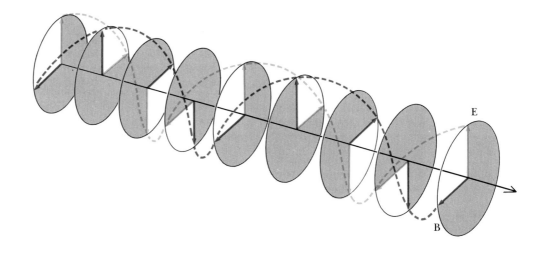

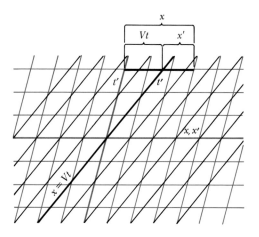

Spacetime grids for two reference-frames in
uniform relative motion.

Let F denote a reference frame in which Newton's laws of motion are valid, and let G denote the coordinate grid associated with this reference frame. On this grid the histories of particles that are at rest are represented by lines x = constant, i.e., lines parallel to the time axis. Lines inclined to the time axis represent the histories of particles moving with constant speed. The history of one such particle, moving with speed V and passing the origin $x = 0$ at time $t = 0$, is represented by the line $x = Vt$ in the bottom right figure on p. 153. Let us now consider a new reference frame F' whose origin coincides with the moving particle. The line $x = Vt$ is the time axis of the grid G' associated with the moving reference frame F', because the time axis of any spacetime grid represents the history of a particle at rest at the point $x = 0$. So far, Einstein still agrees with Galileo and Aristotle. The classical and relativistic paths begin to diverge at the next step.

Where shall we draw the distance axis of the new grid G'? The events represented by points on the new distance axis represent events that have the time coordinate $t' = 0$, where t' denotes the time measured in the moving frame F'. Now, it seemed natural to Aristotle, Galileo, and Newton to suppose that the measurement of time is not tied to any particular reference frame, so that the time-interval between any pair of events has the same value in every (unaccelerated) frame. If we make this assumption and assign the time coordinate $t' = 0$ to the event $x = 0$, $t = 0$, so that the origins of the grids G and G' coincide, then the x'-axis coincides with the x-axis. This is so because events represented by points on the x-axis are simultaneous in frame F'; accordingly, they must all have the time coordinate $t' = 0$, and the locus of these points is, by definition, the x'-axis.

Not only must the x'-axis coincide with the x-axis, but each grid-line $t' = C$ must coincide with the grid-line $t = C$ (see above). The time coordinates of every event are then the same in the two grids: $t' = t$.

All that now remains is to fix the spacing of the grid-lines parallel to the t'-axis. This spacing determines the distance between simultaneous events as measured in the frame F'. More concretely, it determines the length that an observer in frame F' assigns to a rigid rod that has unit length in frame F. In the physics of Galileo and Newton, *the length of a rigid rod is not affected by its motion*. Under this assumption, the grid-lines parallel to the time axis must have the same spacing, measured along lines parallel to the distance axis, in both grids (see facing page). This completes our construction of the grid G'. As we have just seen, the construction depends on just two assumptions: that the time-interval between any two events is the same in both frames; and that the distance between any two simultaneous events is the same in both frames.

We may summarize our conclusions in the following formulas, which connect the coordinates x, t of an arbitrary event, measured in frame F, with the coordinates x', t' of the same event, measured in frame F', which is moving with speed V relative to F:

$$x' = x - Vt, \qquad t' = t. \tag{5.1}$$

The first relation is represented graphically in the figure on the facing page; the second relation expresses the fact that the distance axis and the grid-lines parallel to it coincide in the two grids.

By setting $t = t'$ in the first of Equations 5.1 and transposing, we obtain

$$x = x' + Vt', \qquad t = t'. \tag{5.2}$$

These equations, which tell us how to calculate the coordinates x, t of an arbitrary event from its coordinates x', t', have exactly the same form as Equations 5.1, which tell us how to calculate the pair of coordinates x', t', given x, t. We can obtain Equations 5.2 from Equations 5.1 by interchanging primed and unprimed coordinates and replacing V, the velocity of frame F' relative to frame F, by $-V$, the velocity of frame F relative to frame F'. Thus, *the transformation formulas in Equations 5.1 describe the effects of relative motion in a perfectly symmetric way.*

The Speed of Light and the Principle of Relativity

Consider a particle moving with velocity v relative to a reference frame F. From the lefthand figure on the next page, we see that its velocity v relative to a frame F', which is moving with velocity V relative to F, is $v - V$. This rule is universally valid in classical physics. It applies to waves as well as to particles. For example, if the speed of a sound wave relative to the surface of the Earth is v, an observer traveling with speed V in the same direction should find the speed of the sound wave to be $v - V$. Experience confirms this expectation, even when V is larger than v, as when the observer is aboard a supersonic aircraft.

Does the same rule apply to light waves? In 1887 A. A. Michelson and E. W. Morley reported the outcome of an experiment designed to answer this question. The experiment was based on the following reasoning.

1. There exists a frame of reference F in which the speed of light has the same value c in all directions.

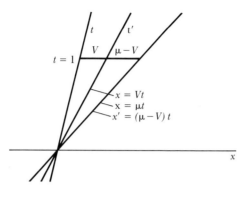

(*Above*) The classical law for the composition of velocities. (*Right*) Albert A. Michaelson (1852–1931), American physicist.

2. Because of its motion around the Sun, the Earth cannot *always* be at rest relative to the frame F just defined. If it happens to be at rest relative to F at one moment, it will certainly be moving relative to F six months later.

3. Suppose that the Earth is moving with speed V in the x-direction, relative to frame F. Then the speed of light traveling in the positive x-direction is $c - V$, and the speed of light traveling in the negative x-direction is $c + V$. In the Michelson-Morley experiment, described in the figure on the facing page, light travels along an arm of length L, is reflected by a mirror, and returns to its starting point. If this arm is parallel to the direction of the Earth's motion relative to frame F, the round-trip time is accordingly

$$\Delta t_{\parallel} = \frac{L}{c - V} + \frac{L}{c + V} = \frac{2L}{c} \frac{1}{1 - V^2/c^2}$$

$$\simeq \left(\frac{2L}{c}\right)\left(1 + \frac{V^2}{c^2}\right). \tag{5.3}$$

4. Light traveling at right angles to the direction of the Earth's motion has speed $\sqrt{c^2 - V^2}$. As shown in the lefthand figure above, its velocity is the resultant of its velocity $\mathbf{c}$ in the frame F and $-\mathbf{V}$, the reflection of the Earth's velocity. Hence, when the arm of length L is perpendicular to the direction of the Earth's motion, the round-trip time is

The Michelson-Morley experiment. Monochromatic light from a source S is split into two beams by a half-silvered mirror M. These are reflected at M_1 and M_2, and parts of them recombine at M to enter the telescope T together. If the path lengths $\overline{MM_1}$ and $\overline{MM_2}$ are nearly equal, and if M_1 and M_2 are not precisely perpendicular to the incident beams, a system of interference fringes (inset) is seen through T.

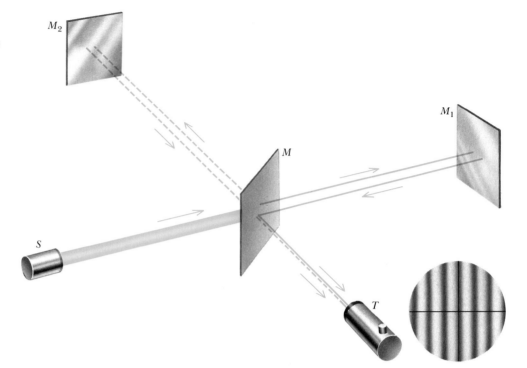

$$\Delta t_\perp = \frac{L}{(c^2 - V^2)^{1/2}} + \frac{L}{(c^2 - V^2)^{1/2}}$$
$$\simeq \left(\frac{2L}{c}\right)\left(\frac{1 + \frac{1}{2}V^2}{c^2}\right). \tag{5.4}$$

The parallel and transverse round-trip times are thus slightly different. The fractional difference is

$$\frac{\Delta t_\| - \Delta t_\perp}{\Delta t_\|} = \frac{\frac{1}{2}V^2}{c^2}.$$

The Earth's orbital speed is about 30 km/sec; the speed of light is 3×10^5 km/sec. Hence the fractional difference between the two round-trip times is about 5×10^{-9}.

5. Michelson thought of a way to detect the tiny fractional difference between the parallel and perpendicular light-travel times. As the above figure shows, monochromatic light is split into

Leonhard Euler (1707–1783),
Swiss mathematician and physicist, in an 1835 engraving.

two beams, which are made to travel back and forth along perpendicular arms before recombining. When they do recombine, they interfere and produce a pattern of bright and dark fringes (see Box on page 154). The experimenter now rotates the entire apparatus through 90° while observing the pattern of fringes through a telescope. Suppose that one of the arms was initially parallel to the direction of the Earth's motion relative to the frame F, in which the speed of light is the same in all directions. As the apparatus rotates, the light-travel time along this arm diminishes from the value given by Equation 5.3 to the slightly smaller value given by Equation 5.4. At the same time, the light-travel time along the other arm increases from the value given by equation 5.4, with L' in place of L (the two arms cannot be *exactly* the same length), to the value given by Equation 5.3, with L' in place of L. Thus the *difference* between the light-travel times along the two arms changes by twice the difference between the values given by Equations 5.3 and 5.4. The fractional change is thus V^2/c^2, or about 10^{-8}. Suppose the arms are 10 meters long. Then the difference between the path lengths of the two beams changes, during the 90° rotation, by $10^{-8} \times 2 \times 10 \times 100$ cm $= 2 \times 10^{-5}$ cm. Michelson and Morley used yellow light with a wavelength of about 6×10^{-5} cm. Thus the change in the difference between the path lengths of the two beams would have been about a third of a wavelength, and such a "large" change would have produced a conspicuous shift in the pattern of interference fringes. Michelson and Morley could detect no change at all in the pattern of interference fringes. They concluded that any change in the difference between the two path lengths, if present at all, had to be less than one sixth of the predicted value.

When an experiment yields an unexpected result, physicists ask themselves: *Should* we have found this result surprising? Were our expectations based on an unjustified assumption or a misleading analogy? In particular, is there any reason why the simple rule $v' = v - V$, which works perfectly well for sound waves and water waves, should not work equally well for light waves? Sound waves are vibrations in a material medium: a gas, a liquid, or a solid. They travel at a definite speed, *relative to the medium*, whose value depends on the nature of the medium and its physical state. For example, the speed of sound in a gas is proportional to the square root of the temperature and inversely proportional to the square root of the mean molecular weight. To calculate the speed of a sound wave in a given reference frame, we must add the speed of the wave relative to the medium to the speed of the medium relative to the reference frame. The same argument applies to water waves, and indeed to any kind of wave in a material medium. We should therefore expect light waves to behave in the same way *if* they too are disturbances in a material medium. The reference frame F in which light travels at the same speed in all directions would then be the frame in which this medium is at rest.

The outstanding theoretical physicists of the seventeenth, eighteenth, and nineteenth centuries all believed in the existence of such a medium, which they called by the name that Greek philosophers had given to the imperishable substance they supposed celestial bodies to be made of: *ether*. Huygens based a sound mathematical theory of the propagation of light in material media on the analogy between light waves and water waves. Newton argued that the propaga-

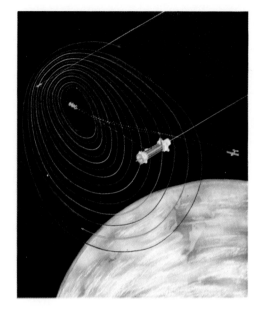

Beryl Langer created this artist's conception of the Spacecraft Array for Michelson Spatial Interferometry (SAMSI) system proposed by Robert Stachnik for deployment from a future Space Station. In this orbiting interferometer, two widely separated telescopes collect light from a star and direct it to a third, centrally located device, which combines the beam to detect and measure interference fringes, as in Michelson's original interferometer. Laser beams are used to monitor and regulate the separations of the components, which vary systematically as the two telescopes move in their spiral orbits. The angular resolution possible with such a system is determined by the maximum separation of the telescopes. When the telescopes are 10 km apart, the angular resolution is 10^{-5} arc second.

ADS 11640 A

Speckle interferometry: a sophisticated technique for processing telescopic images that exploits the phenomenon of interference. The image on the left results from recentering and superimposing 100 short-exposure images of the source ADS 1164OA. The image on the right, which shows that this source is actually a double star whose components have an angular separation of only 1/7 arc sec, results from analyzing the individual short-exposure images (which have a speckled appearance) to recover information contained in the apparently random speckle pattern. This analysis not only resolves the two stars but yields their orientation and relative brightness.

(a)

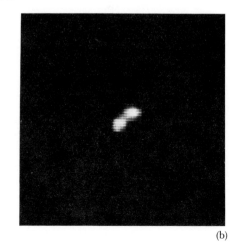

(b)

|⎯⎯⎯⎯⎯⎯⎯⎯⎯⎯⎯⎯|

1 arc sec

tion of heat through an evacuated bell jar demonstrates the existence of the ether, and he conjectured that light excites periodic vibrations in the ether, which transmits them to solid bodies. The great Swiss mathematician Leonhard Euler (1707–1783) conjectured that the ether transmits not only heat and light, but also magnetic and electric forces and gravitation. The physicists who developed the modern wave theory of light during the nineteenth century made the ether a central topic of experimental and theoretical investigation. The ether turned out to have properties unlike those of any other known medium (for example, the property of transmitting *only* transverse vibrations). Moreover, nearly every qualitatively new experiment revealed an unexpected property, forcing physicists to exercise great ingenuity in revising their theoretical models. Even James Clerk Maxwell, whose mathematical theory of electromagnetism accounted for all known properties of light (see Box below) without any assumptions whatever about the physical properties of the ether, believed firmly in its existence. He wrote a long article about it for the *Encyclopaedia Brittanica,* and in 1880 wrote to Michelson suggesting what we now call the Michelson-Morley experiment.

Maxwell's theory

Atoms are made up of electrically charged particles: electrons, which carry equal negative charges, and nuclei, which carry positive charges that are multiples of the electron's charge. The equal and opposite forces that two stationary charged particles exert on each other are, like the gravitational force, inversely proportional to the square of their separation. They are also directly proportional to the product of the charges: like charges repel, unlike charges attract one another. Atoms, molecules, liquids, and solids are held together by attractions of unlike charges.

Just as mass gives rise to a gravitational field, so electric charge gives rise to an *electric field.* The electric field **E** is related to the distribution of electric charge exactly as the gravitational field **g** is related to the distribution of mass. Electric field lines begin in positive charges and end in negative charges. An electric field **E** exerts a force $q\mathbf{E}$ on a charge q.

Moving electric charges produce electric currents, which in turn give rise to *magnetic fields.* The precise relation between current and magnetic field is known as Ampère's law. Conversely, all magnetic fields are produced by electric currents, i.e., moving charges.

When the number of magnetic field lines threading a loop of wire changes with time, a current flows in the wire. This current is driven by an electric field whose value is proportional to the rate of change of the number of field lines threading the loop (Faraday's law of electromagnetic induction). Electric generators are based on this principle.

James Clerk Maxwell, building on earlier work by Carl Friedrich Gauss and George Gabriel Stokes, expanded the language of vectors (whose rudiments are set out in the Appendix of this book) to express Michael Faraday's picture of electric and magnetic field lines and of the way the field lines are related to moving electric charges. He discovered that to obtain a consistent set of mathematical laws, he needed to posit a new source of magnetic fields—that

James Clerk Maxwell (1831–1879), Scottish physicist.

is, a new kind of electric current—proportional to the rate of change of the electric field. Without this new current, which Maxwell dubbed *displacement current,* variable electric currents would cause electric charge to appear or disappear. But Maxwell's postulate had a more spectacular consequence: it implied that varying electric and magnetic fields carrying energy and momentum can propagate in empty space. Maxwell's theory predicts that this time-varying electromagnetic field can be resolved into plane monochromatic waves, each with a definite frequency, traveling in a definite direction with a definite speed; and that this speed, whose value the theory predicts, is independent of the frequency of the wave. Maxwell identified the electromagnetic waves predicted by his theory with light waves. All the predicted properties of Maxwell's electromagnetic waves, including their speed, matched perfectly the experimentally discovered properties of light waves.

Maxwell's theory also predicted that electromagnetic waves are generated by rapidly oscillating electric charge. Heinrich Hertz confirmed this prediction by constructing in his laboratory the first generator and the first receiver of radio waves, which are electromagnetic waves of macroscopic wavelength.

By extending Maxwell's theory in two directions, Einstein founded the two great theories of twentieth-century physics, relativity and quantum theory. The principle of relativity requires the laws of electromagnetism (Maxwell's equations) to take the same form in all unaccelerated frames of reference. The photon hypothesis, published in the same year (1905) as Einstein's first paper on special relativity, stipulates that the absorption and emission of electromagnetic energy and momentum are "quantized" processes.

To explain the negative outcome of the Michelson-Morley experiment, the Irish physicist G. F. Fitzgerald and, independently, the Dutch physicist H. A. Lorentz proposed that a rigid body contracts in the direction of its motion relative to the ether. Comparing Equations 5.3 and 5.4, we see that the round-trip times for the two paths along which the light travels in the Michelson-Morley experiment would be equal if the path length in the direction of the Earth's motion relative to the ether were reduced by the factor $(1 - V^2/c^2)^{1/2}$. In 1904 Lorentz published a generalization of Maxwell's theory in which the Fitzgerald-Lorentz hypothesis appeared as a consequence of more general assumptions. The theory also explained why yet other experiments had failed to detect any motion of the Earth relative to the ether.

Meanwhile, Albert Einstein, unable to find an academic position and working as a clerk in the patent office at Berne, had been thinking along different lines. He found it puzzling that the mathematical laws governing electricity, magnetism, and light (Maxwell's laws, extended by Lorentz to atoms and electrons) define a preferred reference frame (the frame called F in the preceding discussion), whereas Newton's laws of motion and gravitation do not. Why should dynamical phenomena prefer a democracy of unaccelerated reference frames, but electromagnetic phenomena prefer a monarchy? This lack of mathematical harmony weighed more heavily with Einstein than the three-centuries-old consensus about the ether. Einstein's predecessors

Hendrick A. Lorentz (1853–1928), Dutch physicist.

Heinrich R. Hertz (1857–1894), German physicist.

had said, in effect, "There must be an ether, because we can't imagine how there could be vibrations without something that vibrates." Einstein recognized that physical laws are not constrained by what physicists can or cannot imagine. Like Heinrich Hertz, who deduced the existence of radio waves from Maxwell's equations and then made some in his laboratory, Einstein believed that Maxwell's equations *are* his theory. Now, Maxwell's equations do not presuppose or imply the existence of a medium to transmit electromagnetic waves, but they do implicitly define a preferred reference frame: the frame F in which light travels with the same speed in all directions. But if there is no ether to confer privileged status on this reference frame, the laws governing electromagnetic phenomena *ought* not to discriminate between frame F and other unaccelerated reference frames. This is Einstein's generalization of Galileo's principle of relativity: *The laws governing both electromagnetic and mechanical phenomena have the same form in all unaccelerated reference frames.* In particular, the speed of light, which appears (with a different physical interpretation) as a constant coefficient in Maxwell's equations, must have the same value in all directions and in all reference frames.

This conclusion contradicts the simple rule $v' = v - V$, for the principle of relativity requires that $v' = v$ if v, v' represent the speed of light measured in frames F, F' moving with relative velocity V. Einstein concluded that the simple rule must be wrong.

Constructing the Relativistic Spacetime Diagram

Let us go back to our construction of the spacetime diagram shown in the figure on p. 156, which represents the relation between distances, times, and speeds in two unaccelerated reference frames, and try to figure out where we might have gone wrong. In constructing a spacetime grid for frame F (see the bottom left figure on p. 153), we assumed only that space and time are homogeneous, that the grid must not confer preferred status on any point in space or on any moment in time. Next, we identified the line $x = Vt$ with the t'-axis of a spacetime grid for frame F', moving along the x-axis with speed V. This step, too, seems unassailable. But then we made two assumptions: that every event may be assigned the same time coordinate in the two frames, so that the time-interval between any two events is the same in both frames; and that rigid rods have the same length in both frames. Let us now replace these two assumptions, plausible though they are, by the principle of relativity stated at the end of the preceding section.

We can infer at once that both sets of grid-lines of grid G' must be evenly spaced, like the grid-lines of grid G. Otherwise, a clock judged to be running at a constant rate by an observer in frame F would be judged by an observer in frame F' to be running at a variable rate, and marks on a rigid rod judged by an observer in frame F to be equally spaced would be judged by an observer in frame F' to have variable spacing.

Let us now explore the implications of the assumption that the speed of light is the same in both reference frames. For this purpose, we may conveniently replace the time coordinate t by ct, where c is the speed of light, as depicted in the figure on the facing page. Then lines that represent light signals traveling in the positive x-direction are represented on the grid G by one

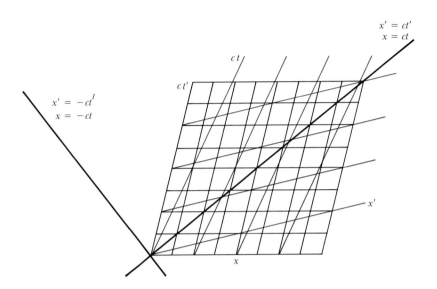

Rhombic grids for two reference-frames in uniform relative motion.

family of diagonals of the rhombic grid-cells; light signals traveling in the negative x-direction are represented by the other family of diagonals. (The equality between adjacent sides of a rhombic grid-cell is expressed by the equation $\Delta x = \pm c \Delta t$. The distance intervals and time-intervals between events in the history of a light signal satisfy one or the other of these two relations, depending on whether the light signal is traveling from left to right or from right to left. Notice that the diagonals of each family are parallel to one another and perpendicular to the diagonals of the other family.) Because grids G and G' must be qualitatively indistinguishable, the two families of diagonals must also be diagonals of the rhombic grid-cells of G'; that is, they must represent the relations $\Delta x' = \pm c \Delta t'$ as well as the relations $\Delta x = \pm c \Delta t$. This conclusion is illustrated above. Notice that the diagram does not change if it is rotated through 180° about any of the light-lines (the lines that represent histories of light signals). Any such rotation leaves the light-lines themselves unchanged but interchanges the two families of grid-lines.

Our last remaining task is to discover how the *size* of a unit cell of the G' grid (a cell with $\Delta x' = c \Delta t' = 1$) is related to the size of a unit cell of the G grid ($\Delta x = c \Delta t = 1$). Let us begin by summarizing our conclusions to this point in algebraic language. To bring out the symmetry between space and time, and to keep the formulas as simple-looking as possible, let us henceforth use the same unit to measure time and distance, and set $c = 1$.* The relation between the grids G and G' discussed above are then expressed algebraically by the equations

* Measuring vertical distances and horizontal distances in the same units enables us to use Pythagoras' formula to express the square of the distance between any two points in space as a sum of squares. Analogously, as we will see later, by measuring time in the same unit as distance and setting the speed of light (which is then a ratio between two distances) equal to unity, we can express the square of the spacetime interval between two events by a formula similar to Pythagoras'. Newton's constant of gravitation is the ratio between a natural unit of mass and an arbitrary, conventional unit; the speed of light is the ratio between two independent conventional units.

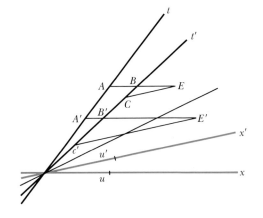

This figure illustrates the argument given below that the quantity γ in Equation 5.5 has the same value for any two events E, E'.

$$x' = \gamma(x - Vt),$$
$$t' = \gamma(t - Vx),$$

(5.5)

where γ is a constant whose value we do not yet know. To prove this assertion, we have only to translate Equations 5.5 into graphical language. Notice, to begin with, that if we interchange x and t, and x' and t', in these equations, the first of them becomes the second, and the second becomes the first; the pair of equations does not change at all. This symmetry of the equations expresses the symmetry of the spacetime diagram mentioned at the end of the preceding paragraph.

Next, let us interpret the equation $x' = \gamma(x - Vt)$. This is slightly tricky, because the *scales* of the two grids, G and G', need not be the same. In fact, as we will see in a moment, we are not free to make them the same; the ratio between the two scales is determined by the principle of relativity. Thus, segments of equal length on the x-axis and the x'-axis do not represent equal intervals of the coordinates x and x'. Let u be the length of a segment $\Delta x = 1$ measured along the x-axis, and u' be the length of a segment $\Delta x' = 1$ measured along the x'-axis. In the figure above, an arbitrary event is denoted by point E, whose x-coordinate is represented by the segment AE. The length of this segment is given by

$$\overline{AE} = x \times u.$$

Similarly,

$$\overline{AB} = Vt \times u, \qquad \overline{BE} = (x - Vt) \times u.$$

Correspondingly, the x'-coordinate of event E is represented by the segment CE, whose length is given by

$$\overline{CE} = x' \times u'.$$

Now, the ratio $\overline{BE}/\overline{CE}$ does not depend on the position of the point E; in the figure, $\overline{BE}/\overline{CE} = \overline{B'E'}/\overline{C'E'}$. Thus the ratio $(x - Vt)u/x'u'$ has the same value at every point on the spacetime diagram, so that $x' = \gamma (x - Vt)$, where γ is some constant. An exactly analogous argument justifies the second of Equations 5.5. The constants of proportionality in the two equations must be equal, because the diagram is symmetric when rotated through 180° about the light-lines $x = \pm t$.

Equations 5.5 are formulas for the coordinates x', t' of an event whose coordinates in frame F are x, t. We may also regard them as a pair of simultaneous equations in which x, t are the unknowns and x', t' are given. To solve these equations, form the combinations $x' + Vt'$, $t' + Vx'$:

$$x' + Vt' = \gamma(1 - V^2)x,$$
$$t' + Vx' = \gamma(1 - V^2)t. \tag{5.6}$$

The principle of relativity requires Equations 5.6 to have the same form as Equations 5.5, with $+V$ in place of $-V$. This will be the case if and only if

$$\gamma(1 - V^2) = \frac{1}{\gamma}$$

or

$$\gamma = (1 - V^2)^{-1/2}. \tag{5.7}$$

The grid G' is now completely determined. To construct its grid-lines, we need to mark the point $x' = 1$ on the x'-axis. On this line, $t' = 0$, so that, by the second of Equations 5.5, $t = Vx$. Setting $t = Vx$ and $x' = 1$ in the first of Equations 5.5 and using Equation 5.7, we find that the x-coordinate of this point $(x' = 1, t' = 0)$ is $x = \gamma$. Its t-coordinate is $V\gamma$. The difference between the squares of these two coordinates is thus

$$x^2 - t^2 = \gamma^2 - V^2\gamma^2 = \gamma^2(1 - V^2) = 1.$$

Thus the point $x' = 1$ on the x'-axis is the point at which this axis meets the hyperbola $x^2 - t^2 = 1$ (see the figure on the next page). Similarly, the point $t' = 1$ on the t'-axis is the point at which this axis meets the hyperbola $t^2 - x^2 = 1$ (in the same figure).

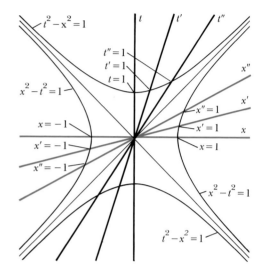

The two families of calibration hyperbolas connect events whose proper distance or proper time separation from the origin have a fixed value.

These results are more general and more important than they may seem at first sight. The unit of time (or distance) is arbitrary, and we may identify any event we choose with the origin of the spacetime diagram. Hence we may rephrase the last sentence of the preceding paragraph as follows: If $\Delta t'$ is the time-interval between two events on the t'-axis (so that $\Delta x' = 0$ for this pair of events), and if Δt, Δx are the time-interval and the space-interval between the same two events, measured in any other frame of reference, then

$$(\Delta t')^2 = (\Delta t)^2 - (\Delta x)^2.$$

The time recorded by a stationary clock—the time we have been denoting by t'—is important enough to have a special name and a special symbol. It is called *proper time* and denoted by τ. With this understanding about the meaning of the symbol τ we can write the preceding equation as

$$(\Delta\tau)^2 = (\Delta t)^2 - (\Delta x)^2.$$

Analogously, we call the length of a rod, measured in a frame of reference in which the rod is at rest, its *proper length*, and denote it by $\Delta\ell$. The proper distance between two events is the distance $\Delta\ell$ measured in a frame of reference in which the events are simultaneous. It is related to the space-interval and the time-interval between the events, measured in any other frame of reference, by the formula

$$(\Delta\ell)^2 = (\Delta x)^2 - (\Delta t)^2.$$

Now consider any two events, separated by a space-interval Δx and a time-interval Δt in some arbitrary frame of reference. We have just seen that the difference between the squares of these intervals, $(\Delta t)^2 - (\Delta x)^2$, has the same value in every frame of reference. If this value is positive, it represents the square of the interval of proper time between the two events, i.e., the square of the time-interval measured by a clock at rest in a frame of reference in which the two events occur at the same place. In this case we say that the interval of spacetime is *timelike*. If the difference is negative, it represents the negative squared proper distance between the events, i.e., the negative of the square of the distance measured in a frame of reference in which the events are simultaneous. In this case we say that the interval of spacetime is *spacelike*. If the difference is zero, so that $\Delta t = \pm \Delta x$, the two events could belong to the history of a particle moving with the speed of light. In this case we say that the interval is *lightlike*.

Proper time and proper length are the relativistic generalizations of classical time and classical length. In classical physics, the time-interval between any two events and the distance between any two *simultaneous* events have the same values in all frames of reference. In Einstein's theory, the spacetime interval between two events is either timelike or spacelike or lightlike, and is equal to the square root of the difference between the squares of the time-interval and the space-interval between the events, measured in any frame of reference. The *invariance* of the spacetime interval—the fact that the value of the difference $(\Delta t)^2 - (\Delta x)^2$ does not depend on the frame of reference in which it is measured—is analogous to the invariance of the distance interval in plane geometry, that is, to the fact that the square of the distance between two points is equal to the sum of the squares of the differences between their x- and y-coordinates, whatever the orientation of the coordinate axes or the position of its origin.

We have found, then, that Einstein's strong form of the principle of relativity can be satisfied if and only if the coordinate grids of unaccelerated reference frames are related in the way specified by Equations 5.5 and 5.7 and represented graphically in the figure on the facing page.

The coordinate transformation defined by Equations 5.5 is called the *Lorentz transformation*, because H. A. Lorentz published similar equations (though with a different physical meaning) in 1904, a year before Einstein's paper appeared. For relative speeds V that are much smaller than the speed of light ($V \ll 1$ in our present units), the Lorentz transformation is indistinguishable from the classical transformation, which preserves the length of rigid rods and the durations of time-intervals. For relative velocities close to the speed of light, however, the phenomena implied by the Lorentz transformation differ strikingly from their classical counterparts, as we will see later in this chapter.

Measurements of Time and Distance in a Moving Reference Frame

Cosmic rays are streams of very energetic particles that enter the Earth's atmosphere from outer space. Some of these particles come from the Sun, others—most of the most energetic ones—come from the Galaxy, and a few may come from intergalactic space. Cosmic-ray particles are thought to be by-products of explosive events, including those responsible for solar

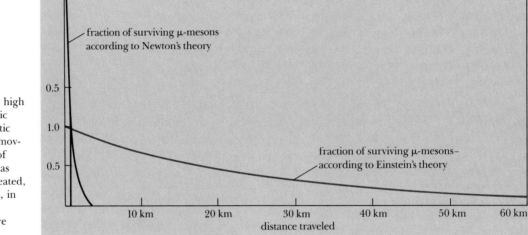

Decay of μ mesons created high in the atmosphere by cosmic rays. Owing to the relativistic time dilation, the mesons, moving with nearly the speed of light, live 50 times as long as they would if they were created, with relatively low energies, in the laboratory. Hence they travel 50 times as far before they decay.

flares, supernovae, and (perhaps) quasars. They are probably accelerated by the magnetic fields implicated in such explosions. When they collide with the nuclei of molecules in the Earth's upper atmosphere, they produce a variety of short-lived particles, including the μ meson. All these short-lived particles can now be produced in high-energy laboratories, and the lifetimes of many of them, including the μ meson, have been measured with great precision. Let t denote the measured lifetime of a μ meson. How far does a μ meson, moving with nearly the speed of light, travel before it decays? According to classical physics, the answer is ct, the product of the meson's speed and its lifetime. In fact, μ mesons produced by cosmic-ray particles in the Earth's upper atmosphere have been found to travel 50 times this distance before decaying (see above). Since (as we will soon see) a particle's speed cannot exceed the speed of light, and since the relation $d = vt$ between distance traveled, speed, and elapsed time is just as valid for particles moving with nearly the speed of light as it is for slow-moving particles, we must conclude that the μ mesons created by cosmic-ray particles actually live 50 times as long as particles of exactly the same kind created in the laboratory.

The transformation theory that we developed in the last section explains why. In the figure on the facing page, the x-coordinate is measured vertically downward from the point in the atmosphere at which a μ meson is born, and the time t is measured from that moment. The history of an exceptionally slow μ meson, with a speed of only $^{24}/_{25}$ the speed of light, is represented by the straight line $x = {}^{24}/_{25}\, t$. This line is the time axis for an observer traveling with the μ meson. Let the segment OT in the figure on the facing page represent the particle's lifetime as measured by such an observer. If the principle of relativity is valid, the outcome of such a measurement cannot depend on whether the *observer* is hurtling toward the Earth with nearly the speed of light or sitting by a monitor in, say, Palo Alto; since the mesons produced in Palo

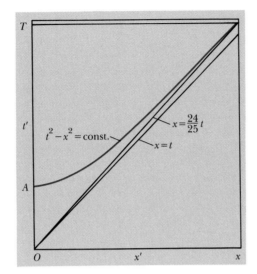

$\overline{OT}$ and $\overline{OA}$ represent the same time-interval, the half-life of a μ-meson, measured in its rest frame. $\overline{OT'}$ represents the meson's half-life measured in a reference frame in which it is moving with $^{24}/_{25}$ the speed of light.

Alto are moving relatively slowly, we may identify the lifetime $\overline{OT}$ with the laboratory value. But the duration of the particle's flight as recorded by an observer on the ground is represented by the segment $\overline{OT'}$. The calibration hyperbola that passes through the point T (see the figure on the facing page) cuts the ground observer's time axis in the point A. The time-interval $\overline{OA}$ measured by an observer on the ground is equal to the time-interval $\overline{OT}$ measured by an observer traveling with the μ meson. Thus $\overline{OA}$ represents the particle's lifetime as measured in the laboratory; and this lifetime is only $^7/_{25}$ of the lifetime $\overline{OT'}$, which, as we have seen, represents the lifetime of the moving particle as recorded by an observer on the ground. The closer the particle's speed comes to the speed of light, the greater the disparity between its lifetime as recorded by an observer on the ground and its "proper" lifetime, the lifetime recorded by a co-moving observer (see above). Notice that the relation between the distance $\overline{OX}$ that the particle travels during its lifetime and the duration of its trip, $\overline{OT'}$, is the normal relation between distance and time in uniform motion: $\overline{OX} = {^{24}/_{25}}\ \overline{OT'}$.

Although the preceding account is somewhat idealized, the experiment is genuine. Its outcome, as well as the outcomes of countless other experiments that test Einstein's theory of unaccelerated motion, agrees perfectly with the theory's predictions. No other theory has been corroborated with such high precision over such a wide range of the relevant parameters.

Let us consider another version of the experiment just discussed. Paula and Petra are twins. On their twentieth birthday, Petra gets married, and Paula boards a spacecraft that is headed for a nearby star. She travels in a straight line for $3\frac{1}{2}$ years at $^{24}/_{25}$ the speed of light and then returns at the same speed, arriving back in time to celebrate her twenty-seventh birthday. She is greeted by Petra, now a matron of forty-five, and by Petra's twin sons Peter and Paul, who have just celebrated their twenty-fourth birthday.

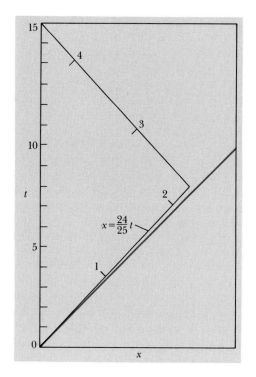

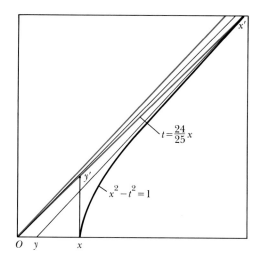

Proper-time intervals along Paula's and Petra's world-lines (spacetime trajectories).

The distance represented by $\overline{OX}$ on the x-axis and $\overline{OX}'$ on the x'-axis are each equal to unity. $\overline{OY}'$ represents the length, measured in the "primed" frame, of a rod of unit length at rest in the "unprimed" frame; $\overline{OY}$ represents the length, measured in the "unprimed" frame, of a rod of unit length at rest in the "primed" frame. In both cases the length of the moving rod is $\frac{7}{25}$ $(=\sqrt{1-(\frac{24}{25})^2})$ the length of the stationary rod. Note that $\overline{OY}:\overline{OX} = \overline{OY}':\overline{OX}'$, and that $X'Y$, which is parallel to the t'-axis, is tangent to the calibration hyperbola at the point X'.

The arithmetic of this example is the same as that of the preceding example. The duration of Paula's trip is longer for Petra than it is for Paula herself, for exactly the same reason that the duration of the μ meson's flight is longer for an observer on the ground than for the particle itself.

People are sometimes troubled by the lack of symmetry between the twins' experiences. "Motion is relative," they say. "From Paula's point of view, it is Petra who has been in motion and who, according to the same calculation, should have aged less than Paula. Since both accounts cannot be true, both must be false; Paula and Petra must be exactly the same age when they are reunited."

This argument rests on the false premise that the twins' experiences are symmetric. It is clear from the figure above, left, that they are not. The lines representing Paula's history in a spacetime diagram form two sides of a triangle; the line representing Petra's history forms the third side. That the elapsed time recorded by a clock whose history is represented by two sides of such a triangle should be shorter than the elapsed time recorded by a clock whose history is represented by the third side is no more mysterious—given the rules we have developed for calculating elapsed times—than that the distance between two points measured along a dogleg should be greater than the straight-line distance.

Relative motion distorts measurements of distance as well as of time. A snapshot of Paula's spacecraft as it whizzes by an observation platform near the edge of the solar system shows a somewhat blurred image only $\frac{7}{25}$ as long as it would have been if the spacecraft had been at rest. The righthand figure on the facing page explains why. The predicted reduction in the length of a rapidly moving rocket is the same as that postulated by Fitzgerald and Lorentz to account for the negative outcome of the Michelson-Morley experiment. Its physical interpretation is different, however. For Fitzgerald and Lorentz, an object moving relative to the ether actually gets shorter, although an observer moving in the same direction at the same speed perceives no change in its length because his measuring rods are shortened by the same factor. In Einstein's theory, the length of a rod—the distance between its endpoints at moments that are simultaneous in the reference frame of the observer—depends only on the relative motion of the rod and the observer (see the figure on page 172).

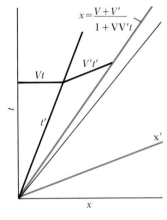

The relativistic law for the composition of velocities.

The Addition of Velocities and the Aberration of Light

Consider a particle moving with velocity **v** in frame F. What is its velocity in a frame F' moving with velocity **V** relative to F? According to the theory of Galileo and Newton, $\mathbf{v}' = \mathbf{v} - \mathbf{V}$. In Einstein's theory the answer must be different, because if the "particle" is a photon, it has the same speed in all reference frames and in all directions. The figure to the left shows graphically that the resultant of two velocities in the same direction can never exceed the velocity of light, and that a particle moving with the speed of light in frame F has the same speed in frame F'.

Let us work out the relation between v, v', and V when all three motions are in the same direction. Let Δx and Δt be the space-interval and the time-interval, respectively, between two events on the line $x = vt$ that represents the history of a moving particle in frame F; and let $\Delta x'$ and $\Delta t'$ by the corresponding intervals between the same two events, viewed in the frame F' moving with speed V relative to the frame F. Then $v = \Delta x/\Delta t$ and $v' = \Delta x'/\Delta t'$. The coordinates (x, t) and (x', t') of each event are connected by the Lorentz transformation (Equations 5.5). If we write down these equations for each of the two events and subtract the second pair of equations from the corresponding equations of the first pair, we obtain

$$\begin{aligned} \Delta x' &= \gamma(\Delta x - V\Delta t), \\ \Delta t' &= \gamma(\Delta t - V\Delta x). \end{aligned} \tag{5.8}$$

Dividing the first of these equations by the second, we find

$$v' = \frac{\Delta x'}{\Delta t'} = \frac{\Delta x - V\Delta t}{\Delta t - V\Delta x}$$

$$= \frac{\Delta x/\Delta t - V}{1 - V\Delta x/\Delta t}$$

$$= \frac{v - V}{1 - Vv}. \tag{5.9}$$

This is Einstein's law for the addition of two parallel velocities. What distinguishes it from the classical law is the denominator $(1 - Vv)$, which ensures that the right side of the formula cannot exceed unity (the speed of light in our present system of units) so long as the numerical values of v and V do not exceed unity. Notice, too, that $v' = 1$ if $v = 1$, whatever the value of V: if a particle has the speed of light in one frame, it has the speed of light in every frame.

How does the relative motion of two reference frames affect the *direction* of a particle's motion? This question is especially interesting when the "particle" is a photon: how does the observer's motion relative to a light source affect the directions of light rays emitted by that source?

In frame F we may let the x-axis coincide with the direction of the velocity $\mathbf{V}$ of frame F' relative to frame F, and we may assume that the velocity $\mathbf{v}$ of the particle whose motion we wish to study is in the x, y-plane. Since the relative motion of the two frames is in the x-direction, the particle's y- and z-coordinates have the same values in the two frames:

$$y' = y, \qquad z' = z. \tag{5.10}$$

Since the particle is moving in the x, y-plane, $z = 0$. It follows from the second of Equations 5.10 that $z' = 0$; in frame F', the particle moves in the x', y'-plane. Suppose that the velocity $\mathbf{v}$ makes an angle θ with the x-axis; that is,

$$\frac{\Delta y}{\Delta x} = \tan\theta, \tag{5.11}$$

where Δx, Δy are coordinate differences between two events in the particle's history. We wish to calculate the angle θ' between the particle's velocity $\mathbf{v}'$ and the x-axis. This angle is related to the coordinate differences $\Delta x'$, $\Delta y'$ in the same way that the angle θ is related to Δx, Δy:

$$\frac{\Delta y'}{\Delta x'} = \tan\theta'. \tag{5.12}$$

The coordinate difference $\Delta x'$ is given by the first of Equations 5.8. By the first of Equations 5.10, $\Delta y' = \Delta y$. Hence

$$\tan \theta' = \frac{\Delta y'}{\Delta x'} = \frac{\Delta y}{\gamma(\Delta x - V\Delta t)} = \frac{\Delta y/\Delta t}{\gamma(\Delta x/\Delta t - V)}$$

$$= \frac{v_y}{\gamma(v_x - V)} = \frac{v \sin \theta}{\gamma(v \cos \theta - V)},$$

(5.13)

The direction of an incoming light ray is affected by the observer's motion.

where v_x and v_y are the components of **v** in the x- and y-directions.

This rather complicated formula simplifies somewhat when the "particle" under consideration is a photon, so that $v = 1$. Now, if a photon's velocity **v** makes an angle θ with the x-axis, the direction from which it appears to be coming (the line of sight) makes an angle $180° - \theta$ with the x-axis (see the figure to the left). Call this angle ϕ. Setting $v = 1$, and bearing in mind that $\cos \phi = \cos (180° - \theta) = -\cos \theta$, and $\tan \phi = \tan (180° - \theta) = -\tan \theta$, we obtain

$$\tan \phi' = \frac{\sin \phi}{\gamma(\cos \phi + V)},$$

(5.14)

which expresses the relation between the angles that the line of sight makes with the direction of relative motion in the two frames.

To see what this relation predicts, let us first look at its limiting form when the relative velocity of the observer and the source is much smaller than the speed of light ($V \ll 1$); this is called the "classical limit." In this limit we may set the quantity γ, given by Equation 5.7, equal to unity; and Equation 5.14 tells us that, because $V \ll 1$, $\tan \phi'$ is nearly equal to $\tan \phi$. That is, the shift in the angle that the line of sight makes with the direction of relative motion, $\Delta\phi = \phi' - \phi$, is small, as we should expect. Setting $\phi' = \phi + \Delta\phi$ in Equation 5.14, and ignoring terms that involve squares and products of the small quantities $\Delta\phi$ and V, we obtain, after a straightforward but somewhat lengthy calculation,

$$\Delta\phi \simeq -V \sin \phi.$$

(5.15)

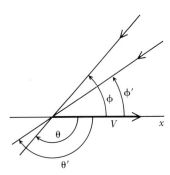

James Bradley (1693–1762). English astronomer.

This formula describes the "aberration of starlight," a phenomenon discovered by James Bradley in 1723: the angle between a star's direction and the direction of its motion relative to the Earth differs from the angle ϕ that would be measured in a frame of reference in which the star is at rest by the small angle $-V \sin \phi$. Now, we may express the star's velocity relative to the Earth as the vector sum of its velocity relative to the Sun and the Sun's velocity relative to the Earth (see the top left figure on the next page). The first component shifts a star's direction by an amount that changes hardly at all in the course of a year (unless the star happens to belong to a close binary). The second component varies periodically with a period of one year. As shown in the top right figure on the next page, and explained in the legend, the star traces out a tiny ellipse on the sky. This ellipse has the same shape as the star's parallactic ellipse (which looks like the Earth's orbital circle, viewed from the star), but the aberrational shift is greatest when the

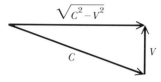

Bradley's explanation of aberration: c represents the velocity of incoming light-particles in the rest-frame, V the velocity of the observer, and $\vec{c} + V$ the velocity of the incoming light-particles as seen by the moving observer.

The aberrational shifts $\overline{SA_1}$, $\overline{SA_2}$, $\overline{SA_3}$ are proportional to, and opposite in direction to, the angular velocities at the points P_1, P_2, P_3 on the star's parallactic ellipse. The two ellipses are similar; their axial ratio is equal to the sine of the star's elevation above the ecliptic (solar latitude). The relative size of the parallactic ellipse is greatly exaggerated in the figure. The star's total displacement is the vector sum of its parallactic and aberrational displacements, e.g., $\overline{SB_2}$.

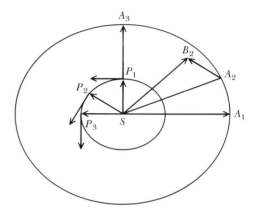

The celestial sphere (a) to a stationary observer, (b) to an observer moving with nearly the speed of light ($\gamma \gg 1$). To the ultrarelativistic observer, all the stars in the sky, except a few directly behind him, are crowded into a circle of radius $^1/_\gamma$ radian.

(a)   (b)

parallactic shift is least, and vice versa. The long axis of the aberrational ellipse is $2V$ radian, where V is the Earth's orbital speed, about 30 km/sec or, 1/10,000 of the speed of light. Since there are about 2×10^5 arc-seconds in one radian, the major semidiameter of the aberrational ellipse is about 20 arc-seconds. By contrast, the major semidiameter of the parallactic ellipse (the star's parallax) is a few tenths of an arc-second for the nearest stars.

Now let's look at the opposite limiting case of Equations 5.14—the "ultrarelativistic limit"—in which the relative velocity V is close to unity and the quantity γ, given by Equation 5.7, is much larger than unity. Suppose that in frame F the sky is uniformly sprinkled with stars. How would the sky look if we were moving relative to F with a speed so close to the speed of light that $\gamma = 100$? We see from Equation 5.14 that the angle ϕ' is very close to 0 unless ϕ is close to 180°.

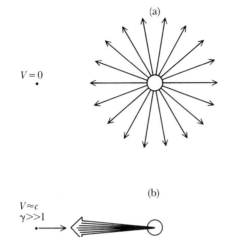

(a)

$V = 0$

(b)

$V \approx c$
$\gamma \gg 1$

A light source that radiates equally in its rest frame (a) sends all its light into a beam of opening angle $1/\gamma$ radian when viewed in a frame in which its velocity is ultrarelativistic ($\gamma \gg$) (b).

Electrons moving with nearly the speed of light in a magnetic field emit radiation in narrow cones centered on the direction of motion.

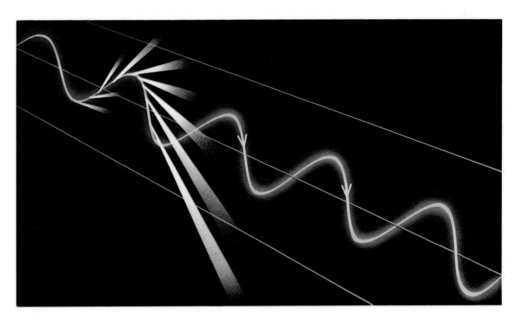

In other words, all the stars in the sky except those almost directly behind us would appear almost directly ahead of us, crowded into a circle of radius $1/\gamma$ radian—a little more than half a degree or twice the radius of the full Moon (see the lower figure on p. 176).

We cannot yet observe the effect just described, but we can observe its precise theoretical equivalent. Consider a light source that, in its own frame of reference, radiates equally in all directions. Suppose we observe this source in another frame F', in which the source is moving fast enough to give the quantity γ defined by Equation 5.7 a value much larger than unity. It follows from Equation 5.14 that in this frame nearly all the light will be focused into a narrow beam, with an opening angle of order $1/\gamma$ centered on the direction of motion (see the top figure on p. 177). In addition, as we will see in the next section, the total power radiated by the source is γ^2 times as great in the frame F' as it is in the source's rest frame. Charged particles accelerated to ultrarelativistic speeds in ring-shaped accelerators called synchrotrons have been observed to radiate in this way. The radiation is confined to a very narrow beam centered on the particles' instantaneous direction of motion. The most powerful astronomical sources of radio-frequency radiation are thought to be giant synchrotrons in which electrons accelerated to ultrarelativistic speeds (by mechanisms not yet completely understood) spiral along magnetic field lines (see the lower figure on p. 177).

Ole Roemer (1644–1710), Danish astronomer.

The Doppler Effect

You have probably noticed the rapid drop in pitch of the warning signal emitted by a passing train or car. The pitch of a tone emitted by an approaching source is higher than it would be if the source were standing still; the pitch of a tone emitted by a receding source is lower.

The difference $\Delta\nu$ between the frequency ν' of an approaching or receding source and its rest frequency ν is given by the formula

$$\frac{\Delta\nu}{\nu} = -\frac{V}{c}, \tag{5.16}$$

where V is the speed at which the source is moving away from the observer (a negative value of V therefore indicates motion toward the observer) and c is the speed of the signal. As we will see in a moment, this formula is valid only for speeds V much smaller than the signal-speed c; but it is valid for all kinds of waves—light waves and water waves as well as sound waves.

The change in the frequency of a periodic signal caused by the relative motion of source and observer is called the Doppler effect, after the Austrian physicist Christian Doppler (1803–1853), but it was discovered and explained in 1675 by the Danish astronomer Ole Roemer (1644–1710). Roemer noticed that the interval between successive eclipses of Jupiter's inner-most satellite Io increases as Jupiter moves from opposition to conjunction, and then decreases as Jupiter moves from conjunction to opposition (see lower figure on p. 179). As Jupiter moves

An engraving of Roemer at work in his study.

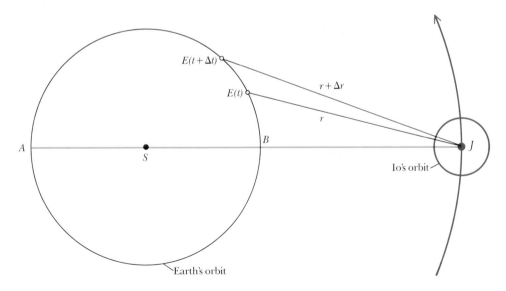

Roemer's experiment, viewed in a rotating frame of reference in which the Sun-Jupiter line SJ has a fixed direction. The light-travel time $\overline{JE}/c$ is least when the Earth is at B (opposition), greatest when the Earth is at A (conjunction). The intervals between successive eclipses of Io by Jupiter increase as the Earth travels from B to A, then decrease as it returns to B.

from opposition to conjunction, the distance between it and the Earth increases. Roemer attributed the progressive lengthening of the interval between successive eclipses during this period to the progressive increase in the time required by light to travel from Jupiter to the Earth. Let T denote Io's orbital period, measured in a frame of reference attached to Jupiter, and T' its orbital period as measured on Earth, the latter being the observed interval between successive eclipses of Io by Jupiter. If the distance between Jupiter and the Earth increases by an amount Δr during one orbital period T, the observed interval T' between successive eclipses is greater than T by the amount $\Delta r/c$, where c is the speed of light:

$$T' = T + \frac{\Delta r}{c}. \tag{5.17}$$

From this formula, using measurements of the accumulated delay in the eclipse times between opposition and conjunction, and a recent estimate of the diameter of the Earth's orbit (Chapter 2), Roemer derived the first estimate of the speed of light.

But Equation 5.17 is also the formula for the Doppler effect. Set $\Delta r = VT$, where V is the radial component of the relative velocity of the Earth and Jupiter. Then Equation 5.17 becomes

$$T' = T(1 + V/c). \tag{5.18}$$

Io's eclipse period is the reciprocal of its eclipse frequency; so we may replace Equation 5.18 by

$$\frac{1}{v'} = \frac{1 + V/c}{v}, \tag{5.19}$$

which is equivalent to Equation 5.16 when $V/c \ll 1$. The same formula evidently connects the emitted and received frequencies of any periodic signal traveling with speed c.

The lefthand figure on the facing page summarizes the classical theory of the Doppler effect for light waves. The oblique parallel lines represent the crests of a light wave traveling in the x-direction. During one second, v crests pass the point $x = 0$, where v is the frequency of the light. The figure shows that during this time-interval, $(1 - V)v$ crests pass an observer moving with velocity V in the x-direction. This number is the frequency v' of the light as recorded by the moving observer. Thus

$$v' = (1 - V)v, \tag{5.20}$$

where we are once again measuring speed in units of the speed of light. If we set $\Delta v = v' - v$,

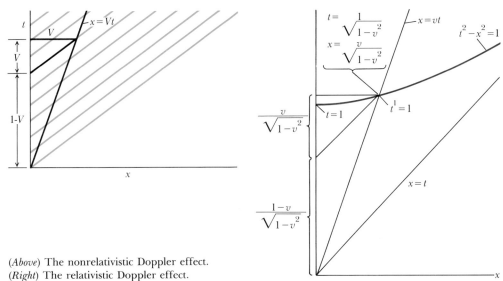

(*Above*) The nonrelativistic Doppler effect.
(*Right*) The relativistic Doppler effect.

we recover Equation 5.16, which summarizes Roemer's explanation of the progressive retardation of Io's eclipses.

Let us do the same calculation using the relativistic spacetime diagram in the righthand figure above. We must now allow for the fact that the stationary and moving observers assign different values to the time-interval between two given events. The figure shows that during a time-interval judged by the moving observer to be 1 second long, the number of crests received is $\gamma(1 - V)\nu$. Hence the moving observer judges the frequency of the light to be

$$\nu' = \gamma(1 - V)\nu = \sqrt{\frac{1 - V}{1 + V}}\,\nu, \tag{5.21}$$

where, as usual, γ is given by Equation 5.7.

The relativistic formula (Equation 5.21) differs from its nonrelativistic counterpart in two important ways.

First, the ratio of the received and emitted frequencies depends only on the relative velocity of the receiver and emitter. If we replace V by $-V$ and interchange ν and ν', the relativistic formula does not change. The nonrelativistic formula (Equation 5.20) does change: the factor $(1 - V)$ on the right side becomes $1/(1 + V)$. The nonrelativistic formula is asymmetric between the receiver and the emitter because, in the classical theory, the speed of light has different values in the two frames.

Second, the relativistic formula (Equation 5.21) predicts that the frequency of a light wave moving directly toward the observer increases without limit as the relative speed approaches the speed of light. The nonrelativistic formula (Equation 5.20) predicts that the frequency of the received light approaches twice the frequency of the emitted light as the relative velocity approaches the speed of light.

The observer's motion relative to a light source affects not only the frequency of the received light but also its intensity. To understand how, let us consider a monochromatic light source emitting photons of energy $h\nu$ (see Box on page 154) at a rate of f per second. The frequency ν' of each received photon is given by Equation 5.21. Moreover, the rate f' at which the receiver judges photons to be emitted is related to the rate f as judged by an observer in the rest frame of the source by the same formula:

$$f' = \sqrt{\frac{1-V}{1+V}}\,f.$$

(5.22)

The apparent brightness of a source is proportional to the rate f' at which photons are emitted and to the energy $h\nu'$ of an individual photon, where h is Planck's constant. The energy per unit time incident on a surface of area A at right angles to the line of sight is given by the formula

$$\text{received power} = f' \cdot h\nu' \cdot \left(\frac{A}{4\pi r^2}\right) \cdot g(\phi).$$

(5.23)

Here r is the instantaneous distance of the source, $A/4\pi r^2$ is the fraction of the surface of a sphere of radius r centered on the source that is covered by the receiving area A, and $g(\phi)$ is a factor that takes account the unequal emission of light in different directions, including the effect of relativistic focusing discussed in the preceding section. For a source approaching the receiver head-on at an ultrarelativistic speed ($\gamma \gg 1$), the focusing factor $g(\phi)$ is of order γ^2. For such a source, $1 + V \simeq \frac{1}{2}\gamma^2$; so the received power is about $\gamma^2 \cdot \gamma^2 = \gamma^4$ times as great as it would be if the source were at rest relative to the receiver at the same distance (as judged by the receiver).

Imagine a spacecraft starting from rest relative to the stars and accelerating at a constant rate equal to the gravitational acceleration at the surface of the Earth. The passengers would experience gravity's familiar pull: they would weigh the same as they did on Earth, and objects would fall toward the stern of the spacecraft exactly as objects near the surface of the Earth fall toward the center of the Earth. As the craft's speed approached the speed of light, the stars would move along great circles toward a point on the celestial sphere directly overhead (the direction of motion). Stars in the upper or forward hemisphere would grow bluer and brighter as they approached the zenith; stars in the lower hemisphere would grow fainter and redder until they reached the horizontal circle, when they would begin to grow brighter and bluer. As the spacecraft continued to accelerate, its speed coming ever closer to the speed of light, the bright patch of starlight centered on the zenith would grow steadily smaller, bluer, and brighter, while everywhere else the sky would grow dark.

Relativistic Energy and Momentum

Newton's theory makes inertia the quintessential property of matter. According to Newton's second law of motion, a body's inertia—its resistance to acceleration—is proportional to its mass. Newton and his successors believed that every particle of matter has a permanent, unchanging mass, and that the mass of a macroscopic body is the sum of the masses of its constituent particles. In a brief note modestly entitled "Does the inertia of a body depend upon its energy-content?" written immediately following his 1905 paper on special relativity, Einstein challenged these beliefs. He argued that a body could gain or lose mass by absorbing or emitting energy. This thesis implies that the mass of any composite system—including atoms and atomic nuclei—cannot be the sum of the masses of its constituent particles, because the motions and interactions of these constituents contribute to the energy of the system and hence to its mass.

Einstein barely discusses the revolutionary implications of his note. In keeping with its modest scale, he devotes one short sentence each to conclusion, discussion, and summary:

> *The mass of a body is a measure of its energy-content; if the energy changes by L, the mass changes in the same sense by L/9 × 10²⁰ [=L/c²], the energy being measured in ergs, and the mass in grams.*
> *It is not impossible that with bodies whose energy-content is variable to a high degree (e.g., with radium salts) the theory may be successfully put to the test.*
> *If the theory corresponds to the facts, radiation conveys inertia between the emitting and absorbing bodies.*

Einstein deduced the equivalence of mass and energy from the following thought experiment. An object at rest in reference frame F emits two photons of frequency ν in opposite directions. By symmetry, the object must be at rest after the photons have been emitted. If the energy of the body was E_0 before it emitted the photons and E_1 afterward,

$$E_0 = E_1 + h\nu + h\nu, \tag{5.24}$$

since, according to Einstein's photon hypothesis (see Box on page 156), the energy of a photon of frequency ν is $h\nu$.

Let us now write down the analog of Equation 5.24 in a reference frame F' moving with velocity V parallel to the direction in which the photons are emitted. Let E_0' denote the energy of the object in the moving frame before the photons are emitted, and E_1' its energy afterward. According to the relativistic formula for the Doppler effect (Equation 5.21), the frequencies of the two photons, as measured in frame F', are given by the formula $\nu' = \gamma(1 \pm V)\,\nu$. Hence,

$$E_0' = E_1' + \gamma(1 - V + 1 + V)\, h\nu = E_1' + 2\gamma h\nu. \tag{5.25}$$

Subtracting Equation 5.24 from Equation 5.25, we obtain

$$E_0' - E_0 = E_1' - E_1 + 2h\nu(\gamma - 1). \tag{5.26}$$

or

$$K_0' = K_1' + 2h\nu(\gamma - 1). \tag{5.27}$$

The differences $(E_0' - E_0) = K_0'$ and $(E_1' - E_1) = K_1'$ represent the *kinetic energies* of the body before and after emission of the photons, because the kinetic energy of a body is the energy attributable to its translational motion, i.e., the difference between its energy in a frame in which it has a given velocity V and its energy in a frame in which it is at rest.

Equation 5.27 states that the emission of the two photons reduces the body's kinetic energy by an amount $2h\nu(\gamma - 1)$. But the emission of the photons caused no change in its *velocity*, for in frame F the body was at rest before it emitted the photons, and it remains at rest after it has emitted them. Since kinetic energy depends only on mass and velocity, and the velocity has not changed, the emission of the photons must have caused a change in the mass. We can verify this conclusion by considering the case $V \ll 1$. Then $K = \frac{1}{2}mV^2$. Also, $\gamma - 1 \approx \frac{1}{2}V^2$. Equation 5.27 then becomes

$$\tfrac{1}{2}m_0 V^2 = \tfrac{1}{2}m_1 V^2 + \tfrac{1}{2}(2h\nu)V^2, \tag{5.28}$$

where m_0 is the mass of the body before emission of the two photons and m_1 its mass afterward. Thus

$$m_0 = m_1 + 2h\nu = m_1 + \Delta E, \tag{5.29}$$

where ΔE is the total energy of the emitted photons, measured in frame F. This completes the proof of Einstein's assertion that if the energy of a body changes by an amount ΔE, its mass changes by an equal amount (or, in conventional units, by $\Delta E/c^2$).

But we have not exhausted the implications of Equation 5.27. Because this relation must hold for all values of V (not just for $V \ll 1$), the kinetic energy K must depend on V in the same way as the quantity $(\gamma - 1)$. And since K is given by the Newtonian formula $K = \frac{1}{2}mV^2$ when $V \ll 1$, the exact formula must be

$$K = m(\gamma - 1). \tag{5.30}$$

The argument leading to Equation 5.29 goes through as before, and we obtain

$$\Delta m = \Delta E, \tag{5.31}$$

where ΔE represents the quantity of energy radiated. We may conveniently measure energy and mass from the same zero-point. Then the total energy E is given by the formula

$$E = K + m = \gamma m = \frac{m}{\sqrt{1 - V^2}} \tag{5.32}$$

or, in conventional units,

$$E = \gamma mc^2. \tag{5.33}$$

In the body's rest frame, $\gamma = 1$, and this equation takes the familiar form $E = mc^2$.

Equation 5.32 shows that the energy of a body of finite rest mass m increases without limit as its velocity approaches that of light. Hence, *a particle of finite rest mass can never be accelerated to the speed of light.* Conversely, if a particle of finite energy is moving with the speed of light, it must have zero rest mass. Besides photons, the only known particles with zero rest mass are neutrinos, which accordingly move with the speed of light.

In Chapter 4 we saw how Huygens used the principle of relativity to derive both the definition of momentum and its conservation law from the theorem that kinetic energy is conserved in elastic collisions. A precisely analogous argument, employing Einstein's law for the composition of velocities (Equation 5.9), yields the relativistic definition of momentum:

$$\mathbf{p} = m\gamma\mathbf{v} = E\mathbf{v}. \tag{5.34}$$

For photons and neutrinos $v = 1$, so that $p = E$.

This concludes our discussion of Einstein's theory of unaccelerated motion. We have seen that Einstein's theory does not merely introduce small corrections to the classical theory. Rather, it predicts that particles moving close to or at the speed of light behave in qualitatively new ways. A photon always has the same speed, even for an observer pursuing it at a speed equal to 0.99 the speed of light. A space traveler could leave the Earth in the year 2000 at the age of 30 and return a century later at the age of 40; but he would have to be shielded from the blinding light of the night sky, which would be concentrated in a tiny disc in the direction of his motion. There can be no rational doubt that these and equally bizarre-seeming predictions of Einstein's theory of unaccelerated motion are valid, because no physical theory—not even New-

Hermann Minkowski (1864–1909) reformulated Einstein's special theory in a mathematical language adapted to its deep structure.

ton's theory of gravitation—has survived more varied and more stringent tests. It is an experimental fact that unstable particles moving at ultrarelativistic speeds live longer than slowly moving particles of the same kind, and by precisely the ratios that Einstein's theory predicts. It is likewise an experimental fact that the radiation emitted by ultrarelativistic charged particles is shifted toward higher frequencies and focused in the forward direction, exactly as Einstein's theory predicts. Even now, almost a century after Einstein published his theory, it seems astonishing that an assumption as simple and plausible as the principle of relativity should have such varied and spectacular consequences.

This chapter has dealt only with Einstein's theory of unaccelerated motion. The complete theory of special relativity treats accelerated motion as well. There are relativistic analogs of Newton's second law of motion, of Maxwell's theory of electromagnetism and electromagnetic radiation, and of Lorentz's theory of electrons. All these theories have their classical counterparts as limiting cases, valid for small relative velocities, and all of them differ markedly, often spectacularly, in the opposite, ultrarelativistic limit. Einstein himself described the most important of these theories in his 1905 paper, and three years later the mathematician Hermann Minkowski succeeded in expressing Einstein's theory in a mathematical language whose rules are adapted to the theory's deep mathematical structure in somewhat the same way that the orthography of a phonetic language is adapted to its pronunciation. When the formulas that express the relativistic generalizations of Newton's and Maxwell's theories are written in Minkowski's mathematical language, they become simpler than their classical counterparts! For example, the classical laws of conservation of momentum and of energy become, in Minkowski's formulation of Einstein's theory, a single law, expressed by a single equation.

Einstein seems never to have regarded special relativity as more than a first step. At first sight, it may not be obvious that anything at all remained to be done, at least outside the realm of atomic and subatomic phenomena. Einstein felt, however, that the principle of relativity, as he had formulated it, was too narrow. It abolished distinctions between unaccelerated frames of reference but left intact the basic Newtonion distinction between accelerated and unaccelerated frames. Accelerated (or unaccelerated) with respect to what? As we saw in Chapter 4, Newton could not answer this question. The Earth is spinning relative to a cosmological reference frame defined by the average motions of distant galaxies. The frames of reference in which Newton's first law of motion holds, frames in which no free particle is accelerated, have no detectable rotation relative to this cosmological frame. That is an empirical finding; and neither Newton's theory nor Einstein's special theory of relativity accounts for it. So Einstein set out to "relativize" accelerated motion: to construct a theory in which accelerated and unaccelerated reference frames would be defined by the distribution of matter and energy on a cosmic scale. In the next two chapters we will retrace his route to that goal.

CHAPTER SIX

EINSTEIN'S THEORY
OF GRAVITATION

The supreme task of the physicist is to arrive at those universal elementary laws from which the cosmos can be built up by pure deduction. There is no logical path to these laws; only intuition, resting on sympathetic understanding of experience, can reach them. Nobody who has really gone deeply into the matter will deny that in practice the world of phenomena uniquely determines the theoretical system, in spite of the fact that there is no logical bridge between phenomena and their theoretical principles; this is what Leibniz described so happily as a "pre-established harmony."

ALBERT EINSTEIN: *Homage to Max Planck*

Einstein's theories of space, time, and gravitation revolutionized the style as well as the substance of theoretical physics. Not only did Einstein invent new physical laws; he also invented a new way of inventing new laws. As the Pythagoreans had recognized that mathematics is a surer guide to physical reality than common sense, so Einstein recognized that symmetry principles are deeper than physical laws. In Chapter 5 we saw how Einstein built the special theory of relativity on the requirement that it be impossible to distinguish between unaccelerated frames of reference. In this chapter we will see how he built a new theory of gravitation from the seemingly innocuous requirement that gravitational forces be indistinguishable from the forces experienced by particles in an accelerated frame of reference.

Is Gravitation a "Real" Force?

Newton's second law of motion states that in an unaccelerated frame of reference, a body's acceleration is proportional to the applied force. The coefficient of proportionality is called the *inertial mass* and is denoted by *m*. Symbolically,

$$\mathbf{F} = m\mathbf{a}, \tag{6.1}$$

where $\mathbf{F}$ is the applied force and $\mathbf{a}$ is the acceleration.

What is an unaccelerated reference frame? It is a frame in which the acceleration of every object is proportional to the applied force. This definition is not as circular as it may seem at first sight. Suppose we measure the accelerations of test objects in a frame F about which we know nothing beforehand. Assume that we know how the various forces that may be acting

(Left) 200-inch Hale telescope and dome.

Albert Einstein (1879–1955).

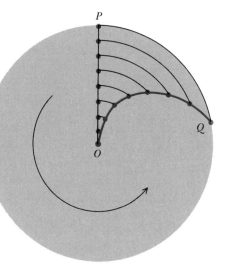

Trajectory of a puck sliding without friction on a spinning ice rink. *OQ* is the trajectory on the rink; *OP*, the trajectory in space. As viewed in the rotating reference frame, the motion appears accelerated.

depend on the positions, velocities, and physical properties of the test objects. We can then calculate the applied forces and decide, by means of appropriate experiments, whether inertial masses can be assigned to all test objects in such a way that the *measured* accelerations are related to the *calculated* forces by Equation 6.1. Usually this will not be possible. But if Newton's theory is correct, we should be able to make all the residual accelerations vanish by choosing a new frame of reference.

For example, suppose we observe that the trajectory of a hockey puck on an ice rink is always an Archimedean spiral,

$$r = V(t - t_0), \qquad \theta = \omega(t - t_0),$$

where r is the radial distance of the puck from its initial position, V is its initial velocity, and θ is its azimuth measured from its initial direction. We could then conclude that the rink was spinning with angular speed $-\omega$.

It is usually not difficult to distinguish between "fictitious" forces (forces resulting from motion of the reference frame) and real forces. The acceleration produced by a real force usually depends on some physical property of the object on which it is acting, and is different for different objects. For example, the acceleration produced by an electric field depends on the electric charge of the accelerated object. Since objects of the same mass can have different electric charges, a given electric field produces different accelerations in different objects. The acceleration produced by a fictitious force, on the other hand, depends not on an object's physical properties but only on its position and velocity.

Now, a body's gravitational acceleration also depends only on its position, and not on its mass or any other physical property. Galileo tested this law in two ways: by dropping objects of

Foucault's experiment at Paris Observatory, demonstrating the Earth's rotation. In a nonrotating reference frame the pendulum vibrates in a fixed plane. Owing to the Earth's rotation, the floor of the Paris Observatory (latitude 49°) rotates relative to such a frame with a period of (sin 49°) × 24 hours.

different weight and composition from a tower, and by observing that the period of a pendulum does not depend on the weight or composition of its bob. Newton observed that Kepler's harmonic law, connecting the orbital diameters and periods of the planets, follows from the inverse-square law of gravitation, if we assume that a planet's gravitational acceleration is independent of its physical properties.

The Eötvös Experiment.

In 1889 R. von Eötvös carried out an experiment to test Galileo's law. Suppose the gravitational acceleration of a piece of wood differed from that of a piece of platinum. Their centrifugal accelerations in a rotating reference frame must be the same. Hence the resultant of the gravitational and centrifugal accelerations cannot have the same value for both objects. At middle latitudes on the surface of the rotating Earth, the resultant accelerations of weights made of wood and platinum will differ in direction and magnitude. If the weights are placed in an Archimedean balance, the wire from which the beam connecting them is suspended experiences a small torque when the weights balance. Eötvös' experiment measured this torque very accurately; it showed that gravitational mass is identical with inertial mass to 1 part in 10^9. An improved version of Eötvös's experiment by R. H. Dicke and his colleagues at Princeton has reduced the margin of error to 1 part in 10^{11}.

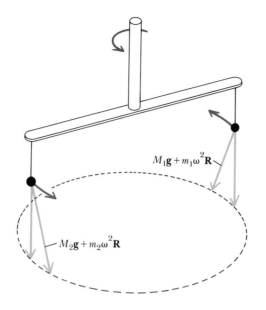

$$M_1\mathbf{g} + m_1\omega^2\mathbf{R}$$

$$M_2\mathbf{g} + m_2\omega^2\mathbf{R}$$

Schematic diagram of the Eötvös experiment.

Galileo's law suggested to Einstein that gravitation is a fictitious force, i.e., that it results from the acceleration of the reference frame in which it is measured. To maintain this view, Einstein had to change Newton's definition of an unaccelerated reference frame. Because the gravitational acceleration vanishes at the origin of any freely falling, nonrotating reference frame, we must regard all freely falling, nonrotating reference frames as unaccelerated; yet any two such frames are, in general, accelerated relative to each other! Later we will see how Einstein resolved this apparent contradiction.

Einstein's way of explaining Galileo's law seems at first sight horribly complicated. No wonder that Ernst Mach, who believed that physical laws are merely devices for summarizing the results of measurements in an economical way, bitterly opposed Einstein's theory. But Einstein believed that his way of looking at gravitation, despite its apparent complexity, was fundamentally simpler, and hence closer to the truth, than Newton's. Moreover, it had testable consequences, which Einstein was able to demonstrate by means of simple thought experiments about the behavior of light in a gravitational field.

How Much Does a Photon Weigh?

Imagine two identically equipped laboratories, one on Earth, the other in a uniformly accelerated spaceship. Einstein's *principle of equivalence* asserts that every experiment will have identical outcomes in the two laboratories, provided the spaceship's acceleration g is equal to the gravitational acceleration at the surface of the Earth.* Consider an object of inertial mass m hanging

---

* This statement of the principle of equivalence is not exact. An experiment that could detect variations in the strength or direction of the gravitational field would have different outcomes in the two reference frames. The outcomes become indistinguishable only in the limit of vanishingly small laboratories. In physicists' jargon, the principle of equivalence is valid locally but not globally.

from a spring balance attached to the ceiling of the space laboratory. (We assume that the spaceship's acceleration is directed from the floor to the ceiling of the laboratory.) The balance registers a force mg. If the object absorbs a photon of energy ΔE, its inertial mass increases by an amount $\Delta E/c^2$ (as we saw in Chapter 5), and the balance registers a force $(m + \Delta E/c^2)g$. The same experiment carried out in the terrestrial laboratory must yield the same result: the balance suspended from the ceiling of the terrestrial laboratory must record a force $(m + \Delta E/c^2)g$. But this force now represents the *weight* of the object. Thus the absorption of a quantity of energy ΔE has increased the object's gravitational mass by an amount $\Delta E/c^2$, the inertial mass associated with this quantity of energy.

This simple argument, together with Einstein's earlier argument that demonstrated the equivalence of energy and inertial mass, shows that *energy, inertial mass, and gravitational mass are one and the same thing;* the three terms are merely different names for a single quantity. In Newton's theory, gravitational and inertial mass are numerically equal but conceptually distinct. Once we accept Einstein's principle of equivalence, the two concepts merge. In all following discussions, therefore, the term "mass" will be used without qualification. Since "mass," in turn, is strictly synonymous with "energy," these two terms will be used interchangeably.

In conventional units the three quantities are connected by the relations $E = mc^2$ and $M = Gm$, where c is the speed of light in empty space and G is Newton's gravitational constant. We can define "natural" units, in which $c = G = 1$, and $E = M = m$. To avoid cluttering up equations with needless cs and Gs, these units, in which time, distance, and mass (or energy) all have the same physical dimension, will be used henceforth.

Einstein's theories teach us that mass (or energy)—not matter—is the indestructible stuff the world is made of. In the fifth century B.C. Leucippus and Democritus speculated that the world is a collection of indestructible particles, and that all change, qualitative as well as quantitative, results from rearrangements of these particles and changes in their motions. Newton believed so strongly in this view that he extended it to light and to the ether, imagining both to consist of tiny material particles. We now know that so-called fundamental particles are not indestructible. A proton and an antiproton may collide and disappear in a flash of light. Energy (mass) is both material and immaterial, continuous and discrete. And energy in all its forms is the source of gravitational fields.

The Gravitational Redshift

When a particle of mass m falls a distance Δz in a uniform gravitational field g, its energy (mass) increases by an amount

$$\Delta m \equiv \Delta E = mg\,\Delta z. \tag{6.2}$$

As we saw in the preceding section, this conclusion must apply to photons as well as material particles. According to Einstein's photon hypothesis, the energy of a photon is proportional to

Periodic frequency shifts of lines in the spectrum of Sirius and its white-dwarf companion (schematic). Because the gravitational potential at the surface of a white dwarf is much larger than the gravitational potential at the surface of a normal star, the gravitational contribution to the white dwarf's frequency shift is much larger than the gravitational contribution to Sirius's frequency shift. The difference between the two gravitational redshifts appears as a difference between the *average* frequency shifts of corresponding lines in the spectra of the two stars.

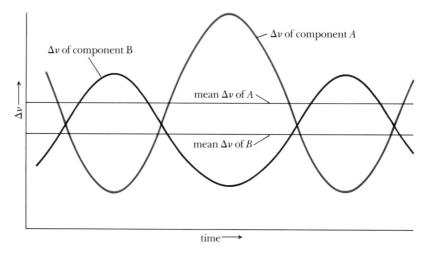

its frequency: $E = h\nu$, where h is Planck's constant. Thus the frequency of a falling photon increases according to the rule

$$\frac{\Delta\nu}{\nu} = g\,\Delta z. \tag{6.3}$$

In conventional units, $g\,\Delta z = g\,\Delta z/c^2$. At the surface of the Earth $g \simeq 10^3$ cm/sec². If $\Delta z = 23$ meters, as in an experiment described in the Box on page 195, the predicted frequency shift $\Delta\nu/\nu$ is about 2.5×10^{-15}.

Photons radiated by a star do work against gravity; hence their frequencies are reduced. In Equations 6.2 and 6.3, we may replace the quantity $g\,\Delta z$ by $\Delta\phi$, where ϕ is the gravitational potential (see p. 126ff in Chapter 4). The resulting equation,

$$\frac{\Delta m}{m} = \frac{\Delta\nu}{\nu} = \Delta\phi, \tag{6.4}$$

is valid for any gravitational field. The fractional redshift for a photon emitted from a star of mass M and radius R is M/R or (in conventional units) Gm/Rc^2. For the Sun, $\Delta\nu/\nu \simeq 2 \times 10^{-6}$; for the companion of Sirius, a so-called white dwarf, $\Delta\nu/\nu \simeq 2 \times 10^{-4}$. A frequency shift $\Delta\nu/\nu$ of order 10^{-4} is not at all difficult to measure. The real difficulty lies in disentangling it from the Doppler shift that arises from the motion of the star as a whole. There is one set of circumstances in which this can be done.

Consider a pair of stars revolving in circular orbits about their common center of mass, and assume for simplicity that the observer lies in the orbital plane. Each star moves toward the observer for half a period, and then away from the observer during the remaining half-period. Each of its spectral lines oscillates in frequency about an average frequency whose deviation

In 1976 Robert F. C. Vessot and collaborators at the Harvard-Smithsonian Center for Astrophysics used a rocket-borne hydrogen maser (above) to measure the Einstein shift. The difference in gravitational potential between the rocket's apogee and perigee gives rise to a predicted Einstein shift

$$\Delta\nu/\nu = 4 \times 10^{-10},$$

more than 100,000 times the predicted Einstein shift in the Pound-Rebka experiment. Because the rocket is in motion, the monochromatic radiation emitted by the hydrogen maser is also redshifted (on the rocket's outward leg) and blue-shifted (on the return leg) by the relativistic Doppler effect (page 178). Vessot devised an ingenious method for electronically subtracting out the first-order (classical) Doppler shift. Thus the experiment measured the sum of the Einstein shift and the part of the Doppler shift that results from relativistic time dilation. These two contributions are approximately equal in magnitude over most of the rocket's flight. The hydrogen maser, a device invented by Norman F. Ramsey of Harvard University, has a frequency stability of 1 part in 10^{14}. The measured relativistic shift agreed with Einstein's theory to 7 parts in 10^5, more than 100 times the accuracy of the Pound-Rebka experiment.

from the rest frequency is the sum of two contributions: a Doppler shift corresponding to the motion of the center of mass, and the star's gravitational redshift (see figure on page 194). Suppose now that one of the two stars is a white dwarf, the other a normal star. The white dwarf's gravitational redshift is 100 times as great as the normal star's. Hence, the difference between the average values of $\Delta\nu/\nu$ for the two stars is nearly equal to the white dwarf's gravitational redshift.

This technique has been used to estimate the gravitational redshifts of several white dwarfs. The results agree with Einstein's prediction.

The Pound-Rebka Experiment and the Vessot Experiment.

In 1960 R. V. Pound and G. A. Rebka, Jr., succeeded in measuring the gravitational redshift in a laboratory experiment at Harvard. Fe^{57} nuclei bound in a solid emit and absorb photons in an exceedingly narrow frequency range ($\Delta\nu/\nu = 3 \times 10^{-13}$) centered on a sharply defined frequency near 3.5×10^{18} Hz (cycles per second). In the Pound-Rebka experiment, a source and an absorber of this accurately monochromatic radiation were separated by a vertical distance of 22.5 m, so that the radiation incident on the absorber was shifted in frequency by an amount $\Delta\nu/\nu = gh/c^2 = 2.5 \times 10^{-15}$. The observed rate of absorption was measurably smaller than the rate that would have been observed in the absence of the gravitational redshift. A suitable Doppler shift introduced by a small, accurately measurable relative motion of the source and the absorber restored the measured absorption rate to its normal value. The Doppler shift needed to restore the absorption rate agreed with the value predicted by Einstein's theory to within 10 percent. Subsequently, the error has been reduced to 1 percent.

Gravitational Time-Dilation; Slowing Down and Bending of Light in a Gravitational Field

In the last section we treated photons as massive (if immaterial) particles, and we concluded that the energy of a photon in a gravitational field depends on its height (or gravitational potential). Since a photon's energy is proportional to its frequency, the frequency of a light wave must also vary with height or gravitational potential. From this conclusion Einstein inferred that *the rates of all physical processes are slowed down in a gravitational field.*

The argument is simple. Suppose that a light source in the ceiling of a high-ceilinged room emits a beam of monochromatic light of frequency ν_2, which is received by a stationary detector on the floor. Since the properties of the light source, the distance between source and receiver, and the gravitational field are all constant in time, the difference $t_1 - t_2$ between the times of reception and emission of a photon or wavecrest must also be constant in time, whatever the shape of the photon's spacetime trajectory. Hence the time-interval separating successive wave-

The time-interval between wavecrests emitted at z_2 is the same as the time-interval between crests received at z_1. Suppose the light is emitted by an atomic transition with a frequency of 10^{15} Hz (cycles per second). Then this time-interval is 10^{-15} sec_2 where sec_2 is the unit of time at z_2. The same atom emits light of the same frequency at z_1; hence $\Delta t_1 = 10^{-15}$ sec_1. That is how we *define* the units of time at different heights. We have found that the frequency of light emitted at z_2 and received at z_1 is multiplied by the factor $(1 + \phi_2)/(1 + \phi_1)$. Hence $\Delta t_1/\Delta t_2 = (1 + \phi_2)/(1 + \phi_1)$.

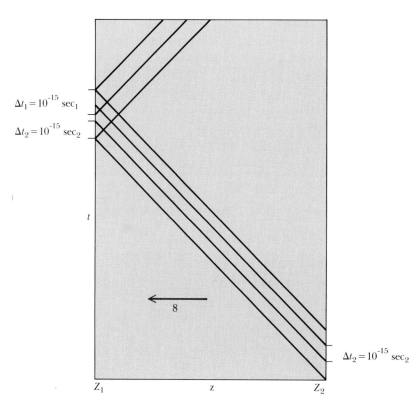

crests must be the same at the source and at the receiver (and at all intermediate points), as illustrated in the spacetime diagram above.

In classical physics and in special relativity, the frequency of a wave in a given reference frame is the reciprocal of the time-interval between successive wavecrests, as measured by a stationary observer in that frame. This relation cannot hold in a gravitational field (or in an accelerated reference frame), because, as we have just seen, the time-interval between successive crests of a light wave propagating in the z-direction is the same at different heights, but its frequency varies with height. Now, we may take as our local unit of time the time-interval between successive crests of the light emitted by a particular atomic transition. In other words, we stipulate that the frequency of light emitted by any given atomic transition is unaffected by the local gravitational potential. Any other idealized clock (frequency standard)—a tiny pendulum, for example—would do just as well, for, by virtue of the principle of equivalence, clocks that go at the same rate in the absence of a gravitational field also go at the same rate in a gravitational field if they are at the same gravitational potential. From our earlier conclusion that the frequency of a falling photon increases, we may therefore infer, with Einstein, that *identical clocks go at different rates at different gravitational potentials.* Only in this way can we reconcile our conclusion that the time-interval between successive crests of a monochromatic light wave does not vary with height and the inference (verified by Pound and Rebka) that its frequency does vary with height.

From Equation 6.4 we can calculate how the unit time-interval defined by a particular atomic transition varies with height. Let Δt_1 and Δt_2 denote the intervals between successive crests of light waves emitted by a given atomic transition at heights z_1 and z_2, respectively. When the light emitted at height z_2 reaches z_1, the interval between successive crests is still Δt_2, as we have seen. An observer at height z_1 will see the light emitted at height z_2 shifted in frequency (as compared with light emitted by the same transition at height z_1). By Equation 6.4,

$$\frac{\nu(z_2 \to z_1) - \nu(z_1 \to z_1)}{\nu(z_1 \to z_1)} = \phi(z_2) - \phi(z_1). \tag{6.5}$$

Here $\nu(z_2 \to z_1)$, the frequency of light emitted at height z_2 and received at height z_1, and $\nu(z_1 \to z_1)$, the frequency of light emitted and received at height z_1, are related to the "standard times" Δt_1 and Δt_2, defined above, by

$$\nu(z_2 \to z_1) = \frac{1}{\Delta t_2}, \qquad \nu(z_1 \to z_1) = \frac{1}{\Delta t_1}. \tag{6.6}$$

Combining Equations 6.5 and 6.6, and writing ϕ_i for $\phi(z_i)$, we obtain

$$\frac{\Delta t_1}{\Delta t_2} = 1 + \phi_2 - \phi_1 \simeq \frac{1 + \phi_2}{(1 + \phi_1)}. \tag{6.7}$$

We have defined the gravitational potential ϕ so that it vanishes in the absence of a gravitational field. Let $\Delta \tau$ denote the interval between successive wavecrests measured in a freely falling reference frame in which the emitting atom is momentarily at rest. Then, if Δt represents the interval between successive crests of light emitted by the same atom when it is at rest at gravitational potential ϕ, we can replace Equation 6.7 by

$$\Delta \tau = (1 + \phi) \Delta t. \tag{6.8}$$

So far we have been talking about stationary clocks in a static gravitational field, but it is not hard to guess how our description should be generalized to apply to moving clocks. In Equation 6.8, $\Delta \tau$ represents the interval between ticks (the emission of successive wavecrests by an atomic clock) as observed in a locally gravitation-free (i.e., freely falling) reference frame. From Chapter 5 we know that in such a frame the interval of proper time between ticks of a moving clock is given by the formula

$$d\tau^2 = dt^2 - d\ell^2, \qquad d\ell^2 = dx^2 + dy^2 + dz^2,$$

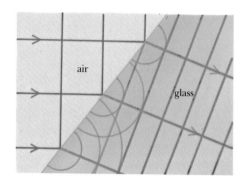

air

glass

Bending of a light ray passing from air to glass. Because light travels slower in glass than in air, successive wavecrests are closer together, and the ray, which is perpendicular to the wavefronts, bends toward the direction perpendicular to the interface. For details of the construction see the figure on page 199.

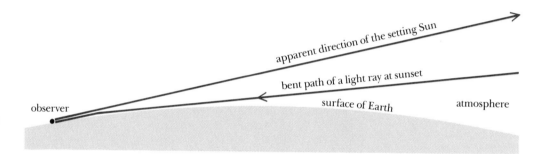

apparent direction of the setting Sun

bent path of a light ray at sunset

observer

surface of Earth

atmosphere

The setting Sun is actually below the horizon, because its rays bend toward the Earth's surface.

where dt, dx, dy, dz represent the coordinate intervals separating nearby events in the history of the moving clock. This suggests that the appropriate generalization of Equation 6.8 for a moving clock is

$$d\tau^2 = [1 + 2\phi(z)]dt^2 - d\ell^2,$$ (6.9)

since $(1 + \phi)^2 \simeq 1 + 2\phi$.

This equation makes some interesting predictions about the behavior of light in a gravitational field. From Chapter 5 we know that the interval of proper time between events in the history of a photon is zero. Since the quantity $d\tau$ has the same meaning in Equation 6.9 as it did in Chapter 5, setting $d\tau$ equal to zero in Equation 6.9 yields a relation that connects the time-intervals and space-intervals between events in the history of a photon in a gravitational field:

$$d\ell^2 = (1 + 2\phi)dt^2 \quad \text{or} \quad \frac{dl}{dt} \equiv c = 1 + \phi(z).$$ (6.10)

Since $\phi \leq 0$, this equation tells us that photons travel more slowly in a gravitational field than in a vacuum. (We will discuss an experimental test of this prediction later.)

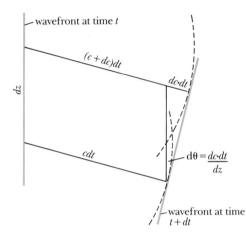

Huygens' construction. From every point on the wavefront at time t draw a sphere of radius $c\,dt$, where c is the local speed of light. The wavefront at time $t + dt$ is tangent to all these spheres.

The slowing down of light in a gravitational field is reminiscent of the slowing down of light in a material medium. Light travels more slowly in air than in a vacuum, and more slowly in glass than in air. Light rays propagating in a medium where the speed of propagation varies with position are bent. A light ray traveling from air into glass or water bends toward the direction perpendicular to the surface that separates the two media. Again, the rays of the setting Sun bend toward the surface of the Earth, because the density of the air diminishes, and the speed of light increases, with height above sea level; hence the setting Sun appears to be higher in the sky than it actually is.

The figure above illustrates a simple geometric construction invented by Huygens to construct the paths of light rays in a medium where the speed of light varies from point to point. From this figure we see that a light ray initially perpendicular to the gravitational field gets bent during a short time-interval dt through an angle $d\theta$ given by

$$d\theta = \frac{[c(z + dz) - c(z)]\,dt}{dz}$$

$$= \frac{[\phi(z + dz) - \phi(z)]\,dt}{dz} = \left(\frac{d\phi}{dz}\right)dt, \tag{6.11}$$

by Equation 6.10. Since $d\phi = -g\,dz$, it follows from Equation 6.11 that the deflection $d\theta$ corresponding to a time-interval dt along the path of the light ray is given by the simple formula

$$d\theta = g\,dt. \tag{6.12}$$

Huygens' construction shows that the light ray is deflected toward the direction of the gravitational field.

We can reach the same conclusion more simply by noticing that a photon, by virtue of its finite mass (= energy), must be accelerated by a gravitational field. Consider a photon moving in the x-direction at right angles to a gravitational field g in the z-direction. During a short time-interval dt it acquires a velocity component $c_z = g\,dt$ at right angles to its initial velocity $c_x = 1$. Hence, during this time-interval its direction changes by an amount $d\theta = c_z/c_x = g\,dt$, in agreement with Equation 6.12.

Let us estimate the time-delay Δt and the deflection $\Delta\theta$ of a photon that just grazes the Sun. The time-delay for a path of length $\Delta\ell$ is, by Equation 6.10,

$$\Delta t = \frac{\Delta\ell}{1+\phi} - \Delta\ell \simeq -\phi\,\Delta\ell;$$

the deflection is

$$\Delta\theta = g\,\Delta t = g\,\Delta\ell.$$

At the surface of the Sun $\phi = -M/R$, $g = -M/R^2$. Thus the total time-delay

$$\Delta t \simeq 2R\phi(R) = 2M = 2GM/c^2 \simeq 10^{-5}\ \text{sec},$$

and the total deflection is

$$\Delta\theta \simeq 2gR \simeq \frac{2M}{R} = \frac{2Gm}{Rc^2} \simeq 1\ \text{arc-sec};$$

so the predicted effects are very small indeed. We will discuss more accurate predictions and their experimental verification later.

Spacetime Curvature and the Motion of Free Particles

Equation 6.8 for the interval of proper time between neighboring events shows that a gravitational field warps spacetime. In the absence of a gravitational field, the interval of proper time is given by Minkowski's formula

$$d\tau^2 = dt^2 - d\ell^2, \tag{6.13}$$

where dt is the time-interval and $d\ell$ the space-interval between the two events. This is just like the Pythagorean formula for the distance between two neighboring points in a four-dimen-

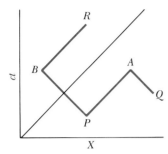

In Minkowskian spacetime any two events, e.g., *P* and *Q* or *P* and *R*, can be connected by a path of zero proper "length" (*PAQ*, *PBR*). The path is made up of segments of photon trajectories.

sional space, except for the minus sign. As we saw in Chapter 5, the minus sign is important. It is responsible for the qualitative distinctions between timelike, spacelike, and lightlike intervals, distinctions that are preserved in all unaccelerated reference frames. Nevertheless, Minkowskian spacetime (the spacetime of special relativity) resembles Euclidean space in an important respect: in both, the graph of a straight line is a straight line.

This statement needs to be explained. For simplicity, let us compare two-dimensional Minkowskian spacetime with two-dimensional Euclidean space. In Euclidean space, we *define* a straight line as the shortest distance between two given points. The advantage of defining straight lines by this property is that it allows us to define straight lines in *any* space, provided we have a definition of distance. For example, the definition implies that straight lines on the surface of a sphere are arcs of great circles (circles whose centers coincide with the center of the sphere).

In Minkowskian spacetime, the Euclidean definition of a straight line must break down, because we can always construct a path of zero "length" connecting any pair of events. In a Minkowskian spacetime diagram, the straight line (in the ordinary Euclidean sense) joining two events separated by a timelike interval is *longer* than the interval of proper time measured along any other path joining the two events. A proof is given below.

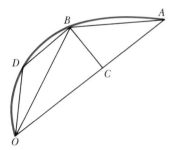

In Minkowskian Spacetime
a Straight Line is the Longest Distance
Between Two Events

Let us first prove the familiar rule "A straight line is the shortest distance between two points" for the Euclidean plane. Let *O*, *A* be any two points. We want to show that the straight line *OA* is shorter than any other line connecting these points. In the figure at the left, choose any point *B* on a given curved line connecting *O* and *A*, and draw *BC* perpendicular to *OA*. By Pythagoras' theorem, $\overline{AB} > \overline{AC}$ and $\overline{OB} > \overline{OC}$. Hence the broken line *ABO* is longer than the straight line *AO*. Similarly, if *D* is a point on the curve *OB*, the length of the broken line *ODB* is greater than the length of the straight line *OB*. Hence the length of the broken line *ODBA* is greater than that of the straight line *OA*. As we add more points to the curve between *O* and *A*, the length of the broken line connecting them increases, and approaches the length of the curve connecting *O* and *A*. Thus the straight line *OA* is the shortest line distance connecting *O* and *A*.

To convince ourselves that a straight line is the shortest distance between two points in a curved space, we must use a slightly different argument. If a curve connecting two given points, *A* and *B*, is not straight, it must be curved in the Euclidean neighborhood of at least one point. Within this neighborhood, select two points *P* and *Q* on the curve, and connect

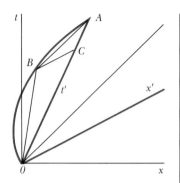

them with a straight line *PQ*. The curve *APQB* is now shorter than the original curve connecting *A* and *B*. We can continue shortening the curve in this way until it is straight in the neighborhood of each of its points, i.e., until it is a straight line. And the method by which we have constructed this straight line shows that it is the shortest distance between the two given points.

If we replace "space" by "spacetime," "point" by "event," and "distance" by "proper time," the preceding considerations apply to timelike curves in spacetime, provided we make one further change: we must also replace "shortest" by "longest." Thus in the figure at the left, the interval of proper time $\tau(OA)$ measured along the timelike straight line OA is greater than the sum $\tau(OB) + \tau(BA)$, where the straight lines OB and BA are both timelike, for we can choose OA to coincide with the time axis. Then if BC is drawn parallel to the corresponding x-axis, we have $\tau^2(OB) = \tau^2(OC) - x^2(BC)$, so that $\tau(OC) > \tau(OB)$. Similarly, $\tau(CA) > \tau(BA)$. Hence $\tau(OA) = \tau(OC) + \tau(CA) > \tau(OB) + \tau(BA)$. The remainder of the argument goes through as before: in Minkowskian spacetime, the proper time measured along a straight timelike line connecting two events is longer than the proper time measured along any other timelike path connecting the same two events. And straight timelike lines in curved spacetime have the same property.

In Minkowskian spacetime, straight timelike lines represent trajectories of free particles. They have equations of the form $x - x_0 = V(t - t_0)$, with $V < 1$. The statement that the interval of proper time between two given events is greatest when it is measured along the straight line connecting the two events is merely a formal statement of the twin "paradox" discussed in Chapter 5. Since the space-traveler's trajectory is represented by a broken line, her elapsed proper time between launch and return must be less than that of her sedentary twin between the same two events.

When there is a gravitational field in the z-direction, the interval of proper time between two neighboring events follows from Equation 6.9:

$$d\tau^2 = [1 + 2\phi(z)]dt^2 - dz^2. \qquad (6.14)$$

This resembles the formula for the distance between two neighboring points on a curved surface. For example, the distance between neighboring points on the surface of a sphere of unit radius is given by

$$d\ell^2 = \sin^2\theta \, d\lambda^2 + d\theta^2, \qquad (6.15)$$

where θ denotes colatitude and λ denotes longitude. The factor $\sin^2\theta$ that multiplies $d\lambda^2$ is analogous to the factor $(1 + 2\phi)$ that multiplies dt^2.

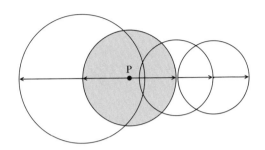

Construction of a straight line on a curved surface.

Although Euclidean geometry does not hold on the surface of a sphere, we can still construct straight lines on it by exploiting the fact that Euclidean geometry is nearly correct in a small region around any given point. (Mathematicians say that the surface of a sphere is *locally Euclidean*.) Thus we can accurately map Kansas City—but not the whole of the United States—on a square grid.

Using the property of local flatness, we can construct a straight line through a given point P in a given direction as follows (see figure above).

1. Surround P by a small circle within which departures from Euclidean geometry are unnoticeable. Through P draw a short segment of a straight line in the given direction.

2. Construct two circles centered on the endpoints of this segment, making the radius of each circle small enough so that the enclosed portion of the surface is approximately flat.

3. Extend the segment through P until it meets these two circles.

By continuing this process, we can extend the line indefinitely in both directions. We can make local deviations from straightness as small as we please by making the radii of the circles small enough. As the radii shrink and their number increases, the constructed curve approaches a perfect straight line.

On the surface of a sphere, this construction yields great circles.

We can carry out an exactly analogous construction in the two-dimensional spacetime defined by Equation 6.14 for the interval of proper time between neighboring events. The analog of a locally flat coordinate grid is a *Minkowskian grid*. A locally Minkowskian reference frame is one in which there is no gravitational acceleration. In such a reference frame, straight timelike trajectories represent motions of free particles. Since a straight line in the curved spacetime defined by Equation 6.14 is made up of tiny, locally straight segments, we conclude that *straight timelike lines in the curved spacetime defined by Equation 6.14 represent the trajectories of particles moving freely in the gravitational field defined by the gravitational potential ϕ.*

This is the conclusion we have been working toward. It describes the motion of particles in a gravitational field without referring to the concepts of force or inertial mass. The spacetime trajectories of particles moving freely in a gravitational field are straight lines in a curved spacetime whose structure is defined by Equation 6.14. Strictly speaking, we have established this result only for a static gravitational field in the z-direction, but we may immediately general-ize the preceding discussion to any gravitational field. We have only to replace the potential $\phi(z)$ by the more general potential $\phi(x, y, z, t)$.

Our geometric way of describing the motion of particles in a gravitational field does not look at all like Newton's algebraic description; yet the two descriptions are mathematically equiva-lent. The problem of finding the path connecting two points in space that minimizes (or maxi-mizes) some integral property of the path, such as its length, was first posed by Galileo, who asked: What is the shape of the path of quickest descent between two points in the Earth's gravitational field? (Galileo guessed, incorrectly, that the path of quickest descent is an arc of a circle.) Some years later the French mathematician Fermat postulated that the path of a light ray in a medium of variable refractive index minimizes the time of propagation between any two points on the path. Jean Bernoulli solved Galileo's problem by using Fermat's principle to find the path of a light ray in a medium where the speed of light varies with height in the same way as does the speed of a particle in a uniform gravitational field. In 1736 Euler showed that all problems of this kind lead to a certain set of differential equations. For the problem of finding the spacetime trajectory that maximizes the proper time between two events in curved spacetime, these differential equations are identical with Newton's equations of motion! The algebraic and geometric descriptions of motion in a gravitational field are mathematically equivalent.

Curved Space: Einstein's Field Equations

If the structure of spacetime were governed by the formula

$$d\tau^2 = (1 + 2\phi)\, dt^2 - (dx^2 + dy^2 + dz^2), \qquad (6.16)$$

space itself would be Euclidean. For the proper distance $d\ell$ between neighboring spacelike events is given by the formula $d\ell^2 = -d\tau^2$ (see Chapter 5, p. 168). Hence, by Equation 6.16, the proper distance between simultaneous events would be given by the Pythagorean formula

$$d\ell^2 = dx^2 + dy^2 + dz^2. \qquad (6.17)$$

A simple thought experiment showed Einstein that this formula cannot be exact.

Imagine a spaceship in the form of a rapidly spinning disc: a flying saucer. Huygens had pointed out that objects in a rotating reference-frame experience an outward gravitational field of magnitude $g = \omega^2 r$, where ω is the angular rate of rotation and r the distance from the axis of rotation.

Before investigating the geometry of the rotating disc, let us notice how the behavior of clocks varies with position on the disc. If the spaceship as a whole is in free fall, a clock at its center goes at the "normal" rate (the rate predicted by special relativity), but, as the following argument shows, an identical clock at the rim runs slower.

The rate of an accelerated clock is the same, at any given instant, as that of an identical clock in the freely falling reference frame in which the accelerated clock is instantaneously at rest. That frame is moving with speed ωr relative to the freely falling frame attached to the center of the spinning disc, and its velocity is perpendicular to the radius. Hence, as we saw in Chapter 5, its rate is slower than the rate of an identical clock at the center of the disc by the relativistic time-dilation factor

$$\gamma = (1 - V^2)^{-1/2} = (1 - \omega^2 r^2)^{-1/2} \simeq 1 + \tfrac{1}{2}\omega^2 r^2.$$

Einstein used an analogous argument to investigate the geometry of the disc. Imagine that many identical rods of length $\Delta\ell$ (as measured in the absence of gravitation) are laid along the rim of the disc and along its diameter. As measured in the freely falling reference frame attached to the center of the disc, the rods along the diameter have the same length as they would have in a nonrotating frame, because their motion is at right angles to their length. But the rods along the circumference have length $\Delta\ell/\gamma$, because of the Lorentz contraction. The rim of the spinning disc coincides with a stationary circle of the same radius. Hence more rods can be fitted along the rim of the spinning disc than along the rim of the same disc when it is not spinning. Therefore, an observer on the disc (for whom all the rods have equal length) will find that the circumference is $\gamma\pi d$, i.e., $(1 + \tfrac{1}{2}\omega^2 r^2)\pi$ times its diameter. Thus, *Euclidean geometry is not valid on the spinning disc: gravitational fields must warp space as well as spacetime.*

To accommodate this conclusion, we must replace Equation 6.16 for the interval of proper time between two neighboring events by one that allows for both the warping of space and the dilation of time. The most general formula of this kind is

$$\begin{aligned}
d\tau^2 = (1 + 2\phi_{00})\,dt^2 &- (1 + 2\phi_{11})\,dx^2 - (1 + 2\phi_{22})\,dy^2 - (1 + 2\phi_{33})\,dz^2 \\
&+ 2\phi_{01}\,dt\,dx + 2\phi_{02}\,dt\,dy + 2\phi_{03}\,dt\,dz \\
&+ 2\phi_{12}\,dx\,dy + 2\phi_{13}\,dx\,dz + 2\phi_{23}\,dy\,dz.
\end{aligned} \tag{6.18}$$

In place of the single gravitational potential ϕ $(= \phi_{00})$, we now have ten gravitational potentials ϕ_{ij}, with $i \leq j = 0, 1, 2, 3$.

What determines these ten potentials? The Newtonian potential ϕ is given by the formula

$$\phi = -\Sigma \frac{M_i}{r_i}.$$

The Einsteinian potentials ϕ_{ij} must also depend on the distribution of energy ($=$ mass) in the universe, but they cannot depend *only* on the distribution of energy. We have already seen that, owing to the relativity of uniform motion, energy and momentum are inseparably linked. Thus a particle that has energy E_0 and zero momentum in its rest frame has energy γE_0 and momentum $-\gamma E_0 \mathbf{V}$ in a reference frame moving with velocity $\mathbf{V}$ relative to the rest frame. The time-dilation factor

$$\gamma = (1 - V^2)^{-1/2} = \frac{dt}{d\tau},$$

since

$$d\tau^2 = dt^2 - d\ell^2 = dt^2 \left[1 - \left(\frac{d\ell}{dt} \right)^2 \right] = dt^2(1 - V^2).$$

Thus the energy and the three components of the momentum are given by the formulas

$$E = E_0 \left(\frac{dt}{d\tau} \right),$$

$$p_x = E_0 \left(\frac{dx}{d\tau} \right),$$

$$p_y = E_0 \left(\frac{dy}{d\tau} \right),$$

$$p_z = E_0 \left(\frac{dz}{d\tau} \right).$$

(6.19)

These formulas show that the energy of a particle is related to its momentum in just the same way that its displacement in time is related to its displacement in space:

$$E : p_x : p_y : p_z = dt : dx : dy : dz.$$

Next, consider a spatial distribution of particles, each of mass m_0 in its rest frame. Let $n(x, y, z, t)$ be the number of particles per unit volume in the neighborhood of the point (x, y, z) at time t, and suppose that all the particles in this neighborhood have the same velocity $\mathbf{v}(x, y, z, t)$. (In a real gas the particles would have random thermal motions in addition to their common velocity $\mathbf{v}$, but for the moment we assume the gas to be so cold that these random motions may be neglected.) The mass density ρ is simply the product of the number density n and the mass $\gamma m_0 = m_0(dt/d\tau)$ of an individual particle. Similarly, the x-component of momentum per unit volume is $nm(dx/d\tau)$. But because of the Lorentz contraction, n, the number of particles per unit volume, has different values in different reference frames. In fact, $n = n_0(dt/d\tau)$, where n_0 is the number density of particles in the local rest frame.*

---

* *Proof:* A box of volume dV contains ndV particles. This number obviously has the same value in every reference frame. The volume dV has different values in different reference frames, but the "four-dimensional volume" $dV\,dt$ has the same value in all reference frames. Since both $n \cdot dV$ and $dt \cdot dV$ have the same value in all reference frames, n must be a constant multiple of dt. In the local rest frame, $n = n_0$ and $dt = d\tau$; so $n = n_0(dt/d\tau)$.

Thus the energy and momentum densities are given by the expressions $n_0 m_0 (dt/d\tau)(dt/d\tau)$, $n_0 m_0 (dt/d\tau)(dx/d\tau)$, $n_0 m_0 (dt/d\tau)(dy/d\tau)$, and $n_0 m_0 (dt/d\tau)(dz/d\tau)$. These four quantities obviously belong to a larger array consisting of the ten quantities $n_0 m_0 (dx_i/d\tau)(dx_j/d\tau)$, where $x_0 = t$, $x_1 = x$, $x_2 = y$, $x_3 = z$. (There are $4 \times 4 = 16$ quantities of this kind, but only ten of them are distinct.) Four of these quantities, the ones we started with, characterize the space density of energy and momentum. The remaining six characterize the flow of momentum. (The flow of energy and the density of momentum are represented by the same three quantities.)

The upshot of this discussion is that, by virtue of the principle of relativity, the spatial distribution of mass is inseparably linked to the spatial distribution and flow of momentum. In a relativistic description of the mass distribution, we must specify ten distinct quantities,

$$T_{ij} \equiv n_0 m_0 \left(\frac{dx_i}{d\tau}\right)\left(\frac{dx_j}{d\tau}\right), \qquad i \le j = 0, 1, 2, 3, \tag{6.20}$$

instead of the single quantity $\rho_0 = n_0 m_0$. This is as it should be: we want the structure of spacetime, which is specified by ten gravitational potentials ϕ_{ij}, to be determined by its contents, which are specified by the ten quantities T_{ij}.

Equation 6.20 for the quantities T^{ij} holds in a locally Minkowskian reference frame, that is, in the absence of a gravitational field. To allow for the effects of gravitation, we have only to stipulate that the interval of proper time $d\tau$ be given by Equation 6.18 instead of by Minkowski's formula $d\tau = dt/\gamma$. This apparently innocuous step brings to light a profound feature of Einstein's theory of gravitation. To describe the distribution and flow of energy and momentum, we must know not only the mass density and velocity in locally Minkowskian reference frames but also how these locally Minkowskian reference frames are connected with one another; that is, we must know the gravitational potentials ϕ_{ij}. These in turn depend (as we will see shortly) on the distribution and flow of energy and momentum. Thus in Einstein's theory *the structure and contents of spacetime cannot be separated* from each other, as they can in Newton's theory. We will return to this important aspect of Einstein's theory later.

Equation 6.20 applies to a cold gas consisting of noninteracting particles. This special case has shown us that the mass density is one of ten connected quantities that jointly characterize the contents of spacetime. In general (and not merely in this special case), these ten quantities describe the distribution and flow of energy and momentum. Einstein postulated that energy and momentum are conserved quantities, as they are in Newton's theory and in special relativity. Thus, the energy contained in an imaginary box changes only by virtue of an inflow or outflow of energy. Similarly, the total momentum in a box can change only if there is an inflow or outflow of momentum. The ten quantities T_{ij} that specify the distribution and flow of energy and momentum therefore satisfy four relations that express the principles of conservation of energy and momentum.

The preceding considerations define the mathematical problem that Einstein had to solve in order to formulate a self-consistent geometric theory of gravitation. To replace the Newtonian formula that connects the gravitational potential ϕ with the spatial distribution of mass, he had to construct a set of ten relations connecting the ten gravitational potentials ϕ_{ij} with the ten quantities T_{ij} that characterize the distribution and flow of energy; and these ten relations had to be consistent with the four relations that express the conservation of energy and momentum. In the limit of weak gravitational fields and small velocities, these ten relations had to reduce to the single Newtonian relation connecting gravitational potential and mass density.

Fortunately, Einstein was able to call upon a powerful theory of curved surfaces and spaces that had been developed in the previous century by Carl Friedrich Gauss and Bernhard Riemann. Using this theory, Einstein constructed the simplest set of ten "field equations" satisfying the conditions just mentioned.

Einstein's field equations look—and are—much more complicated than Newton's, but from a more abstract mathematical standpoint they are exceedingly simple and elegant. They express the simplest connection between the structure and the contents of spacetime that satisfies the conditions that have been described.

We might suppose that, because Einstein's theory has ten gravitational potentials and ten field equations to Newton's one, it is ten times as difficult to apply. In fact, the difficulties are far greater. They arise not from the number of equations but from their nature.

In the first place, Einstein's field equations, unlike Newton's, are *nonlinear*. In Newton's theory the gravitational potential for a collection of point masses is just the sum of the potentials for the individual masses. The gravitational potentials of Einstein's theory do not have this property. We cannot construct the gravitational potentials for two point-masses from the gravitational potentials for a single point-mass. Even today exact solutions to Einstein's equations are known for only a handful of especially simple mass distributions, *not* including a pair of mutually gravitating point-masses.

An equally serious practical difficulty arises from the circular nature of Einstein's field equations. The Einsteinian potentials ϕ_{ij} depend on the quantities T_{ij} that characterize the distribution and flow of energy and momentum. But to specify the quantities T_{ij}, we need to know the structure of spacetime—which is determined by the potentials ϕ_{ij}! This circularity creates even more formidable mathematical difficulties than the nonlinearity.

But it also resolves an old philosophical problem, the conflict between Newton and his supporters, on the one hand, and Leibniz and the British empiricists, Berkeley, Locke, and Hume, on the other, about the nature of space and time. Newton argued that space and time exist absolutely, "without relation to anything external." Leibniz and the British empiricists argued that space and time have no independent reality, that only relations between physical objects and events are real. According to Einstein's theory, Newton's view is wrong, because the structure of spacetime is determined by its contents. But Newton's critics were equally mistaken, because spacetime is no less real or fundamental than matter and motion (or rather, energy and momentum). The structure of spacetime determines its contents no less than its contents determine its structure. Structure and content are interdependent and inseparable aspects of a single physical reality: spacetime-energy.

Experimental Tests of Einstein's Theory

Einstein, like Newton, began by applying his theory to the motion of test particles near an isolated point-mass. In the Newtonian approximation to Einstein's theory, the interval of proper time is given by the formula

$$d\tau^2 = \left(1 - \frac{2M}{r}\right)dt^2 - d\ell^2. \tag{6.21}$$

What would an exact formula for this interval look like?

An isolated point-mass defines a preferred position in space, but no preferred direction; hence we expect $d\tau$ to depend only on the radial coordinate r. Moreover, the mass M (which, in natural units, has the dimension of a length) defines the linear scale of the gravitational field. Accordingly, we expect the coordinate r to occur always in the combination r/M. These considerations suggest than an exact formula for the interval of proper time might have the form

$$d\tau^2 = a(u)dt^2 - b(u)d\ell^2, \tag{6.22}$$

where $u \equiv M/r$, and a and b are functions yet to be determined. If we substitute this formula into Einstein's field equations, we find that they do indeed have a solution of this form, with the functions a and b given by

$$a(u) = \left(\frac{1 - u/2}{1 + u/2}\right)^2,$$
$$b(u) = (1 + u/2)^4, \tag{6.23}$$

where $u \equiv M/r$. This exact solution of Einstein's equations was first obtained by the German astronomer Karl Schwarzschild in 1916.

As we noticed earlier, the ratio $u = M/r$ is very small for ordinary astronomical objects. The Sun's mass in natural units is 1.5 km, and its radius is 7×10^5 km; so $u = 2 \times 10^{-6}$. When u is small, we may use the approximate equality $(1 + u)^n \simeq 1 + nu$, where n is any positive or negative real number, to simplify the preceding formulas for $a(u)$ and $b(u)$:

$$a(u) = 1 - 2u + 2u^2,$$
$$b(u) = 1 + 2u. \tag{6.24}$$

These approximations go one step beyond the Newtonian approximation of Equation 6.21. They are adequate for describing the motion of photons and material particles near the Sun and were used by Einstein for this purpose in 1915, before Schwarzschild found his exact solution.

Einstein's theory predicts that photons emitted by a massive object undergo a gravitational redshift. Led Δt be the interval between ticks of a clock in the gravitational field of a point-mass, and let $\Delta\tau$ be the interval between ticks of an identical clock in the absence of gravitation. The relation between Δt and $\Delta\tau$ is given by Equation 6.22 with $\Delta\ell = 0$, since the space-interval between ticks is zero. Thus

$$\Delta\tau = [g(r)]^{1/2}\Delta t \simeq \left(1 - \frac{M}{r}\right)\Delta t, \tag{6.25}$$

in agreement with our earlier discussion.

Einstein's theory also predicts that photons moving past the Sun are slowed down and deflected. The interval of proper time $d\tau$ between neighboring events in the history of a photon vanishes. From this condition and Equation 6.22, we see that the speed of light $d\ell/dt$ depends on the radial coordinate r but, for a given value of r, has the same value in all directions:

$$c = \frac{d\ell}{dt} = \sqrt{\frac{a(r)}{b(r)}} \simeq 1 - \frac{2M}{r}. \tag{6.26}$$

Comparing this result with Equation 6.10, where $\phi = -M/r$, we see that the predicted change in the speed of light is twice as great as our earlier value. The earlier prediction was based on the Newtonian approximation, in which gravitation makes clocks run slower but does not warp space.

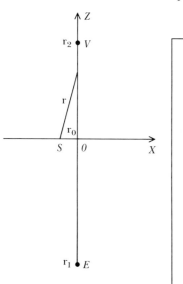

Path of a light-ray passing close to the Sun.

Calculating the Deflection and
Slowing Down of Light by the Sun

For the purpose of the calculation we may assume that the light ray is the straight line EV, passing the center of the Sun, S, at distance r_0. The round-trip time from E to V and back is $2T$, where

$$T = \int_0^{r_2} \frac{dz}{c(z)} + \int_0^{r_1} \frac{dz}{c(z)}.$$

By Equation 6.26, $c(z) = 1 - 2M/r$. Hence

$$\int_0^{r_2} \frac{dz}{c(z)} = \int_0^{r_2} \frac{dz}{1 - \dfrac{2M}{r}} \simeq \int_0^{r_2} \left(1 + \frac{2M}{r}\right) dz$$

$$= r_2 + 2M \int_0^{r_2} \frac{dz}{(r_0^2 + z^2)^{1/2}}$$

$$\simeq r_2 + 2M \log_e \left(\frac{2r_2}{r_0} \right).$$

The round-trip delay is thus

$$\Delta t = 2(T - r_1 - r_2) \simeq 4M \left[\log_e \left(\frac{2r_2}{r_0} \right) + \log_e \left(\frac{2r_1}{r_0} \right) \right].$$

The deflection is given by Equation 6.29 as $d\theta = (dc/dx) \, dt$. Since

$$c = 1 - \frac{2M}{(x^2 + z^2)^{1/2}}, \qquad \frac{dc}{dx} = \frac{2Mx}{r^3} \simeq \frac{2Mr_0}{r^3}.$$

Hence the total deflection between V (the position of a star) and E is

$$\Delta \theta = \int_0^{r_2} + \int_0^{r_1} \left(\frac{dc}{dx} \right) dt = \int_0^{r_2} + \int_0^{r_1} \frac{2Mr_0 \, dz}{(r_0^2 + z^2)^{3/2}}$$

$$\simeq \frac{4M}{r_0}.$$

In 1964 Irwin I. Shapiro of the Massachusetts Institute of Technology proposed an experiment to measure the slowing down of photons passing near the Sun by measuring the time-interval between the emission of a radar signal from the Earth and the reception of the echo from Venus or Mercury near superior conjunction (see the box above). Since photons passing near the Sun are slowed down, the radar echo is delayed. For Venus, the amount of the delay, calculated as explained in the box, is

$$\Delta t \simeq 4M \left[\log_e \left(\frac{2r_V}{R_\odot} \right) + \log_e \left(\frac{2r_E}{R_\odot} \right) \right], \tag{6.27}$$

where r_V denotes Venus' distance from the Sun, r_E the Earth's distance, and $R_\odot$ the Sun's radius. The Sun's mass M, measured in seconds, is $M = 1.5 \text{ km} \div 3 \times 10^5 \text{ km/sec} = 5 \times 10^{-6}$ sec. The Earth's orbital radius is about $215 \, R_\odot$; Venus' orbital radius is about $155 \, R_\odot$. Hence the predicted delay is

$$\Delta t \simeq 4 \times 10^{-6} (\log_e 430 + \log_e 310) \simeq 2.4 \times 10^{-4} \text{ sec.} \tag{6.28}$$

During this period a photon travels only 72 km; so, to measure the delay accurately, we would need to know the distance of a planet's reflecting surface to within a few kilometers. Before Shapiro and his colleagues at MIT's Lincoln Laboratory designed their experiment, planetary distances could not be calculated with an accuracy remotely approaching this requirement; so these scientists could not simply compare measured and calculated values of the round-trip times of radar pulses. Instead, they fitted radar observations taken over many positions of the Earth and Venus in their orbits to a very accurate theoretical description that allowed for the perturbing effects of other planets on the motions of the Earth and Venus, and the shape and reflecting properties of the Venusian surface. Complicated as it was, this description was strongly overdetermined by the data. The experiments yielded estimates of the radar-echo delays from both Venus and Mercury that strongly corroborated Einstein's prediction.

Earlier we used Huygens' (optical) principle to show that the angular deflection of a light ray in a medium where the speed of light varies from point to point is

$$\Delta\theta = \left(\frac{dc}{dx}\right)\Delta t. \qquad (6.29)$$

Here x is the direction indicated in the figure in Box 6.4. To calculate the total deflection, we sum the quantities $\Delta\theta$ given by Equation 6.29 along the (nearly straight) path of the light ray. The result of the calculation, explained in the box, is

$$\Delta\theta \simeq \frac{4M}{r_0}, \qquad (6.30)$$

where r_0 is the distance between the center of the Sun and the light ray. The deflection of a light ray that just grazes the edge of the Sun is $\Delta\theta \simeq 4 \times 1.5$ km $\div 7 \times 10^5$ km $= .85 \times 10^{-5}$ rad $\simeq$ 1.7 arc-sec. In 1919, at the instigation of Eddington, the British government sent expeditions to islands off the coast of Brazil and of West Central Africa to photograph stars near the Sun's edge during a total solar eclipse. The findings agreed with Einstein's prediction to within the experimental error of about 0.2 arc-sec. Data obtained during subsequent total eclipses of the Sun confirmed these early findings but did not improve their accuracy, because local variations in the density of the air near ground level introduce a random error of about 0.2 arc-sec.

At radio frequencies, atmospheric refraction is less serious. Directional measurements with radio interferometers can have errors substantially smaller than 0.1 arc-sec. The quasar 3C279 lies within a quarter of a degree of the Sun's path and is thus occulted by the Sun once a year (in October). By measuring the angular position of 3C279 relative to a nearby quasar, 3C273, radio astronomers have been able to measure the angular shift of 3C279 with an error of only 1 percent. The ratio between the measured deflection and Einstein's prediction is 1.015 ± 0.011.

The most spectacular confirmation of Einstein's prediction that light is bent by a strong gravitational field came in 1979 when Dennis Walsh of Manchester University discovered two quasars whose optical images are separated by only 6 arc-seconds (figure top left, p. 215). Walsh enlisted the aid of Robert Carswell and Ray Weymann of the Kitt Peak National Observatory, who obtained spectra of the two faint quasars and found them to be identical. The three scientists suggested in their discovery paper that both optical images are produced by the same physical object, whose light is bent by an intervening massive galaxy, acting as a "gravitational lens" (see figure top right, p. 215).

But why didn't the intervening galaxy show up on the photographs? Alan N. Stockton of the Institute for Astronomy in Honolulu surmised that its light was masked by one of the quasar images. To test this conjecture he electronically subtracted the bluer quasar image from the redder image, thereby unmasking the intervening massive galaxy (see bottom figure, p. 215).

Why do gravitational lenses produce multiple images? According to Equation 6.30, a light-ray passing a point-mass M at distance r is deflected through an angle $\Delta\theta \simeq 4M/r$. A light-ray passing through an extended spherically symmetric mass distribution at distance r from the center of the distribution is deflected by an amount

$$4M(r)/r$$

where $M(r)$ is the mass within radius (r). The most direct way to study the properties of a lens that bends light rays according to this rule is to build one and look through it. Marc V. Gorenstein of the Harvard-Smithsonian Center for Astrophysics has built plastic lenses that simulate the effects produced by extended galaxies as well as by simple point-masses. Viewed through these lenses, a distant object does indeed split up into replicas of itself, depending on the relative positions of the eye, the object and the lens. The observations of the double quasar and of two additional multiply imaged quasars that have since been found are in agreement with predictions based on Einstein's theory and plausible models of the intervening galaxy.

Finally, Einstein's theory predicts that Kepler's laws of planetary motion are not exact even for a single particle moving around a fixed point-mass. In the Newtonian approximation the orbit is an ellipse with the Sun at one focus. In the next approximation, the orbit depends on the *second-order* contribution to the time dilation [the term $2u^2 = 2(M/r)^2$ in Equation 6.24] and on the *first-order* contribution $2u = 2M/r$ to the space-warp factor $b(r)$ (see Equations 6.24 and 6.22). As we have seen, the orbit is determined by the condition that the spacetime trajectory is a straight line. Using this condition, Einstein found that the only closed orbits are circles. Noncircular orbits approximate ellipses when $M/r \ll 1$, but the axis of the ellipse rotates during each revolution through an angle

$$\Delta\theta = \frac{6\pi M}{a(1 - e^2)}, \tag{6.31}$$

where M is the mass of the Sun, a is the semimajor diameter of the orbit, and e its eccentricity.

This formula predicts that the major axis of Mercury's orbit rotates 43 arc-sec per century.

In his 1915 paper Einstein wrote that this value corresponds "exactly to astronomical observation (Leverrier). For the astronomers have discovered in the motion of the perihelion of this planet, after allowing for disturbances by other planets, an inexplicable remainder of this magnitude."

But the story of the discrepancy and its resolution merits a more detailed account, because it throws an interesting light on the role of "anomalies" in the growth of science.

In the *Principia* Newton proved that any small departure from the inverse-square law of gravitational attraction causes a planet's orbital axis to rotate. Thus the tidal acceleration of the Moon by the Sun causes the Moon's orbital axis to rotate in the direction of the Moon's motion with a period of about nine years. The mutual gravitational attraction of the planets causes their orbital axes to rotate too, though much more slowly. The observed rate of rotation is greatest for Mercury: 532 arc-sec (about a sixth of a degree) per century. In 1845 the French astronomer J. J. Leverrier completed an elaborate calculation showing that Newton's theory actually accounts for all but 35 arc-sec/century of the observed amount. In many branches of science, such a tiny discrepancy between the result of an extremely delicate measurement and a complex theoretical calculation would have been interpreted as a strong confirmation of the theory. But by the middle of the nineteenth century, celestial mechanics had reached such a high degree of precision that Leverrier and other astronomers could regard the discrepancy as real and important. Leverrier suggested that it might be produced by one or more small planets, as yet unobserved, whose orbits lay between Mercury and the Sun.

In 1882 Simon Newcomb recalculated the effects of planetary perturbations on Mercury's orbit and obtained an improved value of 43 arc-sec/century for the residual rate of rotation. Because all efforts to detect planets inside Mercury's orbit had failed, Newcomb conjectured that tiny particles close to the Earth's orbital plane might produce the required perturbation. The presence of such particles had already been inferred from observations of the zodiacal light—a faint glow in the night sky, presumably reflected sunlight, concentrated toward the plane of the Earth's orbit. But Newcomb was unable to construct a theoretical model that would account for both the zodiacal light and the residual rotation without producing additional effects (which had *not* been observed) in the motions of Mercury and Venus. Finally, in 1895, he suggested that perhaps Newton's inverse-square law was inexact at short distances.

But in the following year H. H. Seeliger published a theoretical model that successfully explained the zodiacal light and the residual rotation of Mercury's orbital axis, and had no unpleasant side-effects. Seeliger's model was widely accepted by astronomers. Newcomb himself withdrew his suggestion that Newton's theory was inexact at short distances. By 1915, when Einstein published his calculation, the residual rotation of Mercury's orbital axis had ceased to be an anomaly.

Einstein's prediction caused astronomers to revise their opinion of Seeliger's model. It now appeared artificial and unconvincing. Once Mercury's residual motion was accounted for, the zodiacal light could be explained by simpler and more plausible models. In a sense, Einstein's prediction created, as well as resolved, the anomaly.

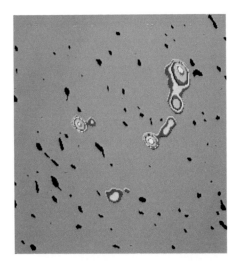

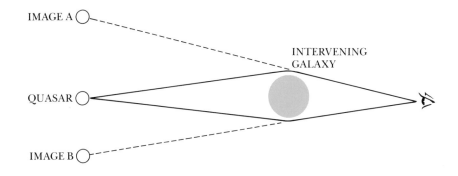

IMAGE A

INTERVENING
GALAXY

QUASAR

IMAGE B

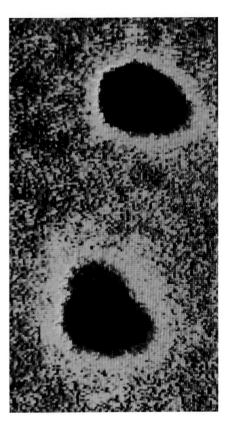

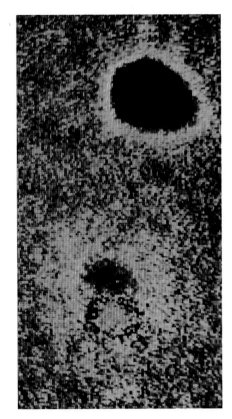

The twin quasars Q0957 + 561 A, B, which appear as multicolored ellipses lined up vertically in the center of the top left figure (a false-color computer-graphic radio map) have identical spectra whose lines are redshifted by identical amounts: $\Delta\lambda/\lambda = 1.4136$. The top right figure shows (schematically) how a massive object (a galaxy or galaxy cluster) lying directly between us and a quasar might produce two images. Careful ray-tracing based on Einstein's theory and a realistic model of the mass distribution in a massive galaxy confirms that compact images of nearly equal brightness can indeed be produced in this way. By electronically subtracting image A from image B, Alan Stockton produced the right-hand display in the bottom figure, in which the gravitational lens, an intervening galaxy previously hidden by image B, now shows up clearly.

Even with the advantage of hindsight it is difficult to fault the astronomical community for not recognizing the residual rotation of Mercury's orbital axis as a genuine anomaly. During the nineteenth century astronomers became aware of small but real discrepancies between the observed motions of the planet Uranus and Newton's theory. In 1846 Leverrier advanced the bold hypothesis that an as-yet-undiscovered planet beyond the orbit of Uranus was responsible for the discrepancies. Using Newton's theory he calculated the unknown planet's position in the sky and communicated his prediction to the astronomer J. G. Galle at the Berlin Observatory. Within 24 hours Galle found the new planet, Neptune, close to the position predicted by Leverrier. (Several months earlier the English mathematician and astronomer J. C. Adams had made nearly the same prediction, using a substantially different method of calculation, but he was less fortunate than Leverrier in persuading observational astronomers to check his prediction.) So astronomers were certainly justified in attributing the residual rotation of Mercury's orbital axis to unobserved matter rather than to a defect in Newton's theory.

The story of Mercury's orbit does not end with Einstein's 1915 calculation. The Sun is slightly flattened by its rotation. Newton's theory shows that the gravitational field of a spheroid departs slightly from the inverse-square law at small distances. Hence the Sun's flattened shape must contribute to the rotation of Mercury's orbital axis. Before we can assert that Einstein's theory explains the residual rotation, we must verify that the solar contribution is negligible.

From the measured rate of rotation at the surface (deduced, for example, from the observed motions of sunspots), we can predict how flattened the Sun would be if it were rotating like a solid body. Calculation shows that the Sun's polar diameter ought to be only 5 km shorter than its equatorial diameter, too little to be important for Mercury's orbit. But what if the interior of the Sun were rotating faster than its outer layers? In 1966 R. H. Dicke and M. Goldenberg carried out a delicate and ingenious experiment to test this possibility by measuring the oblateness of the solar disc photoelectrically. They found that the polar diameter was about 35 km shorter than the equatorial diameter. This degree of oblateness would account for about 10 percent of the residual precession and would spoil the agreement between Einstein's theory and observation.

In 1975, Henry Hill, a former student of Dicke's, carried out an improved version of the Dicke-Goldenberg experiment. He concluded that the oblateness of the Sun *is* consistent with solid-body rotation at the rate indicated by sunspots. But the measurements and the theoretical apparatus needed to interpret them are complex and delicate, and even today we cannot assert with complete confidence that the predicted rate of rotation of Mercury's orbital diameter agrees exactly with astronomical observations.

Neutron Stars and Black Holes

The strength of the gravitational field produced by a mass M at distance R is measured by the ratio M/R, M and R being measured in the same units. At the surface of the Sun, $M/R \simeq 2 \times 10^{-6}$; at the surface of a white dwarf, $M/R \simeq 2 \times 10^{-4}$. As we saw earlier, Einstein's correc-

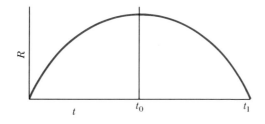

Radius of a cold, nonrotating star according to Newton's theory. The collapse begins at time t_0 when the star is at rest, and ends at time t_1 when it has shrunk to a point-mass of infinite density.

tions to Newton's theory are proportional to M/R. For nearly all kinds of astronomical objects, from planets to superclusters of galaxies, M/R is very small, and predictions based on Newton's theory are accordingly very accurate. The only known astronomical systems for which M/R is not small are *neutron stars* and *black holes*. Here Einstein's theory of gravitation comes into its own; Newton's theory is inadequate even as a first approximation.

White dwarfs, neutron stars, and black holes all represent endpoints of stellar evolution. Young stars derive their energy from thermonuclear reactions that convert hydrogen into helium in the deep interior. When a certain fraction of the hydrogen has been used up, the resulting helium core begins to contract. The star's subsequent evolution depends on whether its mass is less than or greater than a certain critical value known as the Chandrasekhar limit. If the mass lies below this value, the "degeneracy pressure" of the electrons halts the collapse of the helium core before its temperature becomes high enough to ignite thermonuclear reactions that convert helium into carbon.* Meanwhile, the outer layers of the evolving star have been more or less violently ejected. (Planetary nebulae are thought to have been formed in this way.) The helium core, surrounded by a more or less extensive hydrogen envelope, constitutes a white dwarf.

In more massive stars the helium core continues to contract until helium is ignited. The energy liberated by the burning of helium into carbon prevents the core from collapsing further—but only briefly. Once the helium has been consumed, the core resumes its contraction. The temperature increases, and further nuclear reactions occur until all the energy locked in the atomic nuclei has been released. The core is now pure iron, the ultimate nuclear ash, and there is nothing to impede its collapse until its density approaches that of the atomic nucleus itself. The resulting implosion triggers an immense explosion in which the outer layers of the

---

* According to classical physics, the pressure in a perfect gas is proportional to the temperature. According to quantum mechanics, however, the random motions of free electrons in a dense gas do not vanish at zero temperature, and the residual or "zero-point" motions increase with increasing density. In a white dwarf the pressure associated with these zero-point motions keeps the star from collapsing under its own weight. At very much higher densities—those that prevail in a neutron star—the zero-point motions of neutrons and protons play a similar role.

The Crab Nebula. Remains of Supernova of
A.D. 1054.

star are ejected with enormous velocities. These are the explosions that astronomers call *super-novae*.

The fate of the collapsing remnant depends on its mass. If the mass is less than about 2.5 solar masses, the pressure associated with the zero-point motions of the neutrons and protons is large enough to prevent the star from collapsing further under its own weight. Such objects, in which the density of matter equals or even exceeds that prevailing in the atomic nucleus, are called *neutron stars*. Their properties were first studied in the 1930s by J. R. Oppenheimer and G. M. Volkoff.

Degeneracy pressure cannot halt the collapse of a star whose mass exceeds about 2.5 solar masses. Such a freely collapsing object is called a *black hole*. Newton's theory predicts that the radius $R(t)$ of a uniform, uniformly collapsing, spherical star diminishes at an accelerated rate, as shown in the figure on page 217. Thus, according to Newton's theory, the radius of the collapsing star shrinks to zero in a finite time, and the gravitational potential M/R increases without limit.

Einstein's theory tells a different story. The structure of spacetime outside a black hole is described by Equations 6.22 and 6.23. The spacetime trajectory of a photon is given by the condition $d\tau = 0$, from which it follows that the speed of a photon is given by

$$c = \frac{d\ell}{dt} = \left(1 - \frac{M}{2r}\right) \Big/ \left(1 + \frac{M}{2r}\right)^3. \tag{6.32}$$

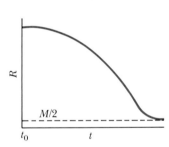

Schematic depiction of the collapse of a cold collapsing star (black hole) according to Einstein's theory.

Thus the speed of a photon diminishes as it approaches the center of a black hole, and vanishes at a finite value of the radial coordinate, $r = M/2$. This means that, for an external observer, a photon falling into a black hole would *never* reach the center; its radial coordinate would always exceed the finite value $M/2$. Since a material particle cannot outstrip a photon, the radius of a black hole must take infinite time to reach the value $r = M/2$. Moreover, photons emitted from the surface of a black hole undergo progressively larger redshifts during the collapse. The energy and frequency of a photon emitted or reflected from the surface approaches zero as the radius $R(t)$ approaches $M/2$. To an external observer, a black hole would at first appear to

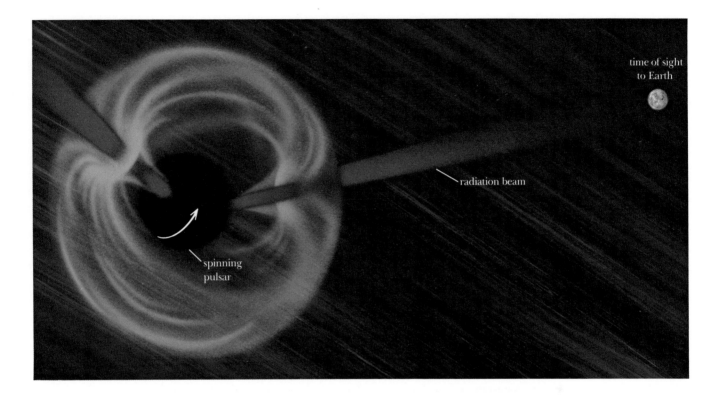

time of sight
to Earth

radiation beam

spinning
pulsar

We can observe only those pulsars whose beams of radiation (however they are caused) happen to cross the line of sight to Earth.

collapse at an accelerating rate. Then, as its radius approached the critical radius $M/2$, the radius would appear to decrease more and more slowly, asymptotically approaching the final value $M/2$.

Neutron stars and black holes have no internal sources of energy; hence they should cool very rapidly. And since they have such small surface areas—a few tens of square kilometers—we should expect them to be exceedingly faint. In fact, thermal radiation emitted by the surface of a neutron star or black hole has not yet been observed. Some neutron stars, however, are powerful emitters of *nonthermal* radiation. These are the *pulsars*, discovered in 1967 by Jocelyn Bell, a graduate student at Cambridge University. Bell had been examining records obtained with equipment designed by Anthony Hewish to record rapid scintillations of radio sources. Among the many records of randomly scintillating sources, she noticed one in which the peaks, though variable in intensity, occurred at precisely equal time-intervals. More accurate observations confirmed the precisely periodic character of the pulses, and a search of the scintillation records turned up two more sources with the same property. Observational and theoretical considerations indicate that pulsars are rapidly spinning neutron stars with extremely strong

Four stellar sources of X-ray emission, shown in false-color displays. *(Upper left)* X-ray emission from a region around Cygnus X-1, companion to a normal star in a binary system (see drawing below). *(Upper right)* Eta Carinae, optically a nebula of the sixth magnitude which in 1843 suddenly brightened by a factor of over 100. Embedded in the nebula are several young stars. *(Lower left)* CTB 109, a supernova remnant whose central star is a pulsar in a binary system. *(Lower right)* SS 433, one of the most remarkable stars known to astronomers. A member of a binary system, it emits two jets that move at speed comparable to the speed of light.

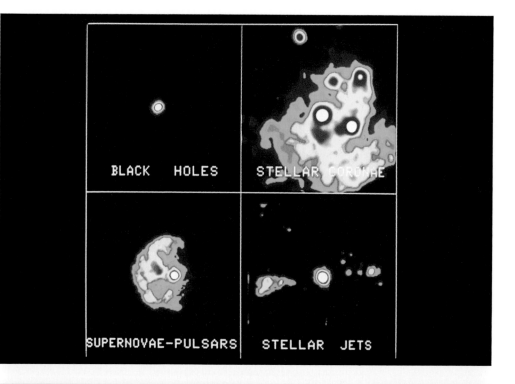

BLACK   HOLES

STELLAR   CORONAE

SUPERNOVAE-PULSARS

STELLAR   JETS

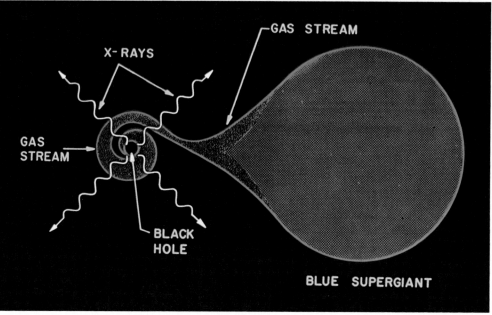

X-RAYS

GAS STREAM

GAS STREAM

BLACK HOLE

BLUE SUPERGIANT

Artist's conception of a black hole in a binary system. Material that flows from the black hole's companion is thought to form a disc around the black hole. This disc is thought to be the source of X-radiation from the system.

magnetic fields. The pulsed radiation is produced by beams emanating from "hot spots" on or near the surface of the spinning star. The precise mechanism of this radiation is still a puzzle.

A few neutron stars have been found to be members of close binary systems. These neutron stars, and no others, are also powerful X-ray sources. Now, in a close binary, one of whose members is a giant or supergiant and the other a dwarf, gas may flow from the distended atmosphere of the giant or supergiant into the gravitational field of the dwarf. Such gas streams in close binaries have long been observed spectroscopically and explained theoretically. If the compact member of the binary is a neutron star (or a black hole), the infalling gas from the other component can be accelerated to very high energies. Ultimately, the kinetic energy acquired by gas molecules falling into the collapsed star will be converted by collisions with other gas molecules into heat and radiation. Rough calculations show that the energy released in this way could easily account for the observed X-radiation from such binaries.

Spectroscopic observations yield information about the masses of stars in a binary system, as we saw in the first section of Chapter 4. In all cases but one, the data are consistent with the assumption that the compact object is a neutron star. The single exception is the X-ray source known as Cygnus X-1, for which the spectroscopic data indicate a mass of about ten solar masses. A compact object this massive could not be a neutron star; it *could* be a black hole. So far no convincing alternative interpretation of the data has been offered.

Black holes are to Einstein's general theory of relativity what ultrarelativistic particles are to his special theory. But whereas the realm of ultrarelativistic particles, high-energy physics, is full of strange phenomena of great importance to experimental physics and observational astronomy, the phenomena associated with black holes are merely strange. I think it likely that important cosmological consequences will eventually flow from the physics of black holes, but for the moment this branch of science is chiefly a playground for theorists. Must we conclude, then, that Einstein's theory of gravitation, though vastly superior to Newton's theory from a theoretical standpoint, tells us little more than Newton's theory about the actual universe? Not at all. In the next chapter we will see that Einstein's theory, unlike Newton's, forms the basis for a self-consistent theory of the universe as a whole; that this theory has many striking and testable predictions; and that it supplies a causal link between freely falling, nonrotating frames of reference and the cosmic distribution of mass and motion.

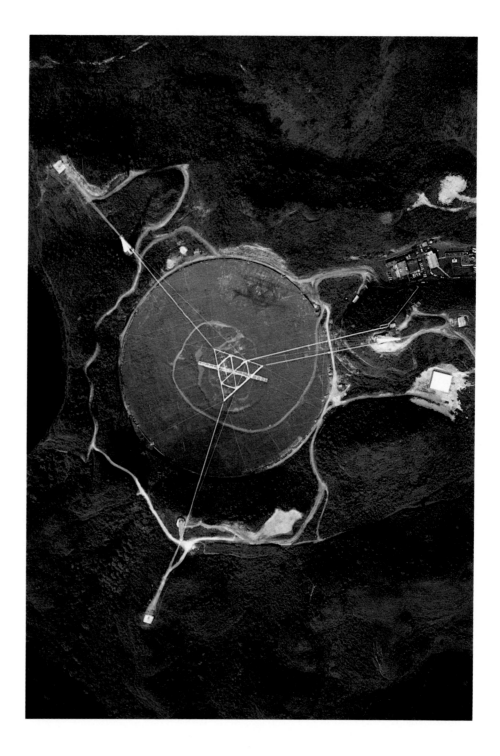

CHAPTER SEVEN

EINSTEIN'S THEORY
AND THE UNIVERSE

We can assert with certitude that the universe is all center, or that the center of the universe is everywhere and the circumference nowhere.
    GIORDANO BRUNO, *Della causa, principio ed uno, V*
    quoted by JORGE LUIS BORGES, in "The Fearful Sphere of Pascal"

In *The Science of Mechanics* Ernst Mach criticized the Newtonian concept of absolute space. A stone whirling at the end of a string experiences a centrifugal force, he said, not because it is accelerated relative to a fictitious absolute space, but because it is accelerated relative to the real universe. And because the acceleration is relative, we could equally well say that the universe is accelerated relative to the stone. The two modes of description must have identical observable consequences, Mach argued, because the distinction between them is merely verbal. Einstein set out to construct a theory of gravitation that not only would abolish the distinction between gravitational mass and inertial mass, but would also link the cosmic distribution of mass and motion to local unaccelerated frames of reference. In Chapter 6 we saw how he accomplished the first objective. In this chapter we will see how Einstein and the Russian mathematician Alexander Friedmann succeeded in constructing a theory of cosmic space and time that, paradoxically, vindicates the deepest insights of both Newton and Mach. In this theory, time, rest, uniform motion, and accelerated motion are no less absolute than in Newton's theory; yet their definitions are tied to the cosmic distribution of mass and motion.

The Cosmological Principle

In a paper entitled "Cosmological Considerations on the General Theory of Relativity," published in 1917, Einstein postulated that *no average property of the cosmic medium defines a preferred place or a preferred direction in space.* Einstein called this assumption the Cosmological Principle.

The Cosmological Principle gives precise expression to a view of the universe held by Newton and Huygens, and, before them, by Giordano Bruno, Lucretius, and the Greek atomists of the fifth century B.C. Einstein saw that he could use this postulate to solve Mach's problem, that is, to show that the cosmic distribution of mass and motion determines local unaccelerated reference frames. We saw in Chapter 6 that Einstein's theory of gravitation makes the structure of spacetime and the distribution of energy (or mass) and momentum mutually dependent; but

(Left) The 1,000-foot radio telescope at Arecibo, Puerto Rico.

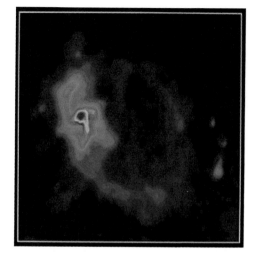

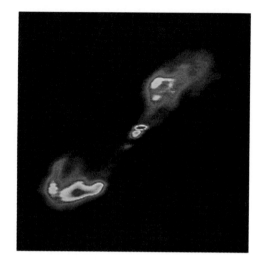

Above left, Sagittarius A. Radio emission from the nucleus of our Galaxy. The red spot is a point source known to be 1000 times smaller than the spot in this picture; it is thought to be the real nucleus of the Galaxy. The spiral shaped region centered on the point source is thermal emission and is a type of structure never seen before. The fainter blue shell is nonthermal and may be a supernova remnant. Above right, Centaurus A (NGC5128). Centaurus A, at a distance of 5 Mpc, is the closest powerful radio galaxy. The general radio jet and the knots within the jet line up remarkably well with the x-ray jet recently found from Einstein Observatory observations. The structure shown here is usually referred to as the "inner-lobes" to differentiate it from the much larger structure seen further out. Both of the observations shown above were made at the Very Large Array (shown below) on the Plains of San Augustín, 50 miles west of Socorro, New Mexico.

it does not guarantee that the cosmic distribution of energy and momentum *determines* the structure of spacetime. Consider, for example, a universe consisting of a single point-mass. A test particle orbiting the point-mass describes a slowly rotating Keplerian ellipse, as discussed in Chapter 6. Rotating with respect to what? Einstein's theory cannot answer this question any more than Newton's theory can: an isolated point-mass cannot by itself define a nonrotating frame of reference. But a cosmic distribution of energy and momentum satisfying the Cosmological Principle both defines and—as we will see at the end of this chapter—determines a standard of rest in the neighborhood of every event. This local standard of rest is functionally indistinguishable from Newton's absolute space.

Cosmologists disagree about the status of the Cosmological Principle. Some regard it as an *approximate description* of the universe, others as an *exact prescription*. This difference reflects a more fundamental disagreement about the relation between cosmology and physics. Physicists distinguish between *laws*, which express the permanent mathematical regularities underlying natural phenomena, and *initial conditions*, which express the contingent aspect of natural phenomena. The nearly elliptical orbits of the planets are necessary features of the architecture of the solar system, because elliptical orbits are a consequence of Newton's law of gravitation. The eccentricities and inclinations of the orbits are contingent features of the architecture. Regularities in such contingent features (the fact that the planetary orbits are nearly circular and nearly coplanar, for example) must be explained by evolutionary processes, not attributed to initial conditions. Analogously, some cosmologists argue, the large-scale uniformity and isotopy of the part of the universe we have so far been able to explore needs to be explained by an evolutionary process, not attributed to an initial condition.

Against this view, we could argue that the conditions that define the universe as a whole differ in kind from those that define ordinary astronomical systems. The properties that define an individual star, such as the Sun, are obviously contingent. We do not ask why the Sun's mass is 1.991×10^{33} gm rather than 1.992×10^{33} gm. The statistical properties of the stellar population that contains the Sun may be less contingent. Although we do not yet have a theory that predicts what fraction of the universe condensed into stars with masses between 0.5 and 1.5 solar masses, this is just the sort of number that a good cosmological theory *ought* to predict.

It seems reasonable to suppose that the universe as a whole is characterized by a definite set of statistical properties, some of which are predictable, at least in principle, and the rest of which must be considered to be axioms, like the laws themselves. I take the Cosmological Principle, which asserts that the universe itself has the same spatial symmetries as the laws that govern its structure and evolution, to be one of the axioms. This admittedly is a strong assumption, but the alternative is not merely unappealing but, to me, unintelligible. If we regard the Cosmological Principle as an approximate description of the part of the universe we have so far been able to observe, what shall we say about the universe as a whole? To refuse to make any assumption at all is to abandon any attempt to interpret cosmological observations or to understand the connection between locally unaccelerated frames of reference and distant matter. To assume that the Cosmological Principle is indeed valid, but only as an approximation, is to assert the

An image-processed optical photo of M82, an active galaxy.

existence of preferred places and directions. That assertion would be a much stronger assumption than the Cosmological Principle, and one for which no supporting evidence has ever been adduced.

Finally, Einstein's theory of gravitation, unlike Newton's, is not just a theory of gravitational interaction. It is also a theory of spacetime, and thus, inevitably, of the universe as a whole. Now, mathematics may be defined as the study of logically possible worlds, but physics, and cosmology as a branch of physics, treat the universe we happen to live in. The Cosmological Principle, or some other postulate of equal power, is an essential component of the theoretical structure that Einstein set out to build. Up until now, no adequate substitute for it has been proposed.

Implications of the Cosmological Principle

What does the Cosmological Principle imply about the structure of spacetime? Because no average property of the cosmic mass distribution defines a preferred place or direction in space, it must be possible to assign space and time coordinates to events in such a way that the average mass density $\bar{\rho}$ depends only on the time coordinate t, and that Equation 6.18, which expresses the interval of proper time between two neighboring events in terms of the coordinate differences and the ten gravitational potentials, does not discriminate between points in space or between directions at a given point.

The simplest formula for the spacetime interval that meets this requirement is Minkowski's formula

$$d\tau^2 = dt^2 - (dx^2 + dy^2 + dz^2). \tag{7.1}$$

However, Einstein's field equations (the differential equations that connect the ten gravitational potentials with the ten quantities that describe the distribution of energy and momentum) show that the mass density in Minkowskian spacetime vanishes identically. Because the average mass density, though small, is finite, cosmic spacetime cannot be Minkowskian. That is, special relativity does not correctly describe a uniform, unbounded distribution of gravitating particles.

Can we find a generalization of Minkowskian spacetime that is consistent with both Einstein's field equations and the Cosmological Principle? In 1917 Einstein published a cosmological theory based on the bold and novel hypothesis that physical space, though uniform and unbounded, is finite—the three-dimensional analog of the two-dimensional surface of a sphere. (We will explore this idea in greater depth in the next section.) Disappointingly, this hypothesis did not, by itself, resolve the abovementioned difficulty: the field equations still had no solution for any finite value of the cosmic mass density. Einstein now took a step that he later described as the greatest mistake of his scientific career. By adding an extra term to his field equations, he succeeded in making them consistent with a finite mass density. Five years later, Alexander Friedmann, a theoretical physicist specializing in dynamical meteorology, thought of a more radical—and, at the same time, more conservative—way to reconcile Einstein's field equations with the Cosmological Principle and the requirement of finite mass density. He showed that the field equations *in their original form* are consistent with the Cosmological Principle and a finite value of the cosmic mass density, *provided that space is not static*. This was a truly innovative idea. Einstein himself did not immediately accept it, and it seems to have escaped the notice of astronomers until Hubble formulated the velocity-distance relation in 1929, seven years after the publication of Friedmann's paper, which has predicted just such a relation.

In the simplest kind of Friedmannian universe, the spacetime interval is given by the formula

$$d\tau^2 = dt^2 - a^2(t)(dx^2 + dy^2 + dz^2), \tag{7.2}$$

where $a(t)$ is some as yet unspecified function of the time. What kind of spacetime does Equation 7.2 represent?

Consider first the behavior of stationary clocks at different points in space. For neighboring events in the history of a stationary clock, $\Delta x = \Delta y = \Delta z = 0$, so that $\Delta\tau = \Delta t$: the interval of proper time between two events in the history of a stationary clock is equal to the coordinate

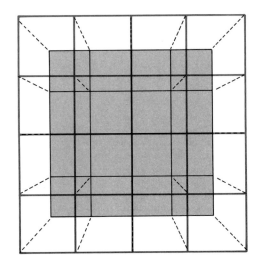

An expanding Cartesian coordinate grid.

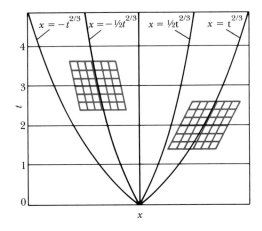

Two-dimensional section of a four-dimensional cosmic spacetime grid. The curves $x = x_0 t^{2/3}$ represent the histories of particles locally at rest in the expanding space. The distance between any two such particles is proportional to the scale factor $a(t) = (t/t_0)^{2/3}$ The colored grids are locally Minkowskian spacetime grids.

time-interval. Thus identical clocks run at identical rates at different points, as the Cosmological Principle requires.

Next, consider the structure of space implied by Equation 7.2. The proper distance ds between neighboring, simultaneous events is given by the formula

$$ds^2 = -d\tau^2 = a^2(t)(dx^2 + dy^2 + dz^2). \tag{7.3}$$

This formula says: to calculate the distance ds between two neighboring points (x, y, z) and $(x + dx, y + dy, z + dz)$, multiply the "coordinate distance," as given by Pythagoras' formula, by the scale factor $a(t)$. Because the distance between neighboring points is given by Pythagoras' formula, the geometry of space is Euclidean. But the distance between any two given points is proportional to the scale factor $a(t)$. If the scale factor $a(t)$ increases with time, Equation 7.3 describes an expanding Cartesian coordinate grid, as illustrated above.

The figure above shows a two-dimensional section of the four-dimensional spacetime grid described by Equation 7.2. The histories of grid points are represented by curves $x = a(t)x_0$ for different values of the constant x_0. The cosmic view must be the same from every grid point; so we may identify any of these curves with the t-axis. Simultaneous events are connected by horizontal lines parallel to the x-axis.

Does Equation 7.2 describe the spacetime structure of a possible universe? We can answer this question, and thereby discover how the scale factor $a(t)$ is related to the mean cosmic mass density $\bar{\rho}(t)$, by the following argument, which avoids actually using Einstein's field equations.

Let us consider an idealized fluid of uniform mass density $\bar{\rho}(t)$, and zero pressure. Consider a small spherical region whose radius is a fixed multiple of the scale factor $a(t)$, so that its mass remains constant as the scale factor changes. What is the acceleration of a fluid particle at the surface of this spherical region? If Newton's theory applied to a uniform, unbounded distribution of mass, we could use Newton's theorem on spherically symmetric mass distributions (see p. 91 in Chapter 3) to answer this question. Although Newton's theory itself cannot be applied in a self-consistent way to a uniform, unbounded distribution of mass, his theorem on spherically symmetric mass distributions holds in Einstein's theory—a fact that provides another illustration of the power of arguments based on symmetry. Thus, *the gravitational field inside any spherical region of a uniform, unbounded fluid is the same as it would be if the region were isolated in empty space.*

Let M and R denote the mass and the radius of our spherical region. We can use Newton's theory to describe the region's behavior if the ratio M/R, with M measured in the same units as R, is much smaller than unity. But M/R is proportional to R^2; so we can make the ratio as small as we please by choosing a sufficiently small value for R. Because Newton's theory is exact in the limit $M/R \to 0$, the resulting equation for the scale factor $a(t)$ is also exact.

The Newtonian theory of a uniform and uniformly expanding (or contracting) self-gravitating sphere is very simple. The radial coordinate of any fluid element is proportional to the scale factor $a(t)$:

$$r(t) = r_0 a(t), \tag{7.4}$$

where r_0 is a constant. Because the scale factor $a(t)$ represents the distance between any two fluid elements, we are free to prescribe its value at some particular moment. We will find it convenient to set the value of the scale factor that corresponds to the present epoch t_0 equal to unity:

$$a(t_0) = 1. \tag{7.5}$$

The constant r_0 in Equation 7.4 then represents the present value of a particle's radial coordinate.

The gravitational acceleration at radius r is $M(r)/r^2$, where $M(r)$ denotes the mass interior to radius r. Thus

$$\frac{d^2 r}{dt^2} = -\frac{M(r)}{r^2}. \tag{7.6}$$

Substituting Equation 7.4 for r in Equation 7.6 and dividing the resulting equation by r_0^3, we obtain an equation that relates the second derivative of the scale factor to the mass density $\bar{\rho}$:

$$\frac{d^2a}{dt^2} = -\frac{4\pi}{3}\,\bar{\rho}\,a \equiv -\frac{\mu}{a^2}, \tag{7.7}$$

where the constant μ denotes the mass within a sphere of radius $a(t)$. Thus the scale factor $a(t)$ satisfies the same equation as the radial coordinate of a particle falling directly toward, or receding directly away from, a fixed mass μ (see the figure and text discussion on p. 217).

To solve Equation 7.7, we begin by multiplying both sides by the first derivative of $a(t)$:

$$\frac{da}{dt}\frac{d^2a}{dt^2} = -\frac{\mu}{a^2}\cdot\frac{da}{dt}$$

or

$$\frac{1}{2}\frac{d}{dt}\left(\frac{da}{dt}\right)^2 = \mu\frac{d}{dt}\left(\frac{1}{a}\right),$$

whence

$$\frac{1}{2}\left(\frac{da}{dt}\right)^2 - \frac{\mu}{a} = E, \tag{7.8}$$

where E is a constant. Er^2 is the Newtonian energy per unit mass of a fluid element at the surface of an expanding (or contracting) sphere of radius $ra(t)$, and $\mu = \frac{1}{3}\pi\bar{\rho}a^3 = \frac{1}{3}\pi\bar{\rho}_0$.

Equation 7.8 describes a particle falling toward, or receding from, a fixed point-mass. At some past or future moment, the particle coincides with the point-mass. Let us choose this moment as the origin of time, so that $a(0) = 0$. The solution of Equation 7.8 is then completely determined by the value of the energy parameter E. Its behavior, illustrated in the figure on the facing page, depends on the sign of E: when E is negative, $a(t)$ represents the radial coordinate of a gravitationally bound particle, which recedes to a maximum distance $-\mu/E$ and then returns to the origin. When E is positive, $a(t)$ represents the radial coordinate of a particle with positive energy; at large distances from the origin, the velocity of the particle approaches the constant value $(2E)^{1/2}$.

The solution for $E = 0$ represents the radial coordinate of a particle with just enough energy to escape. As the particle's radial coordinate a increases, its velocity da/dt approaches zero.

By virtue of the theorem on spherically symmetric mass distributions, Einstein's field equations must give a relation between the scale factor a and the mass density $\bar{\rho}$ that is equivalent to

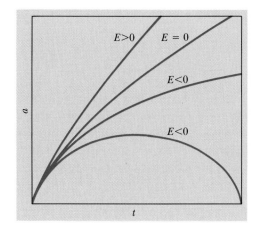

The cosmic scale factor a plotted against time t.
For $E = 0$, $a \propto t^{2/3}$. For $E > 0$, the curve is asymptotic to a straight line, $a \propto t$. The curves for $E < 0$ are cycloids with mirror symmetry about their highest points.

Equation 7.8. But Einstein's field equations must specify a *unique* connection between the scale factor and the mass density when the structure of spacetime is described by Equation 7.2. It follows that the energy parameter E must vanish, because if E were not zero, its value would define a characteristic time and a characteristic distance: the time at which the scale factor takes the value $\mu/|E|$ and the distance that light travels in this time. Because no characteristic time or length appears in Equation 7.2, E must vanish.

Equation 7.8, with $E = 0$, has the following solution, satisfying the additional requirements $a(t_0) = 1$ and $a(0) = 0$:

$$a(t) = \left(\frac{t}{t_0}\right)^{2/3}. \tag{7.9}$$

This solution was used to construct the curve labeled $E = 0$ in the above figure.

By substituting Equation 7.9 into Equation 7.8 and using the definition 7.7 of the constant μ, we obtain an important relation between the mean density $\bar{\rho}$ and the epoch:

$$6\pi\bar{\rho}t^2 = 1, \tag{7.10}$$

where the mass density $\bar{\rho}$ is expressed in natural units.

Finally, the rate of expansion is related to the mean density. Differentiating Equation 7.4, we obtain

$$\frac{dr}{dt} = r_0\frac{da}{dt} = r_0 a \cdot a^{-1}\left(\frac{da}{dt}\right)$$
$$= rH, \tag{7.11}$$

where

$$H = \left(\frac{da}{dt}\right) a^{-1} = \frac{2}{3t}. \tag{7.12}$$

Equation 7.11 states that, at any moment, the relative velocity of two grid points or two fluid particles is proportional to their separation; Equation 7.12 expresses the factor of proportionality, the Hubble parameter H, in terms of the scale factor a. Dividing Equation 7.8 with $E = 0$ by a^2, and using the definition of the constant μ (see Equation 7.7), we obtain the desired relation between the instantaneous values of the Hubble parameter and the cosmic density:

$$H^2 = \frac{8\pi}{3}\bar{\rho}. \tag{7.13}$$

Equations 7.9 to 7.13 contain a complete description of the expanding cosmic medium and the expanding coordinate grid.

The Curvature of Cosmic Space

Euclidean space satisfies the Cosmological Principle: the properties of triangles and other geometric figures are everywhere the same, and nowhere serve to define a preferred direction. Is the converse true? Is a space whose geometric properties do not discriminate between different places or between different directions at a given place necessarily Euclidean?

Let us first consider this question as it applies to two-dimensional spaces. The Euclidean plane is uniform and isotropic: a triangle, or any other figure drawn on a Euclidean plane, will fit anywhere (on the plane) and in any orientation. But the same thing is true of triangles drawn on the surface of a sphere; so the surfaces of spheres are also uniform, isotropic spaces. Are there any other surfaces on which any geometric figure can move with perfect freedom?

Intuition tells us that the answer is no, and mathematicians have indeed proved that no such surface can be constructed in Euclidean space. It *is* possible, however, to construct a self-consistent mathematical description of a two-dimensional space, called the *non-Euclidean* or *hyperbolic plane,* that is neither a Euclidean plane nor the surface of a Euclidean sphere, but on which any geometric figure drawn in one place fits everywhere and in all orientations. Like the Euclidean plane, this space is infinite. Its geometry satisfies all of Euclid's axioms except the axiom of parallels: instead of just one parallel to a line through any point not on the line, as in the Euclidean plane, there are infinitely many.

The invention of non-Euclidean geometry early in the nineteenth century (by Janos Bolyai in Hungary, Carl Friedrich Gauss in Germany, and Nikolai Lobachevsky in Russia) was an

The three inventors of non-Euclidean geometry: Karl Friedrich Gauss (1777–1855), Nikolai Lobachevsky (1793–1856), and Janos Bolyai (1802–1860). Lobachevsky published a detailed account of non-Euclidean geometry in 1829, in a Russian journal. It was not well received by other Russian mathematicians and was largely ignored by non-Russian mathematicians. Janos Bolyai's account appeared two years later, as an appendix to a book on geometry by his father, Wolfgang, a well-known mathematician and a friend of Gauss, the leading mathematician of his day. The elder Bolyai sent a copy of the book to Gauss, who wrote back that he could not praise Janos Bolyai's appendix because "to praise it would amount to praising myself; for the entire content, the path which your son has taken, the results to which he is lead, coincide almost exactly with my own meditations which have occupied my mind for from thirty to thirty-five years." Gauss however, prudently refrained from publishing his meditations on non-Euclidean geometry; they were found among his papers after his death.

important event in the history of Western philosophy, as well as in the history of mathematics. Like Plato and Archimedes, most mathematicians had believed that the axioms, and hence the theorems, of Euclidean geometry are truths about physical space. At the end of the eighteenth century, Immanuel Kant put forward a somewhat different view. He affirmed that space as we perceive it necessarily obeys Euclid's axioms, but he denied that the external world is really extended in space. Our perceptual apparatus imposes a spatial order on external reality; and Euclidean geometry, Kant argued, is built into the human perceptual apparatus. The invention of non-Euclidean geometry showed that the axioms of Euclidean geometry are not necessarily true, in either the Platonic sense or the Kantian sense. They are merely self-consistent, and so are the axioms of non-Euclidean geometry. The statement "The geometry of physical space is X" would nowadays be interpreted to mean "A cosmological theory embodying geometry X gives a more strongly overdetermined account of the observational evidence than do theories embodying alternative geometries."

There are three-dimensional analogs to both spherical and non-Euclidean geometry. The three-dimensional analog of the surface of a sphere is a three-dimensional space that is uniform and isotropic but has finite volume (just as the surface of a sphere has finite area). As on the surface of a sphere, the sum of the angles of a triangle in this space is always greater than 180°. The three-dimensional analog of the hyperbolic plane is uniform, isotropic, and infinite. In this space the sum of the angles of any triangle is less than 180°.

Algebra provides the most direct route to an understanding of non-Euclidean spaces. We may define the surface of a sphere as the set of points whose coordinates x, y, z satisfy the equation

$$x^2 + y^2 + z^2 = R^2, \tag{7.14}$$

where R is the radius of the sphere. Analogously, the three-dimensional surface of a sphere in four-dimensional Euclidean space is populated by points whose coordinates w, x, y, z satisfy the equation

$$w^2 + x^2 + y^2 + z^2 = R^2. \tag{7.15}$$

The geometry of the spherical surface defined by Equation 7.14 is determined by the formula that tells us how to calculate the distance between neighboring points. This is just Pythagoras' formula,

$$d\ell^2 = dx^2 + dy^2 + dz^2. \tag{7.16}$$

Analogously, the geometry of the three-dimensional space defined by Equation 7.15 is determined by the formula that tells us how to calculate the distance between neighboring points in this space. Since we have assumed that the four-dimensional space populated by points whose coordinates are w, x, y, z is Euclidean, the distance between neighboring points is given by the four-dimensional analog of Pythagoras' formula:

$$d\ell^2 = dw^2 + dx^2 + dy^2 + dz^2. \tag{7.17}$$

Is there any other way to construct a surface or a space whose geometric properties do not distinguish between places or between directions at a given place? There is in fact just one more way, discovered around the middle of the eighteenth century by Johann Lambert (1728–1777). Suppose we replace the radius R in Equation 7.14 or 7.15 by a *pure imaginary number iR*, where i stands for the imaginary unit $\sqrt{-1}$. Then the quantity R^2 in Equations 7.14 and 7.15 becomes $-R^2$. But a sum of squares cannot be negative. To avoid this difficulty, we change the sign of one of the terms on the left side, replacing 7.14 by

$$x^2 + y^2 - w^2 = -R^2 \tag{7.18}$$

and 7.15 by

$$x^2 + y^2 + z^2 - w^2 = -R^2. \tag{7.19}$$

These equations may look familiar. If we transform the equation of a circle in the same way, we obtain

$$x^2 - w^2 = -R^2. \tag{7.20}$$

The corresponding equations for the distance between two neighboring points are

$$d\ell^2 = dx^2 + dy^2 - dw^2, \tag{7.21}$$
$$d\ell^2 = dx^2 + dy^2 + dz^2 - dw^2, \tag{7.22}$$
$$d\ell^2 = dx^2 - dw^2. \tag{7.23}$$

It should now be clear that the "embedding space"—that is, the space of dimension $n + 1$ in which we embed the space of dimension n we are really interested in—is mathematically identical with the space that Minkowski invented to formalize Einstein's special theory of relativity! The coordinate w in Equations 7.21 to 7.23 is the analog of the time coordinate; Equation 7.20 has the same form as the equation $t^2 - x^2 = k^2$, which is the locus of points $t' = k, x' = 0$ (see page 168); and Equations 7.18 and 7.19 have similar interpretations in Minkowskian spacetimes of three and four dimensions.

Of course, w is not really a time coordinate. It is a coordinate in a fictitious "embedding space," and it is on a par with the other coordinates in this space. (Remember the symmetry between x and t in Minkowski spacetime diagrams!) Nor is there any *physical* connection between special relativity and the geometry of hyperbolic spaces. It just happens that the geometry of a calibration surface* in Minkowskian spacetime is identical with the geometry of hyperbolic space.

In three-dimensional Euclidean space, rotations of the coordinate axes about a fixed point O do not change the form of the equation that defines the surface of a sphere centered on O. The analogous coordinate transformations in the Minkowskian embedding space, defined by Equation 7.21, are *Lorentz transformations*, which take an equation, such as Equation 7.18, that defines a calibration surface into an equation of the same form but with primed variables in place of the unprimed variables.

The three-dimensional space defined by Equations 7.15 and 7.17 is said to have *positive curvature*, given by $1/R^2$. The three-dimensional space defined by Equations 7.19 and 7.22 is

---

* Such a surface is defined as the locus of events separated from the origin by a fixed interval of proper time or proper distance; the former condition defines a timelike "sphere"; the latter, a spacelike "sphere."

Left, radio picture of radio galaxy 3C430. Right, radio picture of jet of galaxy NGC 6251.

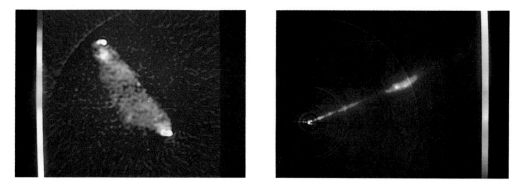

said to have *negative curvature,* given by $-1/R^2$. Euclidean space has zero curvature. It is the common limiting form of spaces of positive and negative curvature as the radius of curvature R increases without limit.

The Cosmological Principle is consistent with the possibility that space is curved. We can allow for this possibility by replacing Equation 7.2 for the interval of proper time between neighboring events in spacetime by

$$d\tau^2 = dt^2 - a^2(t)\,d\ell^2, \tag{7.24}$$

where $d\ell$ is the distance between neighboring points in a curved space. Now, we saw earlier that the scale factor $a(t)$ satisfies an equation that contains an undetermined parameter E. We also saw that if space is Euclidean we must set $E = 0$. In 1922 Friedmann showed that if the structure of spacetime is given by Equation 7.24, then the scale factor satisfies Equation 7.8, and the energy parameter E and the curvature $\pm 1/R^2$ satisfy the relation

$$E = -\frac{1}{2R^2}. \tag{7.25}$$

Thus the curvature of space and the energy parameter have opposite signs.

If the energy parameter is negative, space has positive curvature and hence finite volume; and in this case, as we have seen, the scale factor $a(t)$ increases from zero to a finite maximum and then decreases back to zero. *If space is finite, the duration of an expansion–contraction cycle is also finite.* If the energy parameter is positive (or zero), space has negative (or zero) curvature and infinite volume, and it expands forever.

These are the theoretical possibilities. We will return to the practical problem of deciding between them in Chapter 8.

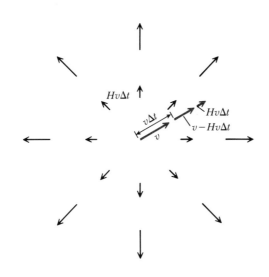

The speed of a skater gliding on an expanding sheet of ice must continually diminish.

Free Particles in an Expanding Universe

Newton's first law of motion, which states that the momentum of a free particle is constant in time, is not valid for free particles in an expanding universe of uniform density. Instead, the momentum of every free particle decreases like the reciprocal of the scale factor $a(t)$:

$$\mathbf{p} \propto \frac{1}{a(t)}. \tag{7.26}$$

According to this law, a free particle does not change its heading, but continually slows down, unless it is moving with the speed of light. Photons and other particles that have zero rest mass must always move with the speed of light, but in an expanding universe their energy and momentum continually decrease.

A free particle cannot change its direction of motion because the cosmic mass distribution and the cosmic expansion nowhere define a preferred direction. To see why a free particle slows down, imagine a skater gliding without friction on a uniformly expanding sheet of ice. Relative to the stationary floor that supports the ice sheet, her velocity must be constant, if we neglect friction. Hence, as shown in the figure above, her speed relative to the ice beneath her skates must continually diminish. Let $v(t)$ denote her speed at time t. During a time interval Δt, she moves a distance $v(t)\Delta t$, arriving at a point that is receding from her position at time t with speed $H(t)v\Delta t$, as we see from the velocity-distance relation (7.11). Thus her speed at time $t + \Delta t$, relative to the ice beneath her skates, is

$$v(t + \Delta t) = v(t) - H(t)v\Delta t. \tag{7.27}$$

Rearranging this equation and replacing H by its definition in terms of the scale factor a, $H = (da/dt)/a$, we obtain

$$\frac{v(t + \Delta t) - v(t)}{v(t)} = -\frac{\left(\frac{da}{dt}\right)\Delta t}{a(t)}$$

or

$$\frac{dv}{v} = -\frac{da}{a}, \tag{7.28}$$

which implies that

$$v \propto \frac{1}{a}. \tag{7.29}$$

That is, the velocity of a skater on an expanding rink decreases in inverse proportion to the scale factor.

To describe the slowing down of an ultrarelativistic skater, we need to replace Equation 7.27 by Einstein's law for the addition of velocities:

$$v(t + \Delta t) = \frac{v(t) - H(t)v\Delta t}{1 - vHv\Delta t}. \tag{7.30}$$

After some rearrangement and simplification, we obtain, in place of Equation 7.28,

$$\frac{d(\gamma v)}{\gamma v} = -\frac{da}{a}, \tag{7.31}$$

where, as in Chapter 5, $\gamma = (1 - v^2)^{-1/2}$. In Chapter 5 we saw that the momentum of a particle moving at any speed is related to its velocity by the formula

$$\mathbf{p} = \gamma m_0 \mathbf{v} = E\mathbf{v}, \tag{7.32}$$

where m_0 is the particle's rest mass and E is its energy. From this formula and Equation 7.31 we obtain, finally,

$$\mathbf{p} \propto \frac{1}{a}. \tag{7.33}$$

That is, the relativistic momentum decreases in inverse proportion to the scale factor. Equation 7.33 includes the nonrelativistic formula 7.29 as a limiting case.

Although the preceding derivation of the law of decreasing momentum does not apply to photons, the law itself does. According to Einstein's photon hypothesis the energy, the momentum, and the frequency of a photon are proportional to one another: $E = p = h\nu$. Hence the frequency of a photon in the expanding universe continually diminishes, and its wavelength, which is equal to the reciprocal of its frequency, measured in natural units, continually increases.

The cosmological redshift is a direct consequence of the diminishing energy and momentum of photons in an expanding space. The redshift is simply related to the change in the scale factor $a(t)$ between the emission and reception of a photon. Let t_1 denote the moment of emission of a photon, t_0 the present epoch; and let us use the labels 1, 0 to distinguish quantities evaluated at these two times: $a(t_1) \equiv a_1$, $\lambda(t_0) \equiv \lambda_0$, and so on. Since the momentum of a photon is inversely proportional both to its wavelength and to the cosmic scale factor,

$$\frac{\lambda_0}{\lambda_1} = \frac{a_0}{a_1}. \tag{7.34}$$

The (fractional) redshift z is defined by

$$z \equiv \frac{\lambda_0 - \lambda_1}{\lambda_1}. \tag{7.35}$$

Earlier we decided to set the present value of the scale factor equal to unity. Hence a photon's redshift z and the cosmic scale factor a_1 corresponding to the moment at which it was emitted satisfy the simple relation

$$1 + z = \frac{1}{a_1}. \tag{7.36}$$

As we look deeper into the past, we see light emitted at progressively smaller values of the scale factor a. Since $a(0) = 0$, photons present at the very beginning of the cosmic expansion would have infinite redshifts. The figure at the top of the next page shows the relations between redshift and scale factor, and (under the assumption that space is Euclidean) between redshift and epoch of emission.

Knowing how free particles behave in an expanding space, we can predict the behavior of an ideal cosmic medium composed of noninteracting particles uniformly distributed in space. Consider first a gas composed of particles moving with speeds much less than the speed of light. The temperature of such a gas is proportional to the average kinetic energy per particle. A slowly moving particle's kinetic energy is proportional, in turn, to the square of its momentum,

$$K = \frac{1}{2}mv^2 = \frac{p^2}{2m}.$$

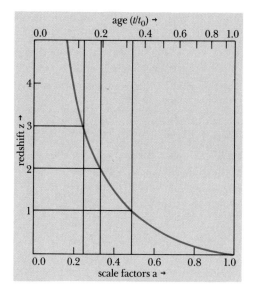

Redshift z of light from a distant source plotted against the scale factor a at the epoch when the light was emitted and against the age of the universe, in units of the present epoch $t_0 \simeq 10^{10}$ years. Light with redshift $z = \Delta\lambda/\lambda = 2$ was emitted when the universe was about ⅕ its present age; light with redshift $z = 3$, when the universe was somewhat more than ¹/₁₀ its present age. Quasars have been observed with redshifts as high as $z \simeq 3.5$.

But the momentum of a free particle varies like the reciprocal of the scale factor. Hence, *the temperature of a nonrelativistic gas decreases like the inverse square of the cosmic scale factor.*

Next, consider a gas composed of ultrarelativistic particles. The kinetic energy, total energy, and momentum of an ultrarelativistic particle are nearly (or, for massless particles, exactly) equal to one another. Hence, *the temperature of an ultrarelativistic gas decreases like the reciprocal of the cosmic scale factor.*

In the next section we will consider a particularly interesting example of an ultrarelativistic gas.

Blackbody Radiation in an Expanding Universe

In the spring of 1964 Arno Penzias and Robert Wilson of the Bell Laboratories were preparing to measure the intensity of continuous radiation from the Galaxy at a wavelength of 20 centimeters (just short of the emission line of neutral hydrogen). Their observation plan was simple in principle:

1. Measure the received signal.

2. Measure the signal arising from noise in the antenna and receiver.

3. Subtract the second signal from the first to obtain the true signal.

Absolute measurements of this kind present great technical difficulties. Penzias and Wilson had decided to undertake them because they had at their disposal an exceptionally noise-free antenna-receiver system, originally designed for communication with deep-space satellites.

The expected signal was very weak, and qualitatively indistinguishable from noise generated by the antenna and receiver; so the noise contributions had to be known very accurately. All but one of them could in fact be reliably measured or calculated. Only the level of noise generated by the antenna itself (a horn reflector with an aperture of 20 feet) was somewhat uncertain.

To reduce this uncertainty, Penzias and Wilson tuned their receiver to a wavelength of 7.3 cm and pointed the antenna at a part of the sky that should have been almost completely dark at that frequency. To their disappointment, the measured signal at this wavelength was much greater than it should have been.

It took the two scientists the better part of a year to convince themselves that neither the antenna nor any other component of the telescope was producing the signal, that it came from without rather than from within. Yet it was a very peculiar signal: it was equally strong in all directions; it did not vary with the position of the Sun; it did not depend on how the antenna was pointed relative to the Earth; and it was unpolarized. It was exactly the sort of signal that would have been produced by a defective resistor. But Penzias and Wilson had ruled out this and similar possibilities. The source of the signal had to be outside the telescope. But where?

They divided the candidate sources into four categories: terrestrial sources, sources in the solar system, Galactic sources, and extragalactic sources. Sources in each of the first three categories would have identified themselves by their directional characteristics. The intensity of a terrestrial source would vary with the orientation of the antenna; the intensity of sources in the solar system would show diurnal or seasonal variation; and the intensity of Galactic sources would vary with Galactic longitude and latitude. Penzias and Wilson could detect none of these variations. The signal was equally intense in all directions and constant in time.

Only one possibility remained: a cosmic distribution of very distant sources. This explanation, however, seemed to have a fatal flaw. All known extragalactic sources are much brighter at wavelengths around 1 meter than at wavelengths of a few centimeters. A cosmic distribution of such sources bright enough to produce the observed signal would have been so bright at meter wavelengths that it would undoubtedly have been noticed earlier by radio astronomers.

Meanwhile, in Princeton, New Jersey, a few miles from the observing site at Holmdel where Penzias and Wilson had set up their telescope, a group headed by Robert Dicke was preparing to search for an isotropic radiation background at a wavelength of 3 cm, using a low-noise antenna-receiver system they had constructed for that specific purpose. For reasons we will discuss in Chapter 8, Dicke hoped to detect blackbody radiation with a temperature of a few degrees Kelvin. Penzias and Wilson had not considered the possibility of such a cosmic radiation field.

A mutual colleague put the two groups in touch. They agreed to submit two letters to the *Astrophysical Journal,* one by Penzias and Wilson describing their measurement, the other by Dicke and the members of his group discussing its theoretical interpretation.

But the nature—indeed, the reality—of the radiation measured by Penzias and Wilson remained in doubt until Dicke's group was able to carry out its own measurement, at a wavelength of 3 cm, a few months later. Their findings fully confirmed those of Penzias and Wilson. At

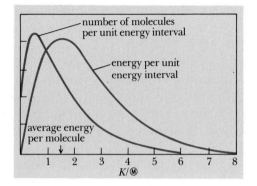

number of molecules
per unit energy interval

energy per unit
energy interval

average energy
per molecule

$K/\text{ⓜ}$

The Maxwell-Boltzmann distribution of molecular kinetic energy. The fraction of molecules with energies in a given energy interval is proportional to the height of the ···· curve at the corresponding value of $K/\text{ⓑ}$ and to the width of the interval. K is the kinetic energy, ⓑ the absolute temperature measured in energy units; $\text{ⓑ} = 1.4 \times 10^{-16}T$ erg if T is measured in degrees Kelvin. The fraction of the energy of the gas carried by molecules with energies in a given narrow range is proportional to the height of the ···· curve at the appropriate energy.

3 cm, as at 7.3 cm, the radiation was equally intense in all directions. Moreover, the measured intensities at both 3 cm and 7.3 cm equaled the predicted intensity of blackbody radiation at a temperature near 3 K, and later observations confirmed the blackbody character of the spectrum in the wavelength range 0.5 cm to nearly 50 cm.

What is blackbody radiation? And why is the discovery of cosmic blackbody radiation considered the most important advance in observational cosmology since Hubble's discovery of the velocity–distance relation? The answers to these two questions are related.

If a quantity of gas is confined in a box whose walls are kept at a constant temperature, the average kinetic energy per molecule approaches a definite limiting value, and the distribution of kinetic energy among the molecules takes on a definite form. In the final, "relaxed" state of the gas, the average kinetic energy per molecule is a fixed multiple of the temperature, and the distribution of kinetic energies is described by a universal function of the ratio K/Θ between the kinetic energy K and the temperature Θ. This universal function, discovered by Ludwig Boltzmann and James Clerk Maxwell, is shown in the figure above.

If the walls are opaque, the box will eventually be filled with radiation that, at every point, is equally intense in all directions and whose energy density is everywhere the same. We may think of this uniform, isotropic radiation field as a gas of photons. The average energy of a photon in the final equilibrium state is a fixed multiple of the temperature of the walls, and the distribution of photon energies is described by a universal function of the ratio $h\nu/\Theta$. This function, whose discovery by Max Planck in 1900 marks the beginning of quantum physics, is shown in the figure on the facing page. The equilibrium radiation is known as *blackbody radiation*. The ratio between the average energy of a photon and the temperature is not the same as the ratio between the average kinetic energy of a molecule and the temperature, nor is Planck's universal distribution function identical with that of Maxwell and Bolzmann. But the most striking difference between the equilibrium distribution of photons and that of massive particles is that the number of photons per unit volume, as well as the average photon energy, is determined by the temperature. It is in fact proportional to the cube of the temperature. Hence, the energy density of blackbody radiation is proportional to the fourth power of the temperature.

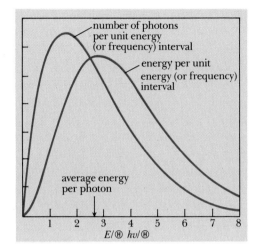

The Planck distribution of photon energy. The fraction of photons with energies in a given energy interval (or with frequencies in a given frequency band) is proportional to the height of the ---- curve at the corresponding value of $E/\oplus$ ($=h\nu/\oplus$) and to the width of the interval (or band). The fraction of the energy density of blackbody radiation in a narrow energy or frequency range is proportional to the height of the ---- curve at the appropriate energy or frequency.

Although the temperature of a gas may vary from place to place, as in the Earth's atmosphere, the local distribution of molecular kinetic energies is usually very close to the equilibrium distribution shown in the figure on page 242, because collisions between molecules redistribute kinetic energy and tend to "randomize" its distribution. The Maxwell-Boltzmann distribution of molecular kinetic energies is maximally random. Photons, by contrast, cannot exchange energy directly. Radiation in an otherwise empty box with perfectly reflecting walls would keep the same spectrum forever. The randomization or *thermalization* of radiation can take place only in the presence of material particles that absorb and reemit photons, and this process is not in general very efficient. The quality of sunlight at the surface of the Earth is not very different, in the visible part of the spectrum, from the quality of sunlight at its source. The Sun's radiating layers have a temperature of around 6,000 K, and the spectrum of sunlight resembles the spectrum of blackbody radiation at this temperature. But if a cubic meter of sunlight at the surface of the Earth were transformed into blackbody radiation with the same total energy, its temperature would be only about 300 K. The process of thermalization would increase the number of photons by a factor of 20, and decrease the average energy of a photon by the same factor.

We are now in a position to understand why the cosmic microwave background is such an interesting phenomenon. The universe as it is today is almost perfectly transparent to microwave and millimeter-wave radiation. Hence, *this radiation could not have been produced under conditions remotely resembling those that prevail in the universe today.* Blackbody radiation could have been produced only when the universe was much more opaque at the relevant wavelengths, and hence much denser, than it is now.

But if the universe had been filled with blackbody radiation at some earlier, denser stage of its history, would the radiation have kept its blackbody character as the universe expanded and cooled? The answer is yes. We have seen that the distribution of photon energies in blackbody radiation is described by a universal function of the ratio $h\nu/\Theta$. We have also seen that, as the universe expands, the frequency ν of a photon varies like $1/a$, the reciprocal of the scale factor. Hence the distribution of photon energies continues to be described by the same universal function, *in which the temperature Θ varies like $1/a$*. In blackbody radiation, the number of photons per unit volume is proportional to the cube of the temperature. As the universe expands, the number of photons per unit volume decreases like $1/a^3$, since a sphere whose radius is proportional to the scale factor always contains the same number of photons. But this rate of decrease is exactly right to maintain the number density of photons appropriate to blackbody radiation, because, as we have just seen, the temperature defined by the spectrum of the radiation decreases like $1/a$. Thus blackbody radiation that has ceased to interact with matter will remain blackbody radiation as the universe expands; its temperature will decrease like the reciprocal of the cosmic scale factor, and its energy density will decrease like the reciprocal of the fourth power of the scale factor.

Deep-Space Measurements

The cosmic expansion is an expansion of space itself, not a systematic recession of galaxies in a static space. At first sight this may seem to be a distinction without a difference. But light behaves differently in expanding and static spaces, and since virtually all cosmological observations rely on light and other forms of electromagnetic radiation (some information also comes from cosmic rays, which are particles), the difference is important.

Suppose that a photon emitted at time t by a distant source is received at the origin at time t_0. In static Euclidean space the source's distance r is equal to the difference $(t_0 - t)$ between the two times (time, as usual, being measured in the same units as distance). In an expanding Euclidean space, the radial coordinate r is related in a more complicated way to the times t, t_0. Let us work out this relation.

The interval of proper time between events in the history of a photon vanishes. Hence it follows from Equation 7.2 that the space-intervals and time-intervals between events in the history of a spherical light wave centered on the origin satisfy the relation

$$0 = dt^2 - a^2(t)\,dr^2 \tag{7.37}$$

or

$$dr = \frac{dt}{a(t)}. \tag{7.38}$$

In a static space the factor $a(t)$, whose present value is unity, would not appear. Integrating this relation between 0 and r, and between the time of emission t and the time of reception t_0, using the formula $a(t) = (t/t_0)^{2/3}$, we obtain

Cluster of Galaxies in Hercules photographed with the 4-meter telescope at Kitt Peak National Observatory, near Tucson, Arizona.

$$r = \int_t^{t_0} \frac{dt}{a(t)} = 3t_0 \left[1 - \left(\frac{t}{t_0} \right)^{1/3} \right].$$

(7.39)

This relation connecting r, t, and t_0 is plotted in the figure on page 246. It coincides with the relation $r = t_0 - t$ when the difference $t_0 - t$ is much smaller than t_0, but as we look back toward the beginning of time ($t = 0$), r approaches, not t_0, but $3t_0$.

The radial *distance* of the source at the moment of emission is $a(t)r$. This quantity is plotted in the figure on page 246. Unlike the radial coordinate r, it does not increase monotonically as we look deeper into the past; it rises to a maximum, then diminishes, approaching zero as $t \rightarrow 0$. Thus sources that are now near the edge of the observable universe were very close to us when they emitted the radiation we now receive.

The velocity V of a source at time t is given by the velocity-distance relation:

$$V(t) = H(t) \cdot a(t)r.$$

(7.40)

V is also plotted. The figure shows (and it is not hard to prove analytically) that the most distant sources, at the moment of emission, were receding at the speed of light ($V = 1$). Sources reced-

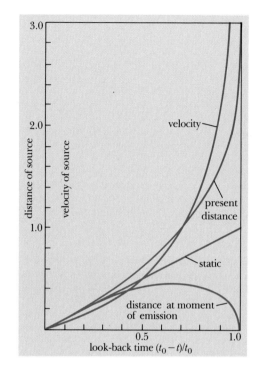

In an expanding space the present distance of a source whose light is now reaching us is greater than it would be in a static space, where distance is proportional to look-back time. On the other hand, the source was closer to us at the moment of emission than it would have been in a static space. Note that the effects of the expansion are insignificant at small distances; and that a source whose light was emitted at the initial moment of the expansion would then have coincided with our present position but would now be at a distance equal to three times the present epoch (3×10^{10} light-years if $t_0 = 10^{10}$ years).

ing with speeds greater than the speed of light were closer to us at the moment of emission. V increases indefinitely as we look deeper into the past. (This result does not contradict special relativity, because special relativity holds only in spacetime regions so small that the effects of the cosmic expansion are negligible within them.)

None of the quantities we have considered so far (the radial coordinate r, the time of emission t, the instantaneous distance ar, the recession speed V) is actually measurable. The simplest measurable properties of a distant source are its redshift z, its angular diameter θ, its apparent brightness l, and its surface brightness l/θ^2. The table on the facing page shows how these quantities are related to one another in an expanding Euclidean universe and in static Euclidean space. Angular diameter, apparent brightness, and surface brightness are plotted against redshift in the figures on pages 247 and 248.

Notice that the formula relating redshift to recession velocity differs from both the nonrelativistic and the special-relativistic formulas, though it coincides with both in the limit of small recession velocities. A source that was receding with the speed of light at the moment of emission has a finite redshift, $z = \frac{5}{4}$. Such a source also has the smallest angular diameter, among sources with a given linear diameter. Sources that were receding faster than light at the moment of emission have larger angular diameters. It may seem strange that light emitted by a source whose speed of recession is greater than the speed of light could ever reach us. At first a

RELATIONS BETWEEN PROPERTIES OF A DISTANT LIGHT SOURCE
IN STATIC AND EXPANDING EUCLIDEAN SPACES

Relation	Expanding space	Static space
Radial coordinate/epoch	$r = 3t_0\left[1 - \left(\dfrac{t}{t_0}\right)^{1/3}\right]$	$r = t_0 - t$
Radial distance/epoch	$R = a(t)r = \left(\dfrac{t}{t_0}\right)^{2/3} r$	$R = r$
Recession velocity/epoch	$V = 2\left[\left(\dfrac{t_0}{t}\right)^{1/3} - 1\right]$	——
Redshift/velocity	$\dfrac{\lambda_0}{\lambda} \equiv 1 + z = \left(1 + \dfrac{1}{2}V\right)^2$	$\dfrac{\lambda_0}{\lambda} = \left(\dfrac{1 + V}{1 - V}\right)^{1/2}$
Redshift/scale factor	$1 + z = \dfrac{1}{a} = \left(\dfrac{t_0}{t}\right)^{2/3}$	——
Angular diameter/distance	$\theta = \dfrac{D}{R}$	$\theta = \dfrac{D}{R}$
Angular diameter/redshift	$\theta = \left(\dfrac{D}{3t_0}\right) \dfrac{(1 + z)^{3/2}}{(1 + z)^{1/2} - 1}$	——
Apparent brightness/distance	$l = \dfrac{a^2 L}{4\pi r^2}$	$l = \dfrac{L}{4\pi r^2}$
Apparent brightness/redshift	$l = \dfrac{1}{36\pi t_0^{\,2}} \dfrac{L(z)}{(1 + z)[(1 + z)^{1/2} - 1]^2}$	——
Surface brightness/redshift	$\dfrac{l}{\theta^2} \propto (1 + z)^{-4}$	$\dfrac{l}{\theta^2} = \text{constant}$

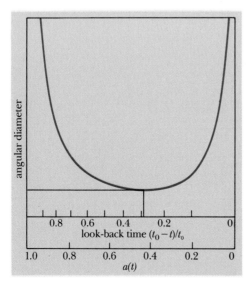

Angular diameter (or parallax) in an expanding Euclidean space, plotted against the cosmic scale-factor $a(t)$ at the moment of emission. A second horizontal scale shows the look-back time $t_0 - t$ in units of the present epoch. The minimum angular diameter in an expanding Euclidean space occurs when $a = 4/9$; earlier sources were closer at the moment of emission; see figure on the facing page.

Apparent faintness and surface brightness in an expanding Euclidean space, plotted against redshift. Apparent faintness is the negative logarithm of the apparent luminosity l. Surface brightness is the logarithm of the apparent luminosity per unit area on the celestial sphere (or per square degree). Astronomers measure apparent faintness in magnitudes, on a scale that makes 5 magnitudes equal to a factor of 100 in apparent luminosity; that is, 5 mag = 2 logarithmic units of apparent faintness. Notice that the surface brightness of a distance source is decreasing very rapidly with increasing redshift just where the effects of redshift on the apparent faintness are beginning to be noticeable.

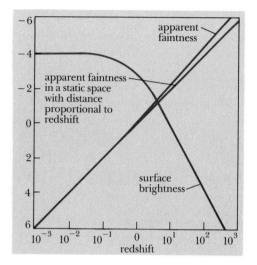

Quasar 3C-279.

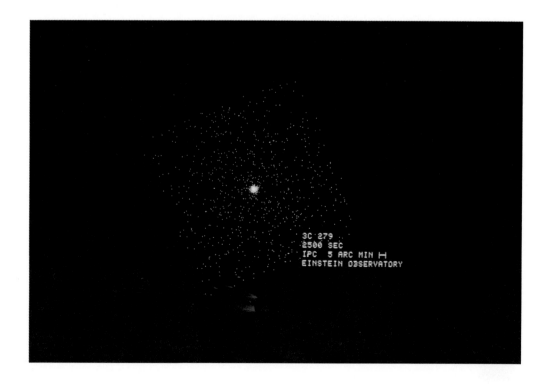

An expanding spherical region whose mass is concentrated at its center and whose mean density remains equal to the density of the surrounding uniform medium.

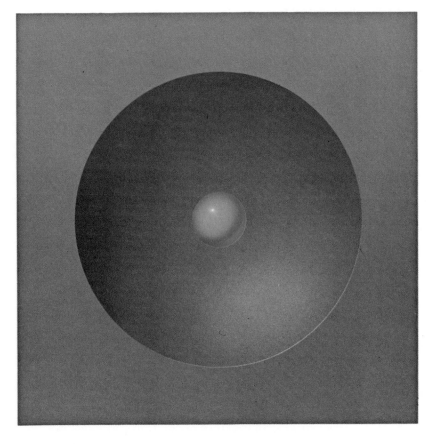

wavefront emitted by such a source in our direction recedes from the origin, but as the expansion slows down, the wavefront retreats ever more slowly, and eventually it begins to advance at an ever-increasing rate toward the origin.

Cosmic Coordinates and Locally Unaccelerated Reference Frames

The idealized models of the universe we have considered so far have been wholly lacking in small-scale structure. We will now discuss a model that is slightly but significantly more realistic. Consider a universe in which the mass density is uniform except in a spherical region whose radius is proportional to the scale factor $a(t)$. Let all the mass that would be needed to make the density in this region equal to the density outside be concentrated in a point at the center of the cavity (see figure above).

Einstein's theory tells us that the gravitational field outside the cavity is the same as it would be if the mass at the center were distributed evenly throughout the cavity. That is, the cavity,

Within the expanding spherical cavity, spacetime is Minkowskian; outside, Friedmannian. The requirement that the spacetime grids join smoothly at the surface of the cavity determines the locally nonrotating frame whose origin coincides with the point-mass at the center of the cavity.

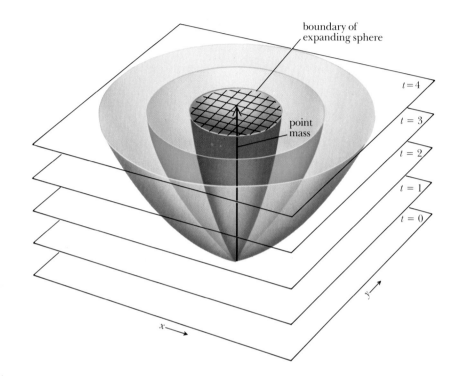

with the point-mass at its center, does not interfere with the expansion of the rest of the universe. Conversely, the rest of the universe does not affect the gravitational field in the cavity; it is exactly the same as it would be if the point-mass were isolated in empty space. Einstein's theory does, however, require that the spacetime grids inside and outside the cavity join up smoothly. This condition implies that the frame of reference in which the gravitational field of the point-mass takes the form given by Schwarzschild's formula (Equations 6.22 and 6.23) is not rotating relative to the rest of the universe (see figure above).

We can replace the point-mass by any reasonably compact self-gravitating system. And we do not need to assume that the mass distribution outside the sphere is perfectly smooth: the gravitational effects of clumpiness extend only to distances comparable to the sizes of the clumps. The key point is that local coordinate grids must fit smoothly into the cosmic coordinate grid, and this requirement links locally unaccelerated reference frames to the cosmic distribution of mass and motion (see figure on the facing page). Thus we can conclude, as before, that the universe as a whole does not influence the dynamics of compact self-gravitating systems, but it does determine the class of reference frames in which Newton's theory is valid in a first approximation. These frames are unaccelerated relative to the cosmic frame whose existence is guaranteed by the Cosmological Principle.

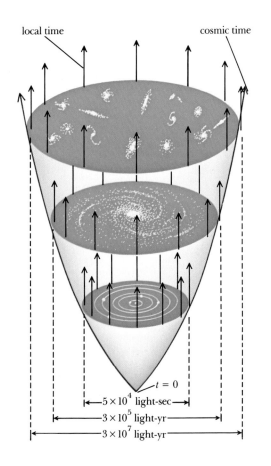

local time   cosmic time

$t = 0$

$\leftarrow 5 \times 10^4$ light-sec$\rightarrow$

$\leftarrow 3 \times 10^5$ light-yr$\rightarrow$

$\leftarrow 3 \times 10^7$ light-yr$\rightarrow$

Cosmic space is an expanding mosaic of nonexpanding tiles, the local spaces in which self-gravitating systems are embedded. Although the "tiles" do not expand, they grow, as progressively larger self-gravitating systems separate out, perhaps in the manner described in Chapter 8.

These considerations show that space is an expanding mosaic whose tiles are static Euclidean spaces, slightly distorted by local concentrations of mass. In the next chapter we will discuss the provenance of the tiles, and related questions.

CHAPTER EIGHT

COSMIC EVOLUTION

As being is to becoming, so is truth to belief. If then, Socrates, amid the many opinions about the gods and the generation of the universe, we are not able to give notions which are altogether and in every respect exact and consistent with one another, do not be surprised.

PLATO, *Timaeus*

What processes have shaped the hierarchy of self-gravitating astronomical systems, and from what beginnings? What is the origin of the cosmic radiation background? These questions differ qualitatively from those we have addressed in earlier chapters. Up until now we have focused attention on physical laws and the ways in which they manifest themselves in the structure and dynamics of astronomical systems. Physical laws account for some of the regularities we observe in the astronomical universe, but not for all of them. For example, Newton's law of gravitation accounts for Kepler's three laws of planetary motion but not for the fact that the orbits of the planets are nearly circular and nearly coplanar. These aspects of the solar system are outcomes of an evolutionary process. Although the process itself is governed by laws we think we understand, its outcome depends on what the primordial solar system was like. Analogously, the laws of physics tell us that the cosmic microwave background must at one time have interacted strongly with matter, but they do not tell us where the photons came from.

Questions about origins are difficult because they are so closely interrelated. To understand why planets move in direct, nearly circular, nearly coplanar orbits, whereas comets move in randomly oriented, highly elongated orbits, we must know what the primordial solar system was like. To learn what the primordial solar system was like we need a theory of star formation, which in turn we can develop only in the context of a theory for the formation and evolution of galaxies. And the problem of the origin of stars and galaxies is, as we will see, inseparable from the problem of the origin of the cosmic microwave background. Thus the strategy "divide and conquer," so fruitful in other branches of science, is useless in cosmogony, the study of cosmic evolution.

Perhaps for this reason, progress in this subject has been slow. There are as yet no strongly overdetermined cosmogonic theories. The most widely accepted view of the early universe, the "standard model" described in the following section, formerly enjoyed a measure of observational support, but, as we will see, now conflicts with the only observational evidence that can be brought to bear on it. Later in the chapter we will consider an alternative theory of the early universe.

(Left) The X-ray telescope of the Einstein Observatory being fitted into the spacecraft. Observations must be made above the earth's atmosphere, which is impermeable to X-rays. At the focus of the telescope a variety of imaging and spectroscopic detectors were available for observations by astronomers throughout the world. Observations ranged in duration from a few thousand seconds to almost a week. The false color photographs in this chapter sample the X-ray sky as seen by the Einstein Observatory.

X-ray image of the unusual object SS433. Emission from the bright central source is seen as well as two jets extending from the central source.

The "standard model" interprets the cosmic microwave background as the remnant of a "primordial fireball." It assumes that the photons that now make up the background were present at the earliest moment in the history of the universe at which current physical theories apply. Because the temperature of a cosmic radiation field is inversely proportional to the cosmic scale factor (as we saw toward the end of Chapter 7), the fireball was very hot. At age 1 second its temperature would have been around 10^{10} K; at age 100 seconds, 10^9 K. At such temperatures all the particles present, including photons, would have had comparable energies; and because there would have been about 10^8 photons for every proton or neutron (as there are now), virtually all the energy of the primordial fireball would have been in the form of radiation.

What is more, protons and electrons would have been held fast by the surrounding radiation, like raisins in an expanding bread dough. Only when the temperature had dropped to about 4,000 K and the protons and electrons combined to form (electrically neutral) hydrogen atoms would the radiation have lost its grip on material particles, allowing them to congregate in self-gravitating clumps.

Now, we saw in Chapter 4 that the mass M of a self-gravitating system, its diameter D, and its internal velocity dispersion $\langle v^2 \rangle$ satisfy the simple relation

$$\langle v^2 \rangle = \frac{M}{D}.$$

Because the velocity dispersion is proportional to the temperature, and the mass and the diameter determine the system's average density, *the mass of a self-gravitating system is determined by its temperature and density.* A piece of the cooling cosmic medium whose mass was smaller than the mass corresponding to the medium's density and temperature could not separate out: its gravitational self-attraction would not be large enough to balance the internal pressure forces. A piece of the cosmic medium whose mass was larger than the mass corresponding to the medium's density and temperature *could* separate out if it were denser than the surrounding medium. At the moment when matter and radiation "decoupled," the least-massive gas clouds that could have separated out would have been about a million times as massive as the Sun. Thus *stars, double and multiple stars, and medium-sized clusters of stars would need to have formed afterward, by some other process.*

Although gas clouds a million or more times as massive as the Sun *could* have separated out when the temperature of the cosmic medium had dropped to a few thousand degrees, this wouldn't actually have happened unless the cosmic medium was already clumpy on the appropriate scales. A smooth cosmic gas at the temperatures and densities that would have prevailed in the "standard model" could never break up into self-gravitating clouds of any size. Thus *the "standard model" must contain a primordial spectrum of large-scale density variations.*

Theorists differ about what sort of primordial density fluctuations it would be reasonable to postulate. Recent theoretical calculations have shown, however, that the assumptions favored by most theorists will not work. For this reason and others to be discussed, there is no longer a consensus among supporters of the fireball hypothesis about what the history of the universe might have been like.

The Cosmic Fireball: A Hypothesis in Search of Evidence

In 1946 George Gamow published a theory for the formation of the chemical elements. He postulated that in the beginning all matter was in the form of neutrons. Occasionally two neutrons would collide to form a deuteron (a nucleus consisting of a proton and a neutron) and an electron. The deuterons formed in this way would then capture neutrons and become tritons (nuclei consisting of one proton and two neutrons), each of which would then capture an additional neutron, and so on, until nuclei with mass numbers of around 250 had been built up. Some nuclei built up in this way would be unstable: they would emit an electron and an antineutrino, a process known as beta decay. A simple calculation, later refined by his colleagues Ralph Alpher and Robert Herman, told Gamow the temperature and density at which these nuclear reactions must have occurred to account for the observed relative abundances of heavy elements.

It turned out that this temperature would have been so high that nearly all the energy would have been in the form of radiation, almost none in the form of material particles. (By contrast, radiation now accounts for only a tiny fraction of the cosmic energy density.) Gamow, Alpher, and Herman remarked that the radiation would cool as the universe expanded, but would remain blackbody radiation (Chapter 7). Alpher and Herman predicted that cosmic space should now be filled with blackbody radiation at a temperature of 5 K.

Astronomers made no effort to detect the predicted radiation, however. Not only would the required measurements have been difficult to make in the 1950s but, more importantly, Gamow's theory had been found to have a fatal flaw.

Enrico Fermi (1901–1959), one of the chief architects of modern subatomic physics.

In 1949 Enrico Fermi and John Turkevich made detailed calculations of the first few steps of Gamow's process. They found that successive neutron captures interspersed with beta decays, under the conditions envisioned by Gamow, would indeed give rise to deuterons, tritons, helium-3, and helium-4. But here the building-up process would come to an abrupt end. Reactions involving neutrons, protons, and nuclei with mass numbers 2, 3, and 4 would not produce significant quantities of heavier nuclei. Gamow had been aware of this bottleneck but had hoped that some way could be found to get around it. The calculations of Fermi and Turkevich dashed this hope.

Gamow's theory also conflicted with observational evidence. By the early 1950s, analyses of stellar spectra had strongly confirmed earlier indications that the abundance ratio of heavy elements to hydrogen is 10 to 100 times greater in the Sun and its neighbors than in stars far from the Galactic plane. If the elements heavier than helium (which was known to be produced by thermonuclear reactions in stellar interiors) were formed in the early universe, their present abundances relative to hydrogen would be practically the same everywhere in the Galaxy. The astronomical observations showed unmistakably that 90 percent or more of the heavy elements in the Sun and in nearby stars must have been nonprimordial.

The fireball hypothesis survived the demise of the theory that gave birth to it. In 1964 two English astrophysicists, Fred Hoyle and Roger Tayler, argued that most of the helium in the universe must have been synthesized near the beginning of the cosmic expansion under conditions similar to those considered by Gamow, Alpher, and Herman. Their argument ran as follows. Stars manufacture helium in their interiors by nuclear reactions that convert four protons (hydrogen nuclei) into one alpha particle (helium nucleus). In each such reaction a little mass disappears (the sum of the individual masses of four protons and two electrons is slightly greater than the mass of a helium nucleus, which contains two protons and two neutrons). Most of the energy represented by the lost mass eventually escapes from stars in the form of visible and invisible radiation. Hoyle and Tayler assumed that the combined luminosity of stars in the Galaxy was never substantially greater than it is now. From the estimated age of the Galaxy, about 10^{10} years, and estimates of its present luminosity, they could then estimate the quantity of energy that has been radiated by all the stars in the Galaxy during its lifetime. Knowing how much radiant energy is released by the reactions that turn hydrogen into helium, they could then calculate the quantity of helium that has been produced in stars during the Galaxy's lifetime. They found that if the stars consisted initially of pure hydrogen there would now be 1 helium atom for every 100 hydrogen atoms. But in 1964, when Hoyle and Tayler published their paper, astronomers believed that there were only 10 hydrogen atoms to every helium atom in all kinds of stars and in the gas between the stars. Thus, Hoyle and Tayler concluded, 90 percent of the helium must have been made before stars came into being, presumably in Gamow's cosmic fireball. Hoyle and Tayler went on to estimate how much helium would be synthesized in the fireball. They reached the important conclusion that for a wide range of initial conditions the expected ratio is indeed 10 hydrogen atoms to 1 helium atom.

(*Left*) The central object in this asymmetric supernova remnant is a pulsar in a binary star system. (*Right*) Puppis a supernova remnant appears in X-rays as a composit of many separate X-ray images.

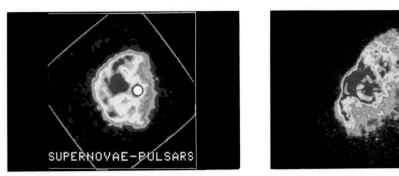

SUPERNOVAE-PULSARS

While Hoyle and Tayler were constructing their theoretical argument, a small group of scientists headed by Robert Dicke at Princeton was mounting a combined experimental and theoretical attack on the same problem. P. G. Roll and D. T. Wilkinson constructed a radiometer that could detect blackbody radiation at a temperature of a few degrees on the Kelvin scale. P. J. E. Peebles worked out theoretical relations similar to those of Hoyle and Tayler, connecting the temperature of the blackbody radiation, the average density of matter in the universe, and the cosmic abundance of helium. Before Roll and Wilkinson could make their first measurement, however, they learned that Arno Penzias and Robert Wilson of the Bell Telephone Laboratories had already carried out a similar measurement (Chapter 7).

Penzias and Wilson announced their discovery in a short letter published in the *Astrophysical Journal* for July 1, 1965. *Preceding* this letter is a longer letter by Dicke and his collaborators, in which they interpret the newly discovered radiation as the remnant of a primordial cosmic fireball. Most astronomers accepted this interpretation instantly. Unlike Kant's island-universe hypothesis, which had to wait 275 years to gain cautious acceptance by the astronomical community, the fireball hypothesis was accorded near-factual status overnight. The attitude that prevailed between 1965 and 1980 is well-expressed by Steven Weinberg in *The First Three Minutes:*

> *Throughout most of the history of modern physics and astronomy, there simply has not existed an adequate observational and theoretical foundation on which to build a history of the early universe. Now, in just the past decade, all this has changed. A theory of the early universe has become so widely accepted that astronomers often call it "the standard model."*

Let us look at some of the astronomical evidence bearing on the "standard model" and the arguments that have been used to interpret it.

Testing the Fireball Hypothesis

The fireball hypothesis has three direct, testable implications. These concern the cosmic abundance of helium, the cosmic abundance of deuterium, and the spectrum of the background

Primordial abundances of helium (this page) and deuterium (p. 259). The "standard model" predicts the dependence of these primordial abundances on the present average cosmic density of matter, ρ. In 1965 the predicted abundances, based on a widely held opinion about the cosmic mass density, agreed with independent estimates of the primordial abundances of helium and deuterium. In 1984, theory and observation no longer agree.

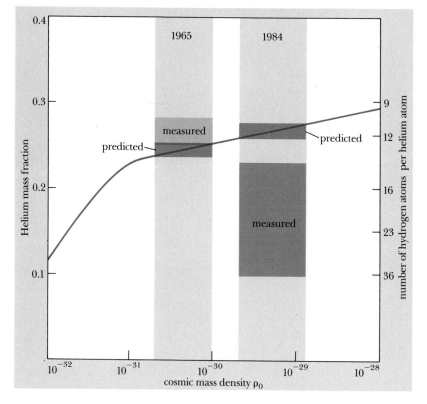

radiation. The predictions depend, in the standard model, on just two parameters: the temperature of the background radiation (which is well-determined by observation to be close to 3 K), and the average density of matter in the universe.

The figures above and right show how the cosmic abundances of helium and deuterium in the standard model depend on the cosmic density of matter. The values of the two relative abundances that were generally accepted in 1965 are consistent with an average density of matter in the universe near 5×10^{-31} gm/cm^3. This value is indeed close to independent estimates of the average density of *luminous* matter.

The third of the standard model's predictions was equally unequivocal: *The spectrum of the radiation background should not deviate significantly from the blackbody (Planck) spectrum.* Blackbody radiation is radiation that has been thoroughly "randomized" by interaction with matter; it consists of photons that have been absorbed and reemitted so many times that they no longer "remember" where they came from or what their original emitter was like. Blackbody radiation is equally intense in all directions and in all places, and its spectrum is described by a universal curve whose formula was first written down by Max Planck in 1900. The fireball would have been pure blackbody radiation, because its photons would have interacted very strongly with

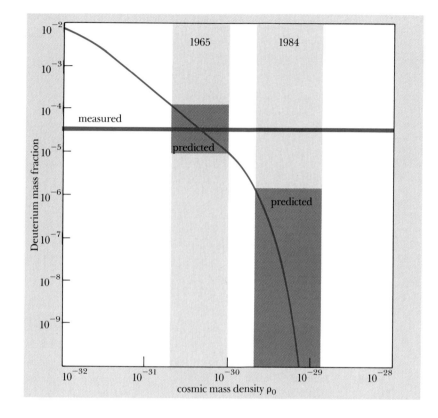

the electrons and other charged particles that were present. After the radiation had ceased to interact strongly with matter, its spectrum would have remained Planckian.

The figure on page 260 shows the Planck curve plotted in two ways, along with measurements of the flux of radiant energy. The measurements recorded were all made at wavelengths between 0.5 cm and 50 cm. The Earth's atmosphere is both quiet and transparent between these wavelengths. At wavelengths shorter than 0.5 cm, so much radiation comes from atmospheric gases that accurate measurements cannot be made from the ground; at wavelengths longer than 50 cm, the background radiation is swamped by radiation from other astronomical sources. Limited though it is, the observable range spans two powers of ten in frequency; and over this range the measurements fit the blackbody curve very well indeed. The fit seems even more impressive if we bear in mind that blackbody radiation is an exceedingly rare phenomenon in nature. The spectra of stars approximate the Planck curve only crudely.

Since 1976, the agreement between the predictions of the standard model and observation has deteriorated, for three main reasons.

First, estimates of the cosmic mass density have increased by factors of 10 or 20. This increase has resulted from improvements in the cosmological distance scale, from increased

The Planck (blackbody) curve plotted in two ways: intensity (or energy density) of the radiation versus reciprocal wavelength ($1/\lambda$) and (in the inset) the logarithm of the intensity (or energy density) versus the logarithm of the reciprocal wavelength. Also shown are the results of microwave measurements made with ground-based radiometers. The log-log plot (inset) is clearly more useful for representing the microwave measurements and comparing them with the Planck relation between intensity and reciprocal wavelength, but it obscures the fact that the bulk of the energy lies at shorter (millimeter) wavelengths. Note that the two curves refer to slightly different temperatures—2.7 K and 3.0 K, respectively.

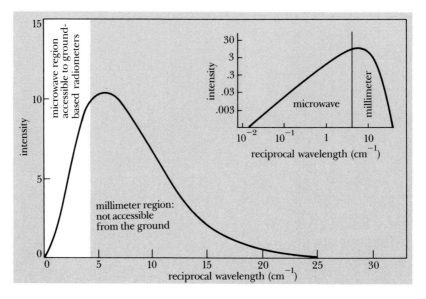

knowledge and understanding of the clustering of galaxies on very large scales, and from new data showing that as much as 90 or even 95 percent of the mass contained in galaxies and galaxy clusters is nonluminous. From the figure on page 259 we can see that current estimates of the cosmic mass density make the predicted abundance of helium only slightly larger than the earlier value. But the predicted abundance of deuterium is now at least a thousand times too small.

Supporters of the fireball hypothesis have suggested that the deuterium that astronomers observe in interstellar gas clouds may not be primordial after all. Where, then, does it come from? Unlike helium and heavier elements, deuterium is consumed rather than synthesized in stellar interiors. It could conceivably be produced in interstellar clouds by shock waves resulting from supernova explosions. However, theoretical calculations have shown that this process would produce anomalously high abundances of other elements, and these have not been observed. Moreover, there is no evidence that the deuterium–hydrogen ratio is greater in the direction of recent supernovae than in other directions. Sherlock Holmes' dictum "When you have eliminated the impossible, whatever remains, no matter how improbable, must be the truth" suggests that interstellar deuterium was manufactured during the early stages of the cosmic expansion.

The second conflict between the standard model and observation stems from improved estimates of the primordial abundance of helium, which indicate that the primordial hydrogen–helium ratio was at least 15 to 1, instead of the predicted ten to one. Several lines of evidence converge on the new, lower value for the primordial helium abundance. Let us consider one of these. Over the years, astrophysicists have constructed increasingly accurate theoretical models of the Sun, embodying libraries of detailed experimental and theoretical data about atomic,

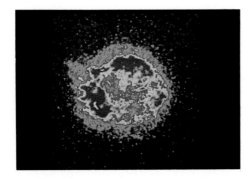

X-ray image of the Cassiopeia A supernova, believed to have exploded about 300 years ago.

radiative, and nuclear processes. The accurately measurable properties of the Sun include its mass, luminosity, effective temperature, age (from the ages of meteorites and Moon rocks), and, most recently, its rate of neutrino emission. The theoretical models that reproduce all these measurements have initial helium abundances of only 1 helium atom to every 20 hydrogen atoms. These models, however, also have very low primordial abundances of heavy elements—less than a tenth of the value inferred from spectroscopic studies of the Sun's outer layers. Astronomers who accept these models suggest that the Sun's outer layers could have been enriched by accretion of interstellar gas rich in heavy elements. Models that equate the primordial abundance of heavy elements to their present surface abundance predict rates of neutrino emission far greater than the observed rates; but even these models have hydrogen–helium ratios of 13 to 1.

Furthermore, not all the helium in such models can be primordial. The Sun is only 4.6 billion years old, less than half the age of the oldest stars in the Galaxy. Its heavy elements, like the heavy elements in all relatively young stars near the plane of the Galaxy, were manufactured in other, more massive stars; these stars exploded, ejecting metal-rich *and helium-rich* gas into the interstellar medium, from which the Sun and other relatively young stars subsequently formed. Astronomers estimate that supernovae eject, on the average, three times as much mass in the form of helium as in the form of heavy elements. If we assume that a fraction Z of the Sun's mass is in the form of heavy elements, then, to calculate the *primordial* helium abundance, we must subtract the quantity $3Z$ from its *initial* helium abundance. When we do this we obtain a value of 16 to 1 for the primordial hydrogen–helium ratio.

The third conflict between the standard model and observation may prove to be the most serious of all. In 1978 D. P. Woody and P. L. Richards of the University of California at Berkeley made the first accurate direct measurements of the cosmic microwave background in the wavelength range that contains most of its energy (see figure on page 260), using a liquid-helium-cooled spectrophotometer carried aloft in a balloon. The measured spectrum deviated significantly from the Planck curve corresponding to the measured energy density (see figure on page 262). (The "effective temperature" of non-blackbody radiation is defined by its energy

Measurements by Woody and Richards with a balloon-borne radiometer of the cosmic background radiation at millimeter wavelengths, compared with the spectrum of blackbody (Planck) radiation with the same energy density. The measured spectrum deviates significantly from the Planck spectrum. It is higher at the peak (between wavelengths of 1 and 2 millimeters), lower at shorter wavelengths. At longer wavelengths (the microwave region) ground-based measurements had previously shown a very good fit with the Planck spectrum for a temperature of about 3 K.

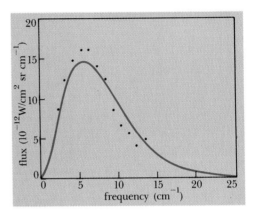

density, represented by the area under the spectral curve. For blackbody radiation, the energy density is a known constant times the fourth power of the temperature. The same formula defines the effective temperature of non-blackbody radiation. The effective temperature calculated from the spectrum measured by Woody and Richards is 3 K.) Before Woody and Richards published their findings, theorists had predicted what deviations from a perfect blackbody spectrum would be observed if the standard model were correct. The predicted deviations were smaller than those found by Woody and Richards and were also qualitatively different.

It is obviously important to repeat the measurements of Woody and Richards. Meanwhile, we may ask, Do the results make sense? Do they fit naturally into *any* theoretical framework? Arthur Stanley Eddington, who founded the modern theory of stellar structure, used to tease his empirically minded colleagues with the maxim, "Don't believe an observation till it is supported by a theory." Penzias and Wilson were perhaps following this advice when, on learning of the fireball hypothesis, they decided that their "excess antenna temperature" might, after all, be a signal from the depths of space and time rather than an obscure fault of their radiometer. Is there any theoretical reason to believe the experimental findings of Woody and Richards? Is there an alternative theoretical framework within which they would make sense?

The Microwave Background as Thermalized Starlight

If the photons that make up the microwave background were not present at the very beginning of the cosmic expansion, they must have been produced by ordinary physical processes. Among the known processes that could conceivably have produced the radiation, the burning of hydrogen into helium in stars is easily the most efficient. The conversion of 1 gm of hydrogen into helium releases 0.007 gm of radiation. Could the microwave background have been produced by stars?

The *present* energy density of starlight in interstellar space is actually comparable to the energy density of the microwave background (about 5×10^{-13} erg cm^3). But the *quality* of star-

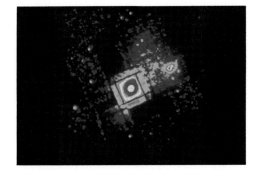

View of the nearest rich cluster of galaxies in the constellation Virgo. This picture is a mosaic of individual X-ray images. The center is dominated by a halo of emission around the galaxy M87. Other galaxies, foreground stars, and background quasars appear as bright spots.

light is radically different from that of the microwave background. Starlight is made up of photons emitted from stellar surfaces whose temperatures range from a few thousands to tens of thousands of degrees Kelvin. The wavelengths of these photons lie mainly in the visible, infrared, and ultraviolet parts of the electromagnetic spectrum. The wavelengths of the photons that make up the microwave background are a thousand times longer. Because a photon's energy is inversely proportional to its wavelength, and the energy density of starlight is comparable to the energy density of the microwave background, the number density of microwave photons in interstellar space is roughly a thousand times as great as the number density of starlight photons. Thus, a theory that makes stars the sources of the microwave background must account not only for the energy density of the radiation but also for its *quality*. That is, it must explain how the starlight came to be *thermalized*—transformed from dilute starlight into blackbody radiation.

How does thermalization take place? Imagine an enclosure with perfectly reflecting walls, filled with dilute starlight. If there is no matter present along with the starlight, the spectrum of the radiation will never change; photons will bounce off the walls, changing direction but not gaining or losing energy; so the number of photons in any given frequency range will remain constant in time. But if a single grain of dust is present it will borrow energy from the radiation field in bills of large denomination (starlight photons) and repay it in small change (photons of much longer wavelength). After many generations of photons have been processed in this way no trace of the original spectrum remains. The radiation is now blackbody radiation at a temperature determined by the original energy density, which remains constant throughout the process of thermalization.

If the universe were filled with dilute starlight and *if it were not expanding*, the same process would occur. Photons would be absorbed and reemitted until eventually their energy distribution had been randomized. But in the actual, expanding universe thermalization is not an inevitable, or even a common, outcome. For example, the radiation now being emitted by stars in our own and other galaxies will never be thermalized. Intergalactic space is so transparent that most of the photons now being emitted will never be absorbed. Starlight could have been thermalized *only* at a much earlier epoch of cosmic evolution, when the universe was opaque.

When was the universe opaque to starlight? Its average density must certainly have been greater than it is now, but how much greater? The answer depends on the form of the matter at that time. Dust grains with diameters comparable to the wavelengths of the thermalized photons catalyze the conversion of interstellar starlight into blackbody radiation most efficiently. Let us assume, then, that the radiation background was thermalized by dust. This assumption puts us in a double bind.

From the measured energy density of the background, we can deduce that the "great flash" cannot have occurred *earlier* than a certain epoch. The argument is simple. As mentioned earlier, 1 gm of hydrogen, when burned into helium, releases 0.007 gm of radiation. Thus if a fraction y of the mass in the universe was converted from hydrogen to helium during the great flash, the ratio between the mass density of radiation and the mass density of matter must have been around $0.007y$ at that time:

$$\left(\frac{\rho_{\text{rad}}}{\rho}\right)_1 = 0.007y, \tag{8.1}$$

where the subscript 1 refers to the epoch of the great flash. The subsequent expansion of the universe has reduced this ratio by the factor $1 + z_1$, where z_1 is the redshift of photons emerging from the great flash: the ratio between the number of photons and the number of material particles in an expanding volume is not affected by the expansion, but the energy of each photon is reduced by the factor $1 + z_1$ (see Chapter 7, p. 239). Thus

$$\left(\frac{\rho_{\text{rad}}}{\rho}\right)_1 = (1 + z_1)\left(\frac{\rho_{\text{rad}}}{\rho}\right)_0, \tag{8.2}$$

where the subscript 0 refers to the present epoch. Combining Equations 8.1 and 8.2, expressing the present mass density ρ_0 as a multiple of the density $\rho* = 2 \times 10^{-29}$ gm/cm^3 that would prevail in a Euclidean universe with $H_0 = 100$ km/sec per megaparsec, and setting $(\rho_{\text{rad}})_0 = 5 \times 10^{-34}$ gm/cm^3, we obtain

$$1 + z_1 \simeq 300y\left(\frac{\rho_0}{\rho*}\right). \tag{8.3}$$

Because the mass fraction y must be less than unity, this approximate equality implies that the redshift z_1 of photons emerging from the great flash must be *smaller* than a certain value that is proportional to the present value of the mass density.

On the other hand, starlight cannot be effectively thermalized unless z is *greater* than a certain critical value. In 1972, Ray Hively and I, with valuable advice from Edward M. Purcell on the electromagnetic properties of small solid grains, derived an inequality that would need to be satisfied by the critical redshift z_1 and the present value of the mass density if the background

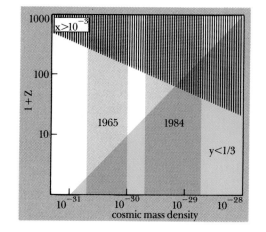

If the microwave background is to be interpreted as thermalized starlight, the redshift z when the light was emitted and the present value of the cosmic mass density must satisfy two inequalities, represented by the horizontally and vertically hatched regions of the diagram. Both inequalities are satisfied in the cross-hatched region, all of whose points correspond to values of the cosmic mass density greater than about 10^{-29} gm/cm^3. This lower bound is consistent with present estimates of the cosmic mass density, but not with estimates widely quoted up to around 1980. If the cosmic mass density is close to 10^{-29} gm/cm^3, the "great flash" would have occurred at redshifts of 50 to 100.

radiation were to be thermalized at long wavelengths. This inequality is

$$(1 + z_1)^{5/2} \left(\frac{\rho_0}{\rho*} \right) x \gtrsim 25, \qquad (8.4)$$

where x is the fraction of the mass that condenses into dust grains.

Inequalities 8.3 and 8.4 are both represented graphically in Figure 8.4. Regions of the redshift-density plane whose points satisfy either inequality are hatched, horizontally for 8.3, vertically for 8.4; so points that satisfy both inequalities lie in the cross-hatched region. (This is a visual way of combining the equations and finding the permitted range of value of z, if there is one.)

Between 1965 and 1979 the ratio $\rho_0/\rho*$ was believed by many astronomers to lie in the range 0.01 to 0.03. This range is represented in the above figure by a strip that does not intersect the cross-hatched area. Therefore, the hypothesis that the microwave background is thermalized starlight seemed untenable. However, more recent and reliable estimates of the cosmic mass density place the ratio $\rho_0/\rho*$ in the range 0.3 to 0.7 (although values as small as 0.1 or as large as 1 cannot yet be ruled out). The permitted range of mass densities is represented in the above figure by a strip that does intersect the cross-hatched area; there is now a small triangle in the redshift-density plane whose points satisfy all three constraints. We conclude that *the microwave background could be starlight produced and thermalized at redshifts of order 100.*

An Early Generation of Massive Stars?

The hypothesis that the microwave background is thermalized starlight is more susceptible to observational refutation than the fireball hypothesis, because it has more testable implications.

Two galaxies, M86 (left) and M84 (right) in the Virgo cluster as seen in X-rays. The X-ray emission from these galaxies comes from gas with a temperature of 5 to 10 million degrees.

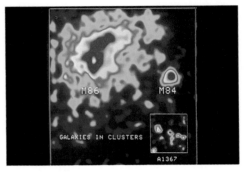

Let us look at some of these implications and at the evidence that can be brought to bear on them.

1. *The radiation background must have been produced by an early generation of stars 10 to 15 times more massive than the Sun.*

From the estimated redshift of the great flash ($z_1 \simeq 100$) and the equation $1 + z_1 = (t_0/t_1)^{2/3}$ that relates z_1 to the corresponding epoch t_1 and to the present epoch t_0 (about 10 billion years) we infer that the great flash occurred some 10 million years after the beginning of the cosmic expansion (in the standard model these two events coincide). The stars that created the great flash must have had lifetimes of this order. Shorter-lived stars would have burned out too quickly; they could not have produced enough radiation to account for the background's present energy density. Longer-lived stars would have emitted most of their radiation when the density of dust was not high enough for it to be thermalized. The theory of stellar evolution tells us that stars with lifetimes of around 10 million years are 10 to 15 times as massive as the Sun.

This inference has two testable implications. Specialists in the theory of stellar evolution have calculated the relative abundances of heavy elements manufactured by massive stars. They have found that the predicted relative abundances depend sensitively on the mass of the star, and that the mix of heavy elements found in the Sun and in other nearby stars must have been synthesized in stars 10 to 15 times as massive as the Sun. Substantially heavier and lighter stars manufacture heavy elements too, but not in the right proportions to account for the abundance ratios inferred from analyses of stellar spectra. Thus the hypothesis that the microwave background is thermalized starlight leads to a correct prediction of the relative abundances of heavy elements. It also predicts, in agreement with observation, that the *relative* abundances of elements heavier than helium should be the same in all stars that, like the Sun, have not yet begun to synthesize these elements.

2. *Elements heavier than helium (and perhaps helium itself) were synthesized by the early generation of massive stars, long before galaxies began to form.*

This prediction conflicts sharply with the following widely held account of how our stellar system evolved. The Galaxy formed as a structureless gas cloud composed of hydrogen and

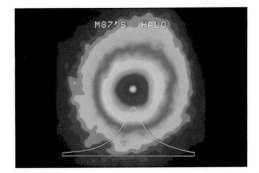

A blow-up of the central region of the Virgo cluster of galaxies shown on page 263. The false color X-ray map shows the distribution around the galaxy M87 which dominates the cluster. The X-ray emitting gas contributes ten percent to the mass of the cluster, about as much as the visible galaxies. The remaining eighty percent of the mass is visible neither in ordinary light nor in X-rays.

helium. During the ensuing 100 million years, this cloud collapsed into a thin disc in which stars began to condense. Massive stars synthesized heavy elements in their interiors and then exploded, ejecting gas rich in heavy elements into the interstellar medium. New stars condensed from the enriched interstellar gas. The most massive of these repeated the cycle, further enriching the interstellar medium. Thus successive generations of stars formed with progressively higher abundances of heavy elements.

By the early 1970s it was possible to confront this speculative account with statistical data derived from spectroscopic analyses of relatively nearby stars in the disc of the Galaxy. The theory did not fit the data at all (see figure on page 268). It predicted far too large a proportion of stars with low abundances of heavy elements. Students of the Galaxy's chemical evolution now agree that a large fraction of the heavy elements must have been synthesized by an early generation of massive, short-lived stars, as the present hypothesis predicts.

3. *The dust grains that thermalized the background radiation must have condensed from gas ejected by the early stars during the explosive phase of their evolution.*

These dust grains must have been made of materials such as carbon, silicates, and iron whose building blocks were synthesized in the massive stars themselves. Could dust grains of the sort that could thermalize starlight efficiently have formed quickly enough? Current theories for the formation of solid grains are consistent with this requirement but are not yet developed enough to answer the question definitively.

4. *The spectrum of the microwave background should deviate from an ideal blackbody spectrum in qualitatively the same way as the spectrum measured by Woody and Richards.*

Because the universe was opaque to starlight during a relatively brief period, it would be surprising if the process of thermalization had gone to completion at all wavelengths, producing a perfect Planck spectrum. How should we expect the spectrum of incompletely thermalized starlight to deviate from the Planck spectrum?

A black body emits blackbody radiation. Let us give the name *emissivity* to the ratio between the power that an object actually radiates when it is maintained at a given temperature and the power that a black body at the same temperature would radiate. A real dust grain radiates more

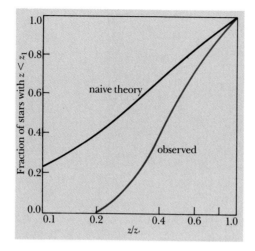

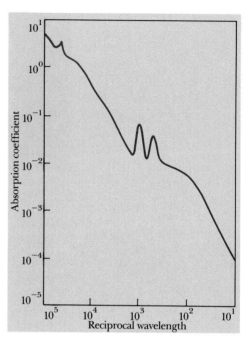

Chemical evidence for an early generation of massive stars. Elements heavier than helium ("metals") are produced in massive stars. If massive stars had always been born at the same rate as they are being born now, more than a fifth of the stars in our neighborhood would have metal abundances less than a tenth the metal abundance of interstellar gas (see curve labeled "Naive theory"). In fact, virtually none of the nearby stars whose spectra have been analyzed have metal abundances as small as one-fifth the metal abundance of interstellar gas. The data indicate that the oldest stars in our neighborhood condensed from gas rich in metals—metals presumably manufactured by the early generation of massive stars.

Emissivity of the commonest kind of interstellar dust grains, against the reciprocal wavelength of the incident radiation. Note that the emissivity declines steeply with increasing wavelength (decreasing reciprocal wavelength). Note also the two peaks at wavelengths of 10 and 20 microns (1 micron = 10^{-6} meter). These peaks are produced by silicates—compounds that are common in terrestrial rocks as well as in interstellar space.

efficiently at some wavelengths than at others: its emissivity varies with wavelength. The right-hand figure above shows how the emissivity of interstellar dust grains varies with wavelengths in the infrared region of the electromagnetic spectrum, where most of the energy of the radiation background would have been concentrated during the period of thermalization. The most conspicuous features of the emissivity spectrum are two broad peaks whose central wavelengths are close to 10 and 20 microns (1 micron = 10^{-6} meter = 10^{-3} millimeter). These peaks are produced by silicates, the molecules that glass and bricks are made of and a common constituent of terrestrial rocks. Astronomical observations in the infrared have shown that silicates are ubiquitous in interstellar space, as they are on Earth; so we may safely assume that they were present in the grains that thermalized the background radiation.

If the background radiation was thermalized by grains with the emissivity shown in the right hand figure on page 268 we would expect it to be more intense than blackbody radiation of the same energy density in the region of the two peaks of the emissivity spectrum, and less intense at shorter and longer wavelengths. The figure on page 262 also shows how the spectrum measured by Woody and Richards would have looked at the epoch of thermalization (i.e., blueshifted by a factor of 100). The measured (and blueshifted) spectrum does indeed rise above the Planck spectrum where the preceding argument suggests that it should. It also dips below the Planck spectrum at shorter wavelengths, as the preceding argument suggests that it should. But it does *not* dip below the Planck spectrum at long wavelengths. This part of the spectrum, which now extends from 0.5 to 50 cm, is the only part of the spectrum that can be, and has been, measured from the ground. As we have seen, it fits a Planck curve very well.

Edward L. Wright of the Massachusetts Institute of Technology and the University of California at Los Angeles found an ingenious solution to this difficulty, following up a suggestion by Edward M. Purcell that Hively and I had quoted in our 1973 paper. Purcell has argued for many years that interstellar grains might be expected to have needlelike, filamentary structures for the same reason that cobwebs have filamentary structure. (Cobwebs, Purcell remarks, are made by electrostatics, not by spiders.) Purcell pointed out that highly elongated grains would be efficient "low-mass antennas for millimeter waves," precisely what are needed to thermalize what is now the microwave region of the cosmic radiation background. In 1981 Wright completed a careful quantitative study of Purcell's idea. He found that a population of graphite needles containing a small fraction of the carbon that would have been available to make them would thermalize what is now the microwave part of the radiation background very effectively. Wright also cited experimental and observational evidence supporting the conjecture that needles with the required properties tend to form under appropriate physical conditions and are actually present in interstellar space.

5. *The present hypothesis accounts for the cosmic X-ray background.*

X-ray astronomy dates from the late 1940s, when scientists at the U. S. Naval Laboratory began to observe the Sun with X-ray detectors carried aloft by balloons and rockets. In 1962 R. Giacconi, H. Gursky, F. Paolini, and B. Rossi of the Massachusetts Institute of Technology and American Science and Engineering used a rocket-borne detector to look for X-rays from the Moon. The detector actually recorded X-rays from a source more powerful than the Sun, in the direction of the constellation Scorpius. Since that accidental discovery of an X-ray source outside the solar system, many discrete sources inside and outside the Galaxy have been found. But early observations also revealed a diffuse background whose intensity was uniform to within a few percent over the entire sky. The integrated intensity of this high-energy background is only 0.02 percent that of the microwave background, and its spectrum is conspicuously nonthermal. Over a range of nearly four decades, the intensity is proportional to a simple power of the wavelength (see figure on page 270).

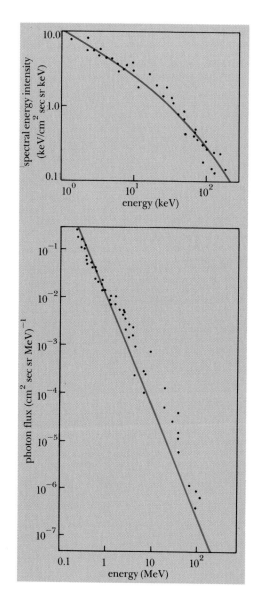

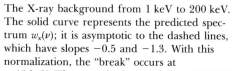

The X-ray background from 1 keV to 200 keV. The solid curve represents the predicted spectrum $w_x(\nu)$; it is asymptotic to the dashed lines, which have slopes -0.5 and -1.3. With this normalization, the "break" occurs at ~ 19 keV. The γ ray background from 0.2 MeV to 200 MeV. The solid curve represents the continuation of $w_x(\nu)$.

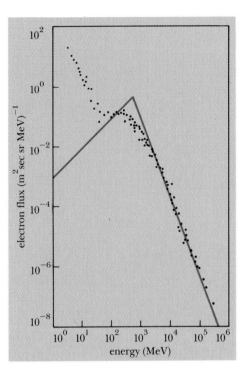

Observed spectrum of cosmic-ray electrons. The dashed line represents the energy spectrum of "primary" electrons—electrons actually produced in the source, rather than by collisions between primary cosmic-ray protons and ambient nuclei.

Arthur H. Compton (1892–1962).

Early in 1976 I suggested to Craig Hogan, then a senior at Harvard, that he look for a mechanism by which the early generation of massive stars that Hively and I had postulated to explain the microwave background might also produce the X- and gamma-ray background. After rejecting several candidate mechanisms, Hogan came up with one that looked promising. Supernovae (massive stars in the explosive stage of their evolution) are known to eject magnetized clouds of ionized gas that contain extremely energetic electrons. Arthur H. Compton had shown in the 1920s that when X-ray photons are scattered by electrons they give up some of their energy and momentum to the electrons. The inverse process can also occur: if the electrons are very energetic, they will lose energy to the photons. Hogan conjectured that the extremely energetic electrons that are a by-product of the supernova explosion would "collide" with ambient ultraviolet photons and promote them to X-ray photons.

The theory of such collisions is well-known, but to test his idea, Hogan and I needed to know how much energy a supernova puts into electrons and how that energy is distributed among them. This information could be extracted from experimental data on cosmic-ray electrons. These extremely energetic particles, detected by instruments mounted on rockets and satellites, are thought to be produced by supernovae in our Galaxy. Their energy spectrum has been accurately measured (see figure above). From the estimated rate of occurrence of supernovae in the Galaxy and calculations that relate the intensity of cosmic rays in interstellar space to the rate at which they are emitted by their sources it was easy to estimate how much energy a typical supernova puts into energetic electrons and how it is distributed. With these data in hand, the X-ray spectrum could be predicted under the single assumption that present-day supernovae resemble their primordial counterparts.

The results of the calculation are represented in the figures on page 270 by the solid line, which is indistinguishable from the best simple curve that can be drawn through the data points. The closeness of the fit is unquestionably accidental. The comparison does, however, test three distinct aspects of the theory. First, the *slope* of the predicted spectrum is determined by the observed slope of the energy spectrum of cosmic-ray electrons and by the theory of Compton scattering. Second, the *amplitude* of the spectrum is determined by the ratio between the total energy emitted by a typical supernova in the form of ultrarelativistic electrons and the energy emitted during the lifetime of the star in the form of radiation. The theory predicts that this ratio is equal to the ratio between the integrated intensities of the X-ray and microwave backgrounds. Finally, the theory successfully predicts the *break* in the spectrum, the energy at which its slope changes. This prediction depends on the observed value of the low-energy cutoff in the spectrum of cosmic-ray electrons and on the epoch of the great flash.

6. *The present hypothesis may explain the enhanced intensity of background gamma-radiation at energies above 1 million electron-volts.* *

The magnetic fields that accelerate electrons in the gases ejected by supernovae also accelerate protons, the principal component of cosmic rays. Collisions between these cosmic-ray protons and ambient gas molecules produce, among other short-lived particles, neutral pions, each of which quickly decays into a pair of high-energy photons. These photons have energies greater than 70 million electron-volts, half the rest energy of the pion. In 1969 Floyd Stecker of the Goddard Space Flight Center suggested that the enhanced intensity of the high-energy background at energies greater than about 1 million electron-volts (see figure on page 270) comes from the decay of neutral pions. To explain the sharp onset of the enhancement, he postulated that there had been a burst of cosmic-ray protons at a redshift of about 100. Photons resulting from such an event would now have energies greater than about .7 MeV. The hypothesis of an early generation of massive stars automatically supplies the initial conditions postulated by Stecker.

7. *If the radiation backgrounds were produced by an early generation of stars, the bulk of the mass in the universe must now reside in nonluminous objects.*

This conclusion follows from our earlier estimate of the mass that must be bound in the early generation of massive stars to account for the quantity and quality of the background radiation;

* The electron-volt is an energy unit commonly used in high-energy physics. It is the energy acquired by an electron in falling through an electrostatic-potential drop of 1 volt; 1 electron-volt (abbreviated eV) equals approximately 1.6×10^{-12} erg.

By measuring the Doppler shift of radiation emitted by cool clouds of atomic hydrogen in a spiral galaxy, viewed edge-on (or nearly edge-on), we can construct the "velocity profile" shown in (b). The velocity V at distance R from the center of the galaxy is related to the mass $M(R)$ within radius R by the formula $V^2 = M(R)/R$. The observation that V is constant at large values of R implies that the mass $M(R)$ is proportional to R. Most of the mass—90 percent or more—lies outside the visible part of the galaxy. It may reside in the remnants of the hypothetical early generation of massive stars.

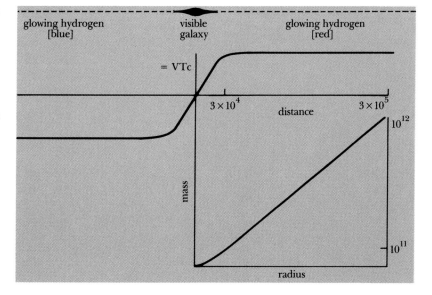

from the secure theoretical inference that the early stars would long since have exhausted their supplies of nuclear fuel; and from the observation that visible stars contain no more than a tenth of the mass that must (according to our hypothesis) have been bound in the early, now-defunct stars.

Modern findings strongly support this conclusion. There are at least two independent lines of evidence.

First, within the last decade we have learned that the objects called "spiral galaxies" are merely the visible cores of systems five to ten times as large and five to ten times as massive. These estimates rest in part on observations at radio wavelengths ($\lambda \simeq 21$ cm) of clouds of atomic hydrogen moving in circular orbits in the central planes of spiral galaxies, like planets around the Sun, at distances far beyond the galaxy's visible limits (see figure above).

The second line of evidence was initiated by the Swiss astrophysicist Fritz Zwicky. In 1933 Zwicky used the virial theorem (Chapter 4) to estimate the mass of a rich cluster of galaxies in the constellation Coma Berenices (figure on p. 113, in Chapter 4) from the measured radial velocities of several member galaxies. He found that the cluster's "virial mass" was about 400 times as large as the sum of the masses of its visible stars. With modern estimates of the cluster's distance Zwicky's estimate of the ratio of nonluminous to luminous mass would be reduced by a factor of 10. Modern applications of the same method to systems of galaxies have confirmed the most striking part of Zwicky's conclusion: that luminous stars make only a minor contribution to the masses of rich galaxy clusters. They have also shown that the ratio of nonluminous mass to luminous mass can differ greatly from cluster to cluster. Finally, modern studies have shown that the ratio of nonluminous mass to luminous mass increases systematically along the

The ratio of non-luminous to luminous mass in self-gravitating systems increases systematically with increasing mass, but at any given mass the estimated values of this ratio span a wide range.

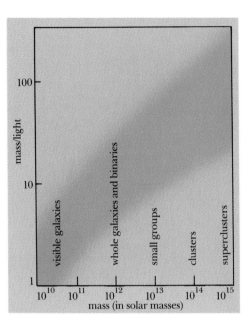

hierarchy of galactic systems; it is smallest for individual galaxies, increases from binaries to multiple systems to small groups in clusters, and is greatest in rich clusters and superclusters (see figure above).

Most supporters of the fireball hypothesis now acknowledge that the bulk of the mass in the universe must be nonluminous. If the nonluminous mass resides in ordinary matter—small solid particles, moon-sized rocks, planets, stars of very low mass, stellar remnants—then it is clear from the figure on page 258 that the primordial abundances of helium and deuterium predicted by the "standard model" cannot be easily reconciled with observation. Some supporters of the fireball hypothesis have postulated that the invisible mass resides not in ordinary matter but in particles called "massive neutrinos" that figure in some speculative theories of elementary particles. But even if we suppose that the mass of the universe resides mainly in these hypothetical particles, the fireball hypothesis still contradicts the astronomical evidence bearing on the primordial abundances of helium and deuterium, because if the cosmic mass density of ordinary matter were low enough to give a primordial hydrogen-to-helium ratio as large as 16 to 1, then the predicted primordial abundance of deuterium would be about 25 times as great as the observed value.

None of the observational evidence that now seems to speak against the fireball hypothesis is absolutely firm. The primordial abundance of helium may prove to be substantially greater than present evidence suggests; interstellar deuterium may prove to be nonprimordial; the measurements of Woody and Richards may have been vitiated by a calibration error; and a unified theory of electromagnetic, weak, and strong interactions may one day predict the existence of neutrinos with just the right masses to be bound in galaxies. All we can say with confidence is that *present* astronomical evidence rather consistently favors the alternative hypothesis that the radiation background is partially thermalized starlight.

The Origin of Astronomical Systems

The cosmic radiation background is not our only clue to the early history of the universe. The most obvious fact about the universe is that the distribution of matter is clumpy rather than smooth; this is an equally important clue.

Newton believed that self-gravitating clumps would form spontaneously in a uniform distribution of gravitating particles, but in this belief he was mistaken. Clumps do not tend to form in a uniform cosmic medium in the way that crystals tend to form in a supercooled liquid. The opposite hypothesis, that self-gravitating systems have always existed, is equally untenable. No present-day astronomical system could have existed at the earliest times, when the cosmic mass density exceeded that which now prevails in the atomic nucleus. If spatial variations in mass density existed at all in the early universe, they must have been tiny ripples in a nearly uniform sea. About this, all cosmologists agree. The hard and controversial questions are, What kind of ripples would produce, 10 billion years later, the kind of structure we now observe? How did these ripples get there in the first place?

Cosmologists also agree that density variations that could evolve into self-gravitating astronomical systems could not have arisen spontaneously at any stage in the evolution of a primordial fireball. The primordial fireball resists spontaneous clumping for much the same reason that air at room temperature and pressure does. Hence if we accept the fireball hypothesis we must postulate that the distribution of matter in the universe was clumpy at the very outset, or at least at the earliest moment when present-day physical laws applied. If the early universe was hot, it must also have been rough. How rough? And in what ways? Since 1965 theorists have worked hard to answer these questions. They have now reached the conclusion that a certain general class of assumptions, regarded by nearly all cosmologists as the most plausible, will *not* work. Detailed computer-assisted calculations based on these assumptions predict structure where none exists (on scales larger than the largest observed scales of galaxy clustering) and fail to predict structure over the vast range of masses and linear scales spanned by actual astronomical systems.

Most cosmologists are still convinced that an explanation for cosmic nonuniformity can be found within the framework of the fireball hypothesis, but some have begun to investigate other possibilities. Cosmologists face some of the same difficulties as historians. They must

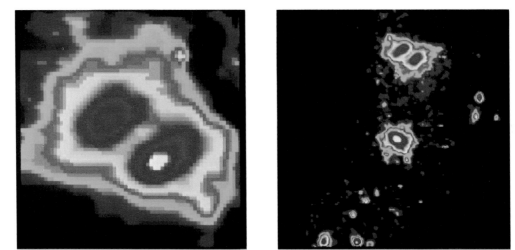

(*Right*) A supercluster of galaxies. Each of the three conspicuous images seen here was made by a cluster of galaxies. The X-rays arise from a hot gas within each cluster at temperatures of 10 to 100 million degrees Kelvin. (*Left*) Detail from the photo on the right. Two mutually gravitating clusters of galaxies embedded in an envelope of X-ray emitting gas.

invent a story that gives coherence and consistency to a heterogeneous collection of facts whose meanings depend on the story itself. Some of the facts become what historian Barbara Tuchman, quoting Pooh-Bah in *The Mikado,* calls "corroborative detail intended to give artistic verisimilitude to an otherwise bald and unconvincing narrative"; other facts become irrelevant; and still others refuse to fit in. And *X*'s corroborative detail may be *Y*'s irrelevancies. Theories of cosmic evolution differ primarily because they were invented to fit different sets of facts. We saw earlier that the fireball hypothesis was put forward to explain the relative abundances of the heavy elements; was revived, after that explanation failed, to account for the cosmic abundance of helium; and was finally used to explain the microwave background. The hypothesis that the microwave background is thermalized starlight has a different genealogy. It was put forward to save an earlier theory for the origin of cosmic nonuniformity, the theory of gravitational clustering. The rest of this chapter deals with that theory and its implications.

We saw in the preceding section that there is now a good deal of evidence for a pregalactic generation of massive stars, but when Ray Hively and I published our paper on the origin of the microwave background in 1973, most astronomers found the idea bizarre and implausible. Conventional wisdom decreed that galaxies came into being first as structureless gas clouds, and that stars later condensed within them. I had begun to question this view of the origin of stars in 1951, after learning that at least half the nearby stars belong to binaries whose members are separated by distances comparable to the diameter of the solar system. We might imagine that double stars form because of encounters between single stars, but closer examination shows that a *triple* encounter is needed (see the figures on page 277). The energy of a binary is negative; the initial energy of the partners in a stellar encounter is positive. A third participant is needed to carry off the excess energy. Triple encounters are exceedingly rare, however, under present conditions. They could not begin to account for the prevalence of binaries. Again, we might

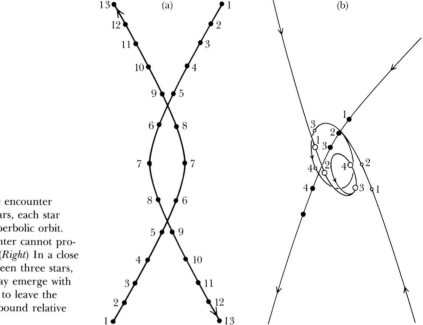

(*Left*) In a close encounter between two stars, each star moves on a hyperbolic orbit. Such an encounter cannot produce a binary. (*Right*) In a close encounter between three stars, one of them may emerge with enough energy to leave the other two in a bound relative orbit.

imagine that binaries have resulted from the fission of rapidly rotating single stars, but this hypothesis, too, runs into insurmountable theoretical and observational difficulties.

Since double stars apparently cannot form under present-day conditions, I began to wonder whether they might not have formed in an earlier, more favorable environment. It occurred to me that binaries and small multiple systems might form very readily if stars condensed directly from the expanding cosmic medium when its average density was comparable to the present average density of the solar system. It seemed plausible that protostars that had just separated from the cosmic medium would tend to clump into self-gravitating systems as the medium continued to expand. This would explain why they are so common today.

But once small clumps of protostars had separated out of the expanding cosmic medium, they would be in the same situation that the newly formed protostars had been in earlier. Gravitation, unlike other forces, is no respecter of scales. If newly formed protostars had indeed congregated in self-gravitating clumps as the universe expanded, then these clumps themselves must have congregated into still larger self-gravitating clumps as the universe continued to expand, and these clumps of clumps into still larger clumps. In this way a hierarchy of self-gravitating systems would be formed step by step, by repetition of a single, scale-independent process, "gravitational clustering." This picture contradicted the then-current, and still widely accepted, view that stars condense from initially structureless, collapsing gas clouds which themselves formed as condensations in larger, initially structureless, collapsing gas clouds.

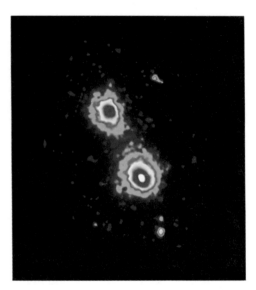

Two rich clusters of galaxies, A399 and A401.

I put forward this picture of hierarchical evolution by gravitational clustering at a meeting of the American Astronomical Society in December 1952, but it remained a picture (rather than a deductive physical theory) for many years. Three main questions needed to be answered:

1. Where do the first clumps come from?

2. Why do newly formed clumps tend to congregate in higher-order clumps as the cosmic medium expands? In other words, what is the physical mechanism of gravitational clustering?

3. What are the quantitative properties of the resulting hierarchy?

The first question was answered by the Soviet astrophysicist Ya. B. Zel'dovich in 1962. Suppose, said Zel'dovich, that the universe began to expand from an initial state at a temperature of absolute zero. Despite its low temperature, the cosmic medium would remain gaseous until its density was about a tenth that of water, when it would solidify. As the medium continued to expand, it would crack, and shatter into fragments. Zel'dovich estimated the size of the fragments by using a sophisticated theory of the growth of cracks in a stressed solid.* The following physical argument allows us to estimate the sizes of the fragments more simply and, probably, more reliably.

The forces responsible for the cohesion of an ordinary solid or liquid bind each atom (or molecule) to its nearest neighbors. The energy needed to create a fragment of surface area S is thus proportional to S and hence to the square of the fragment's radius. But the mass of a fragment is proportional to the cube of its radius; hence the larger the fragments, the less

---

* Zel'dovich abandoned this approach to cosmogony in 1965, after the discovery of the microwave background. Like nearly all his colleagues, he considered the microwave background to be *prima facie* evidence of a primordial fireball.

energy per unit mass is needed to create them. As in a soap bubble or in a system of soap bubbles, the total surface area tends toward the smallest value that is consistent with external constraints. The only obvious constraint on the size of a fragment is that its tensile strength be large enough to keep it from disintegrating. A newly formed fragment is still expanding. The kinetic energy associated with this expansion is almost exactly equal to its gravitational energy, calculated as if the fragment were isolated in empty space. If the fragment is to maintain its integrity, its cohesion energy must exceed its expansion energy; otherwise the fragment will split into at least two parts. Conversely, fragments whose cohesion energy significantly exceeds their expansion energy can, like soap bubbles, reduce their combined surface area by merging. This argument suggests that most of the mass will come to reside in fragments whose cohesion energies are comparable to their expansion energies.

The Mechanism of Gravitational Clustering

After the cosmic medium has shattered into solid fragments, it may be described as a cold, expanding "gas" whose "molecules" attract one another gravitationally. The interactions between "molecules" of the cosmic gas differ, however, from those between molecules in an ordinary gas. Molecular interactions in an ordinary gas are short-range: they come into play only when the distance between two molecules is less than a certain fixed value, usually between 10^{-7} and 10^{-8} cm, which is also the characteristic distance between molecules in the liquid and solid phases. In a "gas" composed of gravitationally interacting "molecules" there is no fixed interaction distance. The range and strength of gravitational interactions are determined by the scale and amplitude of large-scale density variations in the "gas." Because intermolecular forces in an ordinary gas have a fixed and finite range, the gas's macroscopic properties depend on but do not directly influence its microscopic properties. In the cosmic "gas," the "microscopic" and "macroscopic" levels of description cannot be neatly separated; they overlap and mingle. For example, the density of liquid droplets in a vapor depends on the structure of the molecules but not on the density of the vapor in which they condense. By contrast, self-gravitating clumps in the expanding cosmic "gas" can separate out whenever their density is slightly greater than the average density of the "gas," whatever its value.

The cosmic "gas" differs from an ordinary vapor in another important way. Each state of a vapor is characterized by a definite pressure and density (see figure on page 280), and in every normal state a definite fraction of the molecules is in the liquid phase, with the remaining fraction in the gas phase. But there is one exceptional state, represented by the so-called *critical point* in the pressure-density diagram shown in the figure on page 280, which is neither gas nor liquid nor a mixture of gas and liquid. When a vapor approaches this singular state it exhibits a phenomenon known as "critical-point opalescence." It takes on a milky appearance, caused by large-scale density fluctuations. According to the theory we are about to consider, the cosmic "gas" is in a perpetually evolving critical state, and the clumpiness of the astronomical universe is the outcome of instabilities analogous to those that give rise to critical-point opalescence in ordinary vapors.

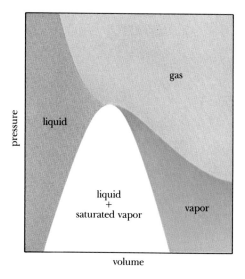

Pressure-volume diagram for an ordinary vapor, showing the ranges of pressure and density (proportional to the reciprocal of the volume) that correspond to the gas and liquid phases. At the critical point the vapor is unstable against large-scale density fluctuations, which give rise to the phenomenon called "critical-point opalescence."

Let us first consider a gas whose molecules are particles of equal mass that attract each other with a force proportional to some negative power β of the interparticle separation. Suppose that such a gas occupies a box of volume V. Equation 8.5, published by Rudolf Clausius in 1870, relates the total kinetic energy K of the molecules, the potential energy U associated with their mutual interactions, the pressure P, and the volume V:

$$2K + (\beta - 1)U = 3PV. \tag{8.5}$$

In an *ideal gas* the molecules do not interact at all; hence $U = 0$ and $2K = 3PV$. The pressure, volume, and temperature of an ideal gas are connected by the ideal-gas law,

$$PV = N\Theta, \tag{8.6}$$

where N is the number of molecules in the box and Θ is the absolute temperature measured in the same unit as energy.* Comparing Equations 8.5 and 8.6, we see that

$$\Theta = \frac{2}{3}\frac{K}{N};$$

---

* The conventional unit of temperature, the degree of the Celsius and Kelvin scales, is $1/100$ the difference in temperature between the boiling and freezing points of water at sea level. The ratio between this unit and the erg is called Boltzmann's constant and is denoted by k_B. Thus $\Theta = k_B T$, where T is the absolute temperature measured in degrees and $k_B \simeq 1.4 \times 10^{-16}$ erg/degree.

that is, the temperature of an ideal gas is equal to two-thirds of the kinetic energy per molecule. We can use the same rule to define the temperature even when the molecules interact.

Next, consider a gas confined to a cylinder fitted with a moving piston. If no energy flows into or out of the gas when its volume undergoes a small change ΔV, the energy $K + U$ changes by an amount equal to the work done by the pressure force on the piston: $\Delta(K + U) = - P \Delta V$, or

$$\frac{d(K + U)}{dt} + P\frac{dV}{dt} = 0. \tag{8.7}$$

Solving Equation 8.5 for P and substituting the result in Equation 8.7, we find

$$\frac{d(K + U)}{dt} + \frac{1}{3}\frac{2K + (\beta - 1)U}{V}\frac{dV}{dt} = 0. \tag{8.8}$$

We are now ready to consider an expanding cosmic "gas" composed of mutually gravitating particles. Let **v** denote the velocity of a particle *relative to the standard of rest at the particle's instantaneous position*. We saw in Chapter 7 that the velocity of a slowly moving free particle is inversely proportional to the cosmic scale factor $a(t)$. Hence the equation of motion for a free particle is not, as in Newtonian mechanics, $d\mathbf{v}/dt = 0$, but

$$\frac{d(a\mathbf{v})}{dt} = 0, \tag{8.9}$$

or

$$\frac{d\mathbf{v}}{dt} = -H\mathbf{v}, \tag{8.10}$$

where $H = \dot{a}/a$, the Hubble expansion parameter.

What is the analog of Newton's second law of motion for particles in the expanding cosmic "gas"? Deviations from the uniformly decelerated motion described by Equation 8.10 result from the *nonuniform* or *fluctuating* component of the cosmic mass density, described by the difference between the actual mass density and the mean density:

$$\rho'(x, y, z, t) \equiv \rho(x, y, z, t) - \bar{\rho}(t). \tag{8.11}$$

Notice that ρ' takes on negative as well as positive values, and that it makes no net contribution to the total mass of a region whose dimensions are large compared with the scale of local irregularities. The fluctuating component of the mass density gives rise to a fluctuating gravitational field, which is related to the mass distribution that produces it exactly as in Newton's theory. Thus the analog of Newton's second law of motion is

$$\frac{d\mathbf{v}}{dt} = -H\mathbf{v} + \mathbf{f},$$

(8.12)

where $\mathbf{f}$ denotes the fluctuating gravitational force per unit mass.

Because regions of positive and negative mass density contribute equally to the field, $\mathbf{f}$ depends almost entirely on the distribution of mass within a sphere whose radius is a few times the largest scale on which the mass density is variable. Contributions to the field from more distant regions cancel out. *The fluctuating gravitational field has finite range.* Therefore it behaves like the intermolecular force in an ordinary gas.

Equation 8.12 plays the same role in a theoretical description of the expanding cosmic "gas" as Newton's second law of motion plays in the theory of an ordinary gas. Carrying the analogy further, William Irvine and I, in 1960, derived an energy equation for the expanding cosmic "gas." This equation has the same form as Equation 8.8, with $\beta = 2$ (because the gravitational interaction obeys the inverse-square law), with K equal to the kinetic energy arising from motion relative to the local standard of rest, and with U equal to the gravitational potential energy arising from the fluctuating component of the mass distribution. K is given explicitly by the formula

$$K = \sum \tfrac{1}{2} m_i v_i^2;$$

(8.13)

the subscript i refers to the ith particle, and the right side is a sum over all the particles in a given (expanding) volume V. To express the potential energy that arises from the fluctuating mass distribution, we divide the volume V into small cells of equal volume ΔV, and we define the "excess mass" m' of a cell in terms of the excess mass density in that cell:

$$m' = (\rho - \overline{\rho})\Delta V \equiv \rho' \Delta V.$$

(8.14)

The gravitational potential energy that arises from the fluctuating part of the mass distribution is then given by the formula (compare Equation 4.14)

$$U = \sum_i \sum_j -\frac{1}{2} \frac{G m_i' m_j'}{r_{ij}}.$$

(8.15)

The right side is a sum that contains one term $-G m_i' m_j'/r_{ij}$ from each distinct pair of cells. Finally, we set $V = a^3$, so that

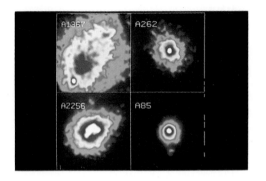

Four clusters of galaxies. The two on the left have broader distributions than those on the right which contain dominant, central galaxies like M87 in the Virgo cluster. The clusters at the top of the figure are less dynamically evolved than the clusters at the bottom as suggested by various indicators of evolution—gas temperature, galaxy velocity dispersion, fraction of spiral galaxies, and presence of small coronae around individual galaxies.

$$\frac{1}{V}\frac{dV}{dt} = \frac{1}{a^3}\frac{da^3}{dt} = \frac{3\dot{a}}{a} = 3H. \tag{8.16}$$

Thus Equation 8.8, the energy equation for the expanding cosmic "gas", takes the form

$$\frac{d}{dt}(K + U) + H(2K + U) = 0, \tag{8.17}$$

with the kinetic energy K given by Equation 8.13 and the potential energy U by Equation 8.15.

The most obvious difference between Equation 8.17 and the analogous equation for an isolated system of mutually gravitating particles is that the total energy $K + U$ of the particles within an expanding volume need not be constant in time. But Equation 8.17 also differs from its classical counterpart in another, equally important way. The potential energy of an isolated system, which figures in the classical equation, is proportional to the *square* of its mass, inversely proportional to its diameter; so if we double the mass of an isolated system, keeping its average density constant, we more than double its potential energy. By contrast, the potential energy U defined by Equation 8.15 is directly proportional to the mass of the expanding region if the diameter of this region is large compared with the largest scale of density variations, because the fluctuating gravitational acceleration $\mathbf{f}$ at any given point depends almost entirely on the fluctuating component of the mass distribution within a sphere whose diameter is a few times the scale of the largest-scale density variations. In this important respect the gravitational interactions that arise from fluctuations in the mass density resemble the interactions between particles in an ordinary gas. And just as in an ordinary gas, the potential energy that arises from these interactions is proportional to the volume of the region to which it refers if the dimensions of the region are large compared with the range of the interactions.

We can express the potential energy U in a form that exhibits its dependence on the volume V of the region and the scale L of local density variations:

$$U = -2\pi G\bar{\rho}\alpha^2 L^2 \cdot \bar{\rho}V. \tag{8.18}$$

Here L is a kind of average scale of density variations, and α represents the relative amplitude of the density fluctuations. This equation shows not only that the potential energy is proportional to the volume, but also that its value is always negative.

What can we infer from Equation 8.17 about the evolution of the expanding "gas"? Suppose, to begin with, that the kinetic energy K is numerically much larger than the potential energy U. According to Equation 8.17, the cosmic "gas" then behaves like an ordinary gas expanding against a moving piston: it simply cools.

Suppose, however, that K and U are more or less equal. An intriguing possibility now arises. Equation 8.17 shows that the cosmic expansion causes the energy $K + U$ to decrease if $2K + U$ is positive. If $2K + U$ were positive and $K + U$ negative, the expansion would cause $K + U$ to become more negative, that is, to increase in numerical value. This could happen only if the potential energy U were to increase numerically, that is, if the mass distribution were to become more clumpy. Thus the growth of clumpiness in a cold, expanding, cosmic "gas" is at least a theoretical possibility. Let us now try to figure out what would actually happen.

At the moment when it is formed, the cosmic "gas" consisting of newly formed solid fragments has zero energy: each fragment has zero peculiar velocity, and the cosmic mass density is uniform. As the fragments separate, a fluctuating gravitational field develops, and this field accelerates the fragments. The pressure, which is proportioned to $2K + U$, becomes positive, causing the energy of a region whose diameter is large compared with the diameter of a typical fragment to decrease and to take on negative values. The following argument suggests that the cosmic "gas" is now unstable against the growth of large-scale density fluctuations.

What makes a gas stable or unstable against the spontaneous growth of density fluctuations? Classical thermodynamics answers this question in a simple way. Consider a small spherical region of the gas. If we compress this region, keeping its temperature constant, and the pressure inside the region increases, then the gas is stable against the growth of density fluctuations on any scale, because the compressed gas bounces back. If, however, the pressure decreases, the gas is unstable against the growth of density fluctuations on all scales. In ordinary gases, a small decrease in the volume at constant temperature always causes the pressure to increase. In a vapor near its critical point, the increase in pressure accompanying a small decrease in volume at constant temperature is very small; at the critical point, it vanishes, and the phenomenon of critical-point opalescence discussed earlier shows that the gas does indeed become unstable against the spontaneous growth of large-scale density fluctuations.

The pressure in the cosmic "gas" as given by Equation 8.5 with $\beta = 2$, and with the potential energy U given by Equation 8.18, is

$$P = \frac{N\Theta}{V} - \frac{2\pi}{3}G\alpha^2\bar{\rho}^2L^2. \tag{8.19}$$

If we compress a region whose diameter is large compared with L, the density ratio α does not change, and the fractional changes of V, $\bar{\rho}$, and L are related by

Optical and X-ray images of the same region of the sky. The optical photograph shows the counterparts of the X-ray source. One is a star and two are quasars.

$$\frac{\Delta V}{V} = -\frac{\Delta\bar{\rho}}{\bar{\rho}} = 3\frac{\Delta L}{L}. \qquad (8.20)$$

With the help of these relations, we obtain from Equation 8.19 the following formula for the change in pressure accompanying a change in volume at constant temperature:

$$\Delta P = \left(\frac{N\Theta}{V} - \frac{4}{3}\cdot\frac{2\pi}{3}G\alpha^2\bar{\rho}^2L^2\right)\left(-\frac{\Delta V}{V}\right)$$
$$= \left(P - \frac{1}{9}\cdot 2\pi G\alpha^2\bar{\rho}^2L^2\right)\left(-\frac{\Delta V}{V}\right). \qquad (8.21)$$

Thus the change in pressure and the change in volume have opposite signs if and only if the pressure exceeds a positive value, given by

$$P_{\mathrm{crit}} = \frac{1}{9}\cdot 2\pi G\alpha^2\bar{\rho}^2L^2. \qquad (8.22)$$

If the pressure were less than this critical value, the "gas" would be unstable against the growth of density fluctuations. The accompanying motions would then cause the pressure to increase beyond the critical value, thereby quenching the instability. On the other hand, the cosmic expansion tends to reduce the pressure, because it attenuates the kinetic contribution to the pressure twice as fast as the gravitational contribution. These opposing tendencies maintain the pressure at a value close to the critical value.

The condition $P = P_{\mathrm{crit}}$, along with Equation 8.19 for the pressure, implies that the kinetic energy and the total energy are fixed multiples of the potential energy:

$$K = -\frac{2}{3}U, \qquad E \equiv K + U = \frac{1}{3}U. \qquad (8.23)$$

Inserting these formulas in the energy equation 8.17 and bearing in mind that $H = \dot{a}/a$, we obtain, finally, the important result that *the kinetic energy, the potential energy, and the total energy of any large coexpanding region all increase at the same rate as the cosmic scale factor:*

$$K, U, E \propto a(t). \tag{8.24}$$

The last result means that the clumpiness of the universe constantly increases. How does it increase? In Equation 8.18 for the potential energy, the product $\bar{\rho}V$ is constant in time, and $\bar{\rho} \propto a^{-3}$; hence the product αL^2 must increase like the fourth power of the cosmic scale factor. Now, when the clustering amplitude is of order unity, clumps of diameter L begin to separate out as self-gravitating systems, and thereafter their energy remains approximately constant. Thus as the universe expands, the clustering amplitude α remains fixed at a value close to unity. The clustering scale L must therefore increase like the square of the cosmic scale factor:

$$L \propto a^2. \tag{8.25}$$

The mass M of clumps of diameter L is proportional to the product of the mean density $\bar{\rho}$, which is inversely proportional to the cube of the cosmic scale factor and to L^3; hence

$$M \propto a^3. \tag{8.26}$$

Equations 8.24 to 8.26 describe the most important aspects of the clustering process. As the universe expands, self-gravitating clusters that encompass larger and larger masses come into being. This process is the reverse of the conventional one, in which a collapsing gas cloud breaks up into fragments which in turn break up into smaller fragments, and so on.

The mass M of clusters just beginning to separate out increases as the cube of the cosmic scale factor; their diameter D increases as the square of the cosmic scale factor, hence as the two-thirds power of the mass; their energy per unit mass, ε, which is proportional to M/D, increases like the cosmic scale factor itself, hence as the cube root of the mass:

$$D \propto M^{2/3}, \quad \varepsilon \propto \frac{M}{D} \propto M^{1/3}. \tag{8.27}$$

To calculate the coefficients of proportionality in these relations, we need to know the value of ε at the onset of the clustering process.

At the moment when the solid cosmic medium breaks up into fragments, the kinetic energy K and the potential energy U of a region whose diameter is large compared with the diameter of a typical fragment are both zero. As the fragments separate, a fluctuating gravitational field develops, which accelerates them. The energy $K + U$, decreasing from its initial value of

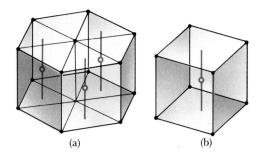

(a) (b)

In solid molecular hydrogen, each lattice site in the close-packed hexagonal structure (*left*) is occupied by a hydrogen molecule. In solid metallic hydrogen, each lattice site in the body-centered cubic lattice (*right*) is occupied by a proton; electrons wander freely through the lattice.

zero, now assumes negative values, causing the "gas" of solid fragments to become unstable against clumping. I estimate that at the onset of this instability, the energy per unit mass $(K + U)/\bar{\rho}V$, which is the quantity we denoted by ε, is given by

$$\varepsilon_1 = -\frac{1}{3}QR_f^2\left(\frac{Gm_f}{R^3}\right), \qquad (8.28)$$

where m_f is the mass of a typical fragment, R is the radius of a sphere of average mass m_f, and R_f is the radius of a typical fragment. R_f is related to the mass m_f and the density ρ^* of the solid medium at the moment of fragmentation by the equation

$$m_f = \frac{4\pi}{3}R_f^3\rho^*. \qquad (8.29)$$

Q is a small number, whose value is difficult to estimate but probably lies between $\frac{1}{10}$ and $\frac{1}{100}$.

As we saw earlier, the gravitational potential energy per unit mass of a typical fragment should be comparable to the cohesion energy per unit mass of the solid medium:

$$\frac{3}{5}\frac{Gm_f}{R_f} \simeq \varepsilon^*, \qquad (8.30)$$

where ε^* denotes the cohesion energy per unit mass. We can use Equations 8.29 and 8.30 to estimate the mass and radius of a typical fragment if we know the density ρ^* and cohesion energy per unit mass ε^* of the solid medium.

Zel'dovich, in his 1963 paper, assumed that the early cold universe, consisting almost entirely of hydrogen, would solidify as a molecular lattice (see figure (a) above). In 1970, Ray Hively, then a graduate student at Harvard, carried out detailed quantum-mechanical calculations indicating that hydrogen can solidify not only as a molecular lattice but also as a metal with properties similar to those of metallic sodium. When Hively wrote his dissertation, the existence of a stable metallic phase of hydrogen was in doubt. Since then, physicists have succeeded in producing metallic hydrogen for short periods, and the discovery that Jupiter has a strong magnetic field has led planetary astronomers to postulate that it has a metallic hydrogen core. Hively's calculations suggested that an initially cold universe composed of hydrogen would solidify and shatter in the metallic phase, but they did not rule out the possibility that the universe might solidify in the molecular form, or solidify first in the metallic form and then undergo a phase transition to the molecular form. The zero-pressure density and cohesion energy per unit mass of the metallic phase, as estimated by Hively, are

$$\rho^* = 0.65 \text{ gm/cm}^3, \qquad \varepsilon^* = 4 \times 10^{11} \text{ erg/gm} \qquad \text{(metallic).} \qquad (8.31)$$

The corresponding values for the molecular phase are

$$\rho^* = 0.089 \text{ gm/cm}^3, \qquad \varepsilon^* = 4 \times 10^9 \text{ erg/gm} \qquad \text{(molecular).} \qquad (8.32)$$

The fragment masses calculated from Equations 8.29 and 8.30 are

$$m_f = \begin{cases} 2 \times 10^{28} \text{ gm} = & 10^{-5} \text{ M}_\odot & \text{(metallic)} \\[2mm] 5 \times 10^{25} \text{ gm} = 2.5 \times 10^{-8} \text{ M}_\odot & \text{(molecular).} \end{cases} \qquad (8.33)$$

The molecular-hydrogen fragments are comparable in mass to Jupiter's Galilean satellites and the satellites of Saturn. The metallic-hydrogen fragments are less massive than the giant planets by one to two orders of magnitude.

   With the help of Equation 8.30, we can rewrite Equation 8.28 for the initial clustering energy per unit mass in the form

$$\varepsilon_1 = -\frac{5}{9} Q \left(\frac{R_f}{R} \right)^3 \varepsilon^*. \qquad (8.34)$$

The ratio R_f/R probably lies between $\frac{1}{2}$ and $\frac{1}{3}$, and Q probably lies between $\frac{1}{10}$ and $\frac{1}{100}$.

   The figure on page 289 shows binding energy $(-\varepsilon)$ plotted against mass (m) for self-gravitating systems ranging in size from the Jupiter and Saturn systems to rich galaxy clusters. Also shown is the theoretical relation $\varepsilon/\varepsilon_1 = (m/m_f)^{1/3}$, with ε_1 given by Equation 8.34. The plotted points are well represented over the entire range of masses—18 powers of ten—by the theoretical relation based on the prediction that the primordial fragments are metallic hydrogen with density and cohesion energy given by Hively's calculations.

   The theory of gravitational clustering that I have just sketched rests on the assumption that space is infinite and Euclidean. As we saw earlier, direct estimates of the cosmic mass density, based on the assumption that the ratio of nonluminous to luminous matter is the same in the

A logarithmic plot of binding energy per unit mass against mass, according to the theory of gravitational clustering discussed in the text. Data points refer to representative members of their classes.

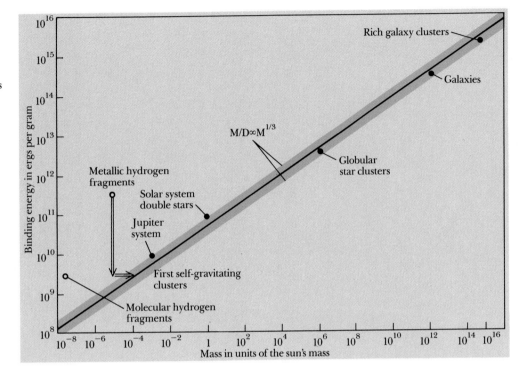

universe as a whole as in the Local Supercluster, yield values that are half or less than half the value appropriate to zero spatial curvature. The theory of gravitational clustering allows us to estimate the curvature of space more directly, without making an assumption about the cosmic ratio of luminous to nonluminous matter.

If the curvature of space is negative—that is, if the cosmic mass density is smaller than the value corresponding to zero spatial curvature—then every spherical region, regarded as a Newtonian system, has positive energy. It is easy to modify the theory of gravitational clustering to allow for this new feature. The binding energy per unit mass $-\varepsilon$ of a cluster formed when the cosmic scale factor has the value a must be multiplied by the factor

$$1 - 3(\Omega_0^{-1} - 1)a, \tag{8.35}$$

where Ω_0 is the ratio between the present value of the cosmic mass density and the value appropriate to a universe of zero spatial curvature. Now, the average density of the Local Supercluster is probably not much more than twice the average cosmic density; hence this system either formed very recently or is now forming. Yet its binding energy per unit mass fits the theoretical relation derived for a universe of zero spatial curvature (see figure above). This suggests that the factor in 8.35 is close to unity, and hence that Ω_0^{-1} is close to unity. The simplest

EINSTEIN OBSERVATORY
QSO 0420-388 Z=3.1
5 ARC-MIN: ⊢―――⊣

X-ray image of a quasar with a redshift $z = 3.1$. The light from this object was emitted when the universe was about one-eight its present age.

possibility that is consistent with the present theory and with astronomical observations is $\Omega_0 = 1$: the universe has zero spatial curvature. This implies that space is infinite, that the universe expands forever, and that the process of clustering continues forever. If $\Omega_0 < 1$, space is infinite and the universe expands forever, but the clustering process ends when the factor (8.35) vanishes, that is, when

$$a^{-1} = 3(\Omega^{-1}_0 - 1).$$

The Visible Universe and the Invisible Universe

The theory of the microwave background discussed early in this chapter rests on the assumption that a large fraction of the mass of the universe condensed in an early generation of massive stars. The theory of gravitational clustering described in the preceding sections is consistent with this assumption (whereas the fireball hypothesis is not), but it leaves open the questions why the bulk of the mass in the universe should have condensed in stars, and why these stars should be ten times as massive as the Sun, rather than, say, a hundred times or a tenth as massive. Also, we have not yet explained the curious relation between the visible universe and the invisible universe. Visible matter makes up only a small fraction of the mass of the universe and is concentrated in relatively compact, flattened systems that form the cores of much larger spherical systems composed of invisible matter. In this section we will see how these gaps might be filled.

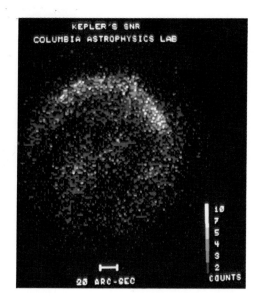

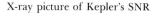

X-ray picture of Kepler's SNR

The members of the first clusters are fragments of solid hydrogen. The binding energy per unit mass of these clusters is much smaller than the cohesion energy per unit mass of the fragments; hence collisions between the solid fragments are unlikely to disrupt them. On the other hand, the clusters themselves are likely to be disrupted when they congregate into larger clusters, because the binding energy per unit mass increases with mass during the clustering process. Thus, progressively more massive and more tightly bound clusters of solid fragments will come into being. Eventually the binding energy per unit mass of a newly formed cluster will be almost equal to the binding energy per unit mass of the fragments that compose it. At this stage of the process, collisions between the fragments convert a large fraction of their mass into gas.

Gas clouds resist tidal disruption. They radiate away energy, becoming more tightly bound as they do so; hence it is plausible that a large fraction of the mass in the universe condensed into protostars that survived subsequent stages of the clustering process. In principle it should be possible to predict the masses and mass distribution of these protostars. When this calculation is done, it will provide a stringent test of the present theory.

Let us assume, then, that 90 percent, say, of the clusters formed at some definite stage of the clustering process are not disrupted at the next stage, but evolve into massive stars; and that the remaining 10 percent are disrupted at the next clustering stage. Clusters formed at this stage will then be of two kinds: clusters of protostars, and clumpy gas clouds with masses about a hundred times greater than that of the Sun. Some of these massive clouds may themselves evolve into stars. Let us assume, however, that the internal dynamics of some of these clouds prevent them from condensing into single, massive protostars. (I have argued in the technical

X-ray picture of Quasar PKS2216-038

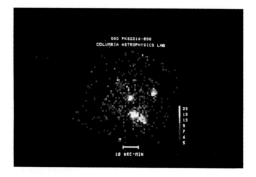

literature that magnetic fields and the transfer of angular momentum dominate the "internal dynamics" of such gas clouds.) What then?

The formation of massive gas clouds causes the evolutionary process to branch. The main branch is the one we have already discussed. It gave rise to the radiation background and to a hierarchy of self-gravitating systems composed mainly of no-longer-luminous matter. Could the second branch, which proceeds from the massive gas clouds, have given rise to the luminous component of galaxies?

Gas clouds participate in the clustering process along with stars and other objects, but the outcome of the process for them is different from that for stars. Clouds tend to stick together when they collide, transforming the kinetic energy of their relative motion into thermal energy that is subsequently radiated away. Thus we may expect clustering to give rise to *cloud complexes*. Because these objects are much more massive than stars, they settle toward the centers of their parent clusters, for the same reason that heavy particles tend to settle to the bottom of a liquid suspension. This explains why the stars that eventually form in the clouds are concentrated in the cores of much larger nonluminous systems.

Because clouds and cloud complexes dissipate kinetic energy but not angular momentum in collisions, we expect them to form more or less highly flattened systems. Thus we can understand why massive gas clouds are now found either in extended discs or in compact regions at the centers of galaxies.

To complete this highly schematic overview of cosmic evolution, let us consider one more speculative notion. Astronomical observations tell us that stars range in age from 10 billion years to less than a million years. Evidence accumulated during the past decade strongly suggests that in our own Galaxy, stars are born in compact cloud complexes of a million or more solar masses, and that these massive cloud complexes may be a billion or more years old. This last finding conflicts sharply with conventional ideas about the origin and evolution of stellar systems but fits in nicely with the picture described here. If we assume that the clouds themselves are old and have approximately the same age, as the present theory predicts and as modern observations strongly suggest, they must have *different gestation periods for stars:* the time that elapses between the formation of a cloud and the birth of stars in it must cover a wide range of values amounting perhaps to several billion years in some clouds.

Both observational evidence and theoretical arguments indicate that, as Walter Baade once remarked, star formation is "a contagious disease." Once stars begin to form, the process spreads rapidly. The onset of star formation transforms a cloud into a cluster of stars, which heat the gas that fails to condense into stars. The heated gas then expands and escapes, becoming part of the interstellar medium. We may think of clouds that have comparable gestation periods as members of a *galactic subsystem* characterized by their common gestation period. The age of a *stellar* subsystem is thus the difference $t_0 - t$ between the age of the universe, t_0, and the subsystem's gestation period, t. We would therefore expect to find a correlation between the ages of galactic subsystems and their degree of flattening. The youngest stellar subsystems, which had the longest gestation periods, should be flattest; the oldest, the least flattened. Such a correlation is in fact observed.

Let us suppose, for the sake of concreteness, that the gestation periods of clouds in a given galaxy have an exponential frequency distribution. The fraction of clouds with gestation periods between t and $t + \Delta t$ is then equal to $e^{-t/T} \Delta t/T$, where T is the mean gestation period. It is conceivable that the mean gestation period would differ from galaxy to galaxy. What would be the observational consequences of such a variation?

Let T_{disc} denote the shortest time required by a system of clouds to form a disc by means of dissipative encounters. Observational evidence suggests that $T_{disc} \simeq 5 \times 10^9$ years. The mean gestation period T is either small compared with T_{disc}, comparable to T_{disc}, or large compared with T_{disc}. Let us consider briefly the observable consequences of these three possibilities.

If $T \ll T_{disc}$, the average age of nonprimordial stars (stars born in clouds) will be comparable to the age of the universe, and these stars will form slightly flattened systems. We may identify such systems with elliptical galaxies.

If $T \simeq T_{disc}$, comparable numbers of stars form in spheroidal subsystems and in a disc. Most of the disc's light will come from stars more than 10^8 years old. We may identify such systems with "early" spirals, that is, spirals of type Sa and Sb in Hubble's classification.

Finally, if $T \gg T_{disc}$, most of the nonprimordial stars form in a disc. With increasing values of T, the spheroidal subsystem diminishes in mass relative to the disc, and the average age of the stars in the disc decreases. We may identify these systems with "late" spirals and irregulars.

Retrospect and Prospect

The main phenomena that need to be explained by a theory of cosmic evolution are the existence of self-gravitating systems; gravitational clustering and its spectrum (the observed relation between the masses and diameters of self-gravitating systems); the role of stars and galaxies in the hierarchy of self-gravitating systems; the preponderance of invisible mass; the phenomenon of stellar populations; the cosmic abundances of the chemical elements; the X-ray and gamma-ray backgrounds; and, last but not least, the cosmic microwave background. Ever since its discovery in 1965, the microwave background has rightly been regarded as the single most important clue to the physical state of the early universe. Most cosmologists interpret the microwave background as the remnant of a primordial fireball. We have seen that this interpretation, though simple and natural in itself, runs into difficulties. Each of its three direct predictions (which concern the cosmic abundances of helium and deuterium and the shape of the spectrum at millimeter wavelengths) conflicts with present astronomical evidence. In addition, theorists have not yet been able to reconcile the fireball hypothesis in a convincing way with the existence of cosmic nonuniformity.

The hypothesis of a cold initial state was put forward originally to explain the clumpiness of the cosmic mass distribution. According to this hypothesis, the universe began to expand from a state of uniform mass density at temperature zero. Though cold, the expanding cosmic medium remained gaseous until its density dropped to a value close to that of water, when it solidified. As the medium continued to expand, it shattered into fragments of subplanetary mass. The cold, expanding "gas" composed of solid fragments was unstable against the spontaneous growth of clumps in the same way that an ordinary vapor at its critical point is unstable against the growth of large-scale density fluctuations. The instability led to clumpiness on progressively larger scales, causing progressively more massive self-gravitating systems to separate out. A crude theory of this process predicts a linear relation between the logarithm of the mass and the logarithm of the diameter of newly formed self-gravitating systems. The slope of the predicted relation agrees well with observation, and so does the intercept, if, as theoretical calculations by Ray Hively suggest, the cosmic medium solidified as metallic hydrogen. The fact that the estimated mass and diameter of the Local Supercluster fit the theoretical relation implies that the cosmic mass density is not much smaller than the value appropriate to zero spatial curvature.

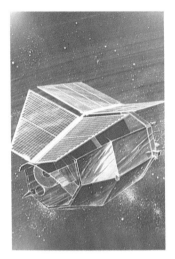

Artist's conception of the Einstein Observatory in space.

The theory of gravitational clustering in an initially cold universe is consistent with the hypothesis that the bulk of the matter in the universe condensed into massive stars that produced a great flash of light when the universe was around 10 million years old. This light, partially thermalized by solid grains formed from elements synthesized in the interiors of the same stars, became the microwave background. This theory for the origin of the background is supported by measurements of the shape of the spectrum at millimeter wavelengths. The stars that gave rise to the microwave and millimeter-wave background would also have produced, during the explosive phase of their evolution, ultrarelativistic electrons and protons, which in turn produced the observed X-ray and gamma-ray background by mechanisms discussed earlier in this chapter.

Much of the matter that did not condense into the early generation of massive stars would have formed massive, clumpy gas clouds. Ensuing stages of gravitational clustering would have given rise to cloud complexes, forming relatively compact and flattened subsystems of clusters whose main constituents are the remnants of the early, massive stars. Finally, these subsystems would have evolved into visible galaxies.

This, very schematically, is how an initially cold and structureless universe might have evolved. Although this "alternative history" seems to fit observational data better than do histories based on the fireball hypothesis, it needs more work. Its predictions need to be made more precise, less elastic. The observational picture, too, needs to be clarified. As the theoretical and observational pictures come into sharper focus, they may turn out not to be the same picture after all. Almost certainly, the picture sketched here will need to be modified in important ways. But that is why we make models: to suggest calculations, experiments, and observations that will cause us to revise or abandon the model. Before we can learn what the world is like we must imagine what it might be like.

In this chapter we have considered three main periods of cosmic history: early, middle, and modern. During the early period, the cosmic medium was smooth and structureless, or nearly so; during the middle period, the hierarchy of self-gravitating astronomical systems came into being; during the modern period, the structure of the universe has changed only on the largest scales. But the questions that many theorists find most exciting concern a period of cosmic history not mentioned here at all: the exceedingly brief period at the very beginning of the cosmic expansion when particles were so tightly packed that present physical laws did not apply. The very early universe is a frontier where particle physics and cosmology meet and join. It is a testing ground for unified theories of the fundamental forces of nature; and these theories, in turn, could revolutionize our ideas about the early universe. They may rescue the fireball hypothesis, or they may replace it by a hypothesis as novel and fruitful as Friedmann's hypothesis of expanding space.

Yet we can be confident that future discoveries and theories will enrich, rather than destroy, the universe that Pythagoras, Aristarchus, Kepler, Newton, and Einstein have taught us to see: a universe no less harmonious than that of Pythagoras and Plato, but whose harmonies lie in mathematical laws; a universe no less perfect than that of Aristotle, but whose perfection lies in abstract symmetries; a universe in which the vast emptiness of intergalactic space is filled with a soft glow, carrying a message from the depths of time that we cannot yet fully decipher; a universe with a beginning in time but with no beginning or end in space; a universe that may expand forever, or that, just possibly, may one day stop expanding and begin to contract. It is a very different universe from that imagined by those brave souls who first dared to ask, What is the world *really* like? But I think it would not have displeased them.

Kelvin

APPENDIX

VECTOR ARITHMETIC

A displacement from a point A to a point B along a given straight line is defined by its *magnitude,* the distance $\overline{AB}$, measured in some specified unit, and by its *direction* or *sense,* whether B lies to the right or to the left of A. Thus we may represent displacements by numbers: positive numbers represent displacements to the right; negative numbers, displacements to the left. Conversely, every positive and negative number represents a definite displacement. We will denote the displacement from A to B by the symbol $\overrightarrow{AB}$. Notice that the displacement $\overrightarrow{AB}$ does not depend on the position of A, but only on the length and direction of the displacement itself (figure below).

Analogously, we may represent a displacement in a plane by a *pair* of signed numbers (a, b), where the first specifies the magnitude and sense of the displacement in the x-direction, and the second the magnitude and sense of the displacement in the y-direction. In a city whose streets run east–west and north–south, forming a square grid, we can get from one streetcorner A to

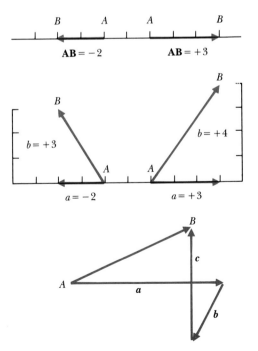

(Left) Idealized portion of the Universe. The geometric figure is Newton's figure for his proof that Kepler's law of elliptic motion implies the inverse square law of gravitation.

Displacements in one, two, and three dimensions.

another streetcorner B by walking a blocks east and b blocks north, in either order. (We stipulate that walking -3 blocks east means walking 3 blocks west, walking -5 blocks north means walking 5 blocks south.) We represent the displacement itself by an arrow drawn from A to B, and denote it by $\overrightarrow{AB}$. Its length $\overline{AB}$ is given by Pythagoras' theorem,

$$\overline{AB}^2 = a^2 + b^2,$$

and its direction is defined by the ratio b/a.

Analogously, a displacement $\overrightarrow{AB}$ in three-dimensional space may be defined either by its *components* (east–west, north–south, up–down) or by its magnitude $\overline{AB}$ and its direction (specified, for example, by the two ratios a/b, a/c).

When we travel from one point to another point, by whatever route, our displacement is defined by its two endpoints and their order. Suppose we travel from A to K and then from K to B. Our net or resultant displacement is described by $\overrightarrow{AB}$ (see figure at left). Thus we obtain the following rule for combining displacements. *To combine two displacements, place the arrows representing them end to tip. The arrow that joins the free end to the free tip represents the resultant displacement.* The figure shows that the resultant of two displacements does not depend on the order in which they are combined.

The mathematical quantity that we have denoted by $\overrightarrow{AB}$ is called a *vector*. The relation between vectors and displacements in space is exactly the same as that between signed numbers and displacements along a line. And just as displacements along a line are a special case of displacements in space, so signed numbers are a special case of vectors. Vectors are generalized numbers, endowed with direction as well as magnitude.

We may conveniently represent the statement "displacements $\overrightarrow{AK}$ and $\overrightarrow{KB}$ combine to form the resultant $\overrightarrow{AB}$" by the equation

$$\overrightarrow{AK} + \overrightarrow{KB} = \overrightarrow{AB}$$

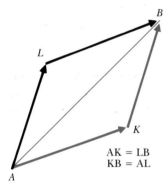

AK = LB
KB = AL

as seen in the figure at the left. Of course, the plus sign does not mean the same thing in this equation as it does in the equation $2 + 3 = 5$. (Its meaning is given by the preceding italicized rule.) But "vector addition" contains ordinary addition as a special case. If $\overrightarrow{AK}$ and $\overrightarrow{KB}$ are parallel, the *magnitude* of their resultant $\overrightarrow{AB}$, which we denote by $\overline{AB}$, is the ordinary sum of the magnitudes of $\overrightarrow{AK}$ and $\overrightarrow{KB}$: $\overline{AK} + \overline{KB} = \overline{AB}$. Otherwise, $\overline{AK} + \overline{KB} > \overline{AB}$.

The preceding rule implies that $\overrightarrow{AB} - \overrightarrow{BA}$ is a vector of zero length, represented by the point A (figure on page 299). The same rule tells us that the resultant of any displacement $\overrightarrow{AB}$ and the displacement of zero length is the displacement $\overrightarrow{AB}$. Thus the displacement of zero length plays the same role in vector addition as the number 0 in ordinary addition. The displacement of zero length is denoted by 0 (though a more consistent notation would be $\vec{0}$). Thus we write $\overrightarrow{AB} + \overrightarrow{BA} = 0$.

The displacement of zero length enables us to define negative displacements, just as the number 0 allows us to define negative numbers. The formula $\overrightarrow{AB} = -\overrightarrow{BA}$ means the same thing as the formula $\overrightarrow{AB} + \overrightarrow{BA} = 0$. Analogously, the relation $\overrightarrow{BC'} + \overrightarrow{C'C} = \overrightarrow{BC}$ means the same thing as the relation $\overrightarrow{C'C} = \overrightarrow{BC} - \overrightarrow{BC'}$.

We may generalize ordinary multiplication by defining the product of a positive or negative number k and a displacement $\overrightarrow{AB}$ as the displacement whose magnitude (length) is equal to $k \cdot \overrightarrow{AB}$, and whose direction is the same as that of $\overrightarrow{AB}$. Thus $\overrightarrow{AB} + \overrightarrow{AB} = 2\overrightarrow{AB}$, $m\overrightarrow{AB} + n\overrightarrow{AB} = (m + n)\overrightarrow{AB}$, and so on. In other words, $k\mathbf{a}$ is a vector with the same direction as $\mathbf{a}$ and with magnitude ka. Here and in what follows we use boldface letters to represent vectors,

$$\mathbf{a} = \overrightarrow{AB},$$

and we represent the magnitude of a vector by the corresponding letter in italic type.

There are two ways to multiply a vector by a vector. The first is the *scalar* or *dot product,*

$$\mathbf{a} \cdot \mathbf{b} = ab \cos \theta, \tag{1}$$

where θ is the angle between $\mathbf{a}$ and $\mathbf{b}$. The work done by a force $\mathbf{F}$ on a particle that undergoes a displacement $\Delta\mathbf{r}$ is

$$\Delta W = \mathbf{F} \cdot \Delta\mathbf{r} = F\Delta s \cos \theta, \tag{2}$$

where $\Delta s \equiv |\Delta\mathbf{r}|$, the length of the displacement. The gravitational potential ϕ is related to the gravitational force $\mathbf{F}$ by a similar relation:

$$\Delta\phi = -\mathbf{F} \cdot \Delta\mathbf{r}. \tag{3}$$

To represent the relation between an applied force $\mathbf{F}$, the radius vector $\mathbf{r}$ drawn from an immovable point (the center of rotation) to the point where the force is applied, and the resulting torque $\mathbf{K}$, we need the *vector* or *cross product:*

$$\mathbf{K} = \mathbf{r} \times \mathbf{F}. \tag{4}$$

The magnitude of $\mathbf{K}$ is the area of the parallelogram defined by $\mathbf{r}$ and $\mathbf{F}$:

$$K = rF \sin \theta \tag{5}$$

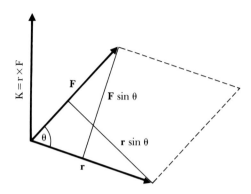

The relation between **r**, **F**, and **K** = **r** × **F**.

(see figure above). The direction of **K** is perpendicular to the plane defined by **r** and **F**, and **K** points in the direction that a right-handed screw would be driven by a rotation that carries **r** into **F** through the angle θ, the smaller of the two angles defined by **r** and **F**. This rule is illustrated in the righthand figure above.

The product symbolized by Equation 4 behaves in some ways like the product of two ordinary numbers. Thus **a** × (**b** + **c**) = **a** × **b** + **a** × **c**, and if k is an ordinary number, k(**a** × **b**) = **a** × (k**b**) + (k**a**) × **b**. From these rules we can derive the "chain rule" for the rate of change of a product **a** × **b**:

$$\frac{d}{dt}(\mathbf{a} \times \mathbf{b}) = \lim \frac{(\mathbf{a} + \Delta\mathbf{a}) \times (\mathbf{b} + \Delta\mathbf{b}) - \mathbf{a} \times \mathbf{b}}{\Delta t}$$

$$= \left(\frac{d\mathbf{a}}{dt}\right) \times \mathbf{b} + \mathbf{a} \times \left(\frac{d\mathbf{b}}{dt}\right). \tag{6}$$

On the other hand, the "vector product" is neither commutative nor associative:

$$\boldsymbol{a} \times \boldsymbol{b} = -\boldsymbol{b} \times \boldsymbol{a} \tag{7}$$

and

$$\boldsymbol{a} \times (\boldsymbol{b} \times \boldsymbol{c}) \neq (\boldsymbol{a} \times \boldsymbol{b}) \times \boldsymbol{c}. \tag{8}$$

A torque tends to produce rotation. By setting the force **F** = $d\mathbf{p}/dt$ in the definition of the torque (Equation 4) and using the chain rule (Equation 6), we discover exactly what it is that an applied torque tends to change:

$$\mathbf{K} \equiv \mathbf{r} \times \mathbf{F} = \mathbf{r} \times \left(\frac{d\mathbf{p}}{dt}\right)$$

$$= \frac{d}{dt}(\mathbf{r} \times \mathbf{p}) - \left(\frac{d\mathbf{r}}{dt}\right) \times \mathbf{p}$$

$$= \frac{d}{dt}(\mathbf{r} \times \mathbf{p}) \equiv \frac{d\mathbf{J}}{dt}, \tag{9}$$

since $d\mathbf{r}/dt = \mathbf{v} = \mathbf{p}/m$ and $\mathbf{p} \times \mathbf{p} = 0$, by Equation 7. The quantity

$$\mathbf{J} = \mathbf{r} \times \mathbf{p} = m(\mathbf{r} \times \mathbf{v}) \tag{10}$$

is called *angular momentum*. Equation 9 equates the rate of change of a particle's angular momentum to the applied torque. As noted in the table on page 128, this rule is analogous to Newton's second law of motion, which equates the rate of change of a particle's momentum to the applied force.

GUIDE TO FURTHER READING

CHAPTER ONE

Einstein, A., 1954. *Ideas and Opinions.* New York: Bonanza Books.

A collection of short essays, articles, talks, and prefaces, including "The Method of Theoretical Physics," "Physics and Reality," and "Geometry and Experience," in which Einstein explains his view of scientific knowledge.

Feynman, R.P., 1967. *The Character of Physical Law.* Cambridge: M.I.T. Press.

A series of popular lectures in which one of this century's most original theoretical physicists explains some of the deepest concepts in classical and contemporary physics in a simple, clear, and lively way.

Galilei, Galileo, 1623. *Il Saggiatore (The Assayer).* Translated by Stillman Drake, 1957, in *Discoveries and Opinions of Galileo.* New York: Doubleday.

Along with other writings of Galileo, perhaps the best and clearest introduction to the philosophy of science.

Copernicus, N., 1543. *On the Revolutions of the Heavenly Spheres.*

Translations of the most interesting sections of this book, along with well-chosen extracts from many other writers, including Plato, Aristotle, Lucretius, Ptolemy, Newton, Kant, and Hubble may be found in Munitz, M.K., 1957, *Theories of the Universe.* New York: The Free Press.

Stebbins, G.L., 1969. *The Basis of Progressive Evolution.* Chapel Hill: University of North Carolina Press.

An elegant and philosophically sophisticated account of biological evolution and its relation to cultural evolution, by one of the leading evolutionary biologists of our day.

Koyré, A., 1957. *From the Closed World to the Infinite Universe.* Baltimore: The Johns Hopkins Press.

A scholarly yet highly readable study of the interaction between scientific, philosophical, and religious ideas about space and the extent of the Universe from Nicholas of Cusa in the fifteenth century to Newton and Leibniz at the end of the seventeenth century.

Koestler, A., 1959. *The Sleepwalkers.* New York: Macmillan.

A lively and unorthodox account of the astronomical revolution and the men who made it by a famous novelist who held strong views on this (and most other) subjects. One section of the book, *The Watershed,* which is available separately [Garden City, N.Y.: Anchor Books], is a fine biography of Kepler that concentrates on his scientific achievements but does not neglect other aspects of his life and personality.

Kuhn, T., 1970. *The Structure of Scientific Revolutions.* Chicago: University of Chicago Press (second edition).

Kuhn argues that "scientific revolutions" are not stages in a progression toward a true picture of the world but merely shifts from one "paradigm" to another, "incommensurable" paradigm. His argument rests on the tacit (and incorrect) assumption that verbal (as opposed to purely mathematical) concepts are an essential ingredient of physical theories.

Quine, W.V.O., 1952. "Two Dogmas of Empiricism" in *From a Logical Point of View.* Cambridge: Harvard University Press (second edition, 1980).

This famous essay elaborates a view of the relation between natural science and experience similar to Einstein's and to the view illustrated by the figure on page 20 of this book. Quine, however, asserts that science is "extremely underdetermined by experience," whereas I claim that it has a growing, strongly overdetermined core. These two views may not be as different as they seem, because I regard mathematically equivalent forms of a given theory as the same theory in different guises while Quine may perhaps regard them as distinct theories.

CHAPTER TWO

Galileo, G., 1967 (first publ. 1632). *Dialogue on the Two Chief World Systems: Ptolemaic and Copernican.* transl. by Stillman Drake. Berkeley and Los Angeles: University of California Press.
Perhaps the best introduction to modern science ever written.

Koyré, A., 1973. *The Astronomical Revolution.* Ithaca, N.Y.: Cornell University Press.
Contains an especially insightful discussion of Kepler's scientific contributions to the astronomical revolution.

Heath, T., 1913. *Aristarchus of Samos.* London: Clarendon Press.
A thorough and readable account of Aristarchus' achievement in its historical context.

Dreyer, J.L.E., 1953 (first publ. 1906). *A History of Astronomy from Thales to Kepler.* New York: Dover (second edition).
A clear, authoritative account that deals adequately with the mathematical aspects of the subject but presupposes some technical knowledge of classical astronomy.

CHAPTER THREE

Galileo, G., 1954 (first publ. 1638). *Dialogues on Two New Sciences.* transl. H. Crew and A. de Salvio. New York: Dover.
The first dialogue expounds Galileo's theory of size and scale, the second his theory of motion.

Newton, I., 1962 (first edition publ. 1686). *The Mathematical Principles of Natural Philosophy.* transl. A. Motte and F. Cajori. Berkeley and Los Angeles: University of California Press (two volumes).
Rewarding for the insight it gives into Newton's philosophical, mathematical, and scientific thinking, even if one skims the (still) difficult proofs.

Mach, E., 1960 (first publ. 1883). *The Science of Mechanics.* transl. T.J. McCormack. LaSalle, Illinois: Open Court.
A history of mechanics with a strong positivist bias.

CHAPTER FOUR

Kant, I., 1969 (first publ. 1755). *Universal Natural History and Theory of the Heavens.* Ann Arbor: University of Michigan Press.
An exuberant, highly readable account of a theory that lives up to the title.

Hubble, E., 1958 (first publ. 1936). *The Realm of the Nebulae.* New York: Dover.
An excellent popular introduction to observational cosmology by the astronomer whose discoveries initiated the modern era.

Abell, G., 1982. *Exploration of the Universe.* Philadelphia: Saunders.
A well-balanced, complete, and clearly written modern textbook.

CHAPTER FIVE

Einstein, A., 1952 (first publ. 1916). *Relativity: the Special and General Theory.* New York: Crown.
A very clear, brief, nonmathematical account of the two theories' main ideas.

Lorentz, H.A., Einstein, A., Minkowski, H., Weyl, H., 1923. *The Principle of Relativity.* New York: Dover.
A very well-selected collection of original papers on special and general relativity, translated by W. Perrett and G.B. Jeffery. The introductory sections of Einstein's papers are clear and nontechnical.

Einstein, A., 1953. *The Meaning of Relativity.* Princeton: Princeton University Press (fifth edition).
A considerably more demanding discussion than *Relativity: the Special and General Theory.*

CHAPTERS SIX AND SEVEN

Einstein, A. Articles in *The Principle of Relativity* (cited above). The 1911 paper, "The Influence of Gravitation on the Propagation of Light," is especially rewarding, because it uses very elementary mathematics to arrive at deep and surprising conclusions.

CHAPTER EIGHT

Silk, J., 1980. *The Big Bang*. New York: W.H. Freeman and Co.
A popular account of modern discoveries and ideas by a proponent of the fireball hypothesis.

Weinberg, S., 1977. *The First Three Minutes*. New York: Basic Books.
A readable account of the discovery of the cosmic microwave background and the "standard model" of the early universe by a leading student of the very early universe.

CREDITS

INDEX